AS & A2 Physics

Exam Board: AQA A

Complete Revision and Practice

AS-Level Contents

A2-Level Contents

Unit 5: Option C — "Applied Physics" isn't covered in this book.

Editors:
Amy Boutal, Sarah Hilton, Julie Wakeling, Sarah Williams

Contributors
Tony Alldridge, Jane Cartwright, Peter Cecil, Peter Clarke, Mark A. Edwards, D. Kamya, Barbara Mascetti, John Myers, Zoe Nye, Moira Steven, Andy Williams, Tony Winzor

Proofreaders:
Ian Francis, Glenn Rogers

Published by CGP

ISBN: 978 1 84762 418 5

With thanks to Jan Greenway for the copyright research.

Groovy website: www.cgpbooks.co.uk
Jolly bits of clipart from CorelDRAW®
Printed by Elanders Ltd, Newcastle upon Tyne.

Based on the classic CGP style created by Richard Parsons.

AS-Level Physics

Exam Board: AQA A

The Scientific Process

'How Science Works' is all about the scientific process — how we develop and test scientific ideas. It's what scientists do all day, every day (well, except at coffee time — never come between a scientist and their coffee).

Scientists Come Up with Theories — Then Test Them...

Science tries to explain **how** and **why** things happen — it **answers questions**. It's all about seeking and gaining **knowledge** about the world around us. Scientists do this by **asking** questions and **suggesting** answers and then **testing** them, to see if they're correct — this is the **scientific process**.

1) **Ask** a question — make an **observation** and ask **why or how** it happens. E.g. what is the nature of light?
2) **Suggest** an answer, or part of an answer, by forming:
   - a **theory** (a possible **explanation** of the observations) e.g. light is a wave.
   - a **model** (a **simplified picture** of what's physically going on)
3) Make a **prediction** or **hypothesis** — a **specific testable statement**, based on the theory, about what will happen in a test situation. E.g. light should interfere and diffract.
4) Carry out a **test** — to provide **evidence** that will support the prediction, or help disprove it. E.g. Young's double-slit experiment.

The evidence supported Quentin's Theory of Flammable Burps.

A theory is only scientific if it can be tested.

...Then They Tell Everyone About Their Results...

The results are **published** — scientists need to let others know about their work. Scientists publish their results in **scientific journals**. These are just like normal magazines, only they contain **scientific reports** (called papers) instead of the latest celebrity gossip.

1) Scientific reports are similar to the **lab write-ups** you do in school. And just as a lab write-up is **reviewed** (marked) by your teacher, reports in scientific journals undergo **peer review** before they're published.
2) The report is sent out to **peers** — other scientists that are experts in the **same area**. They examine the data and results, and if they think that the conclusion is reasonable it's **published**. This makes sure that work published in scientific journals is of a **good standard**.
3) But peer review **can't guarantee** the science is **correct** — other scientists still need to **reproduce** it.
4) Sometimes **mistakes** are made and bad work is published. Peer review **isn't perfect** but it's probably the best way for scientists to self-regulate their work and to publish **quality reports**.

...Then Other Scientists Will Test the Theory Too

Other scientists read the published theories and results, and try to **test the theory** themselves. This involves:

- Repeating the **exact same experiments**.
- Using the theory to make **new predictions** and then testing them with **new experiments**.

If the Evidence Supports a Theory, It's Accepted — for Now

1) If all the experiments in all the world provide evidence to back it up, the theory is thought of as **scientific 'fact'** (for now).
2) But they never become **totally undisputable** fact. Scientific **breakthroughs or advances** could provide new ways to question and test the theory, which could lead to **new evidence** that **conflicts** with the current evidence. Then the testing starts all over again...

And this, my friend, is the **tentative nature of scientific knowledge** — it's always **changing** and **evolving**.

The Scientific Process

So scientists need evidence to back up their theories. They get it by carrying out experiments, and when that's not possible they carry out studies. But why bother with science at all? We want to know as much as possible so we can use it to try and improve our lives (and because we're nosey).

Evidence *Comes From* ***Controlled Lab Experiments...***

1) Results from **controlled experiments** in **laboratories** are **great**.
2) A lab is the easiest place to **control variables** so that they're all **kept constant** (except for the one you're investigating).

For example, finding the resistance of a piece of material by altering the voltage across the material and measuring the current flowing through it (see p. 26). All other variables need to be kept the same, e.g. the dimensions of the piece of material being tested, as they may also affect its resistance.

... That You can Draw ***Meaningful Conclusions*** *From*

1) You always need to make your experiments as **controlled** as possible so you can be confident that any effects you see are linked to the variable you're changing.
2) If you do find a relationship, you need to be careful what you conclude. You need to decide whether the effect you're seeing is **caused** by changing a variable, or whether the two are just **correlated**.

"Right Geoff, you can start the experiment now... I've stopped time..."

Society ***Makes Decisions*** *Based on* ***Scientific Evidence***

1) Lots of scientific work eventually leads to **important discoveries** or breakthroughs that could **benefit humankind**.
2) These results are **used by society** (that's you, me and everyone else) to **make decisions** — about the way we live, what we eat, what we drive, etc.
3) All sections of society use scientific evidence to make decisions, e.g. politicians use it to devise policies and individuals use science to make decisions about their own lives.

Other factors can **influence** decisions about science or the way science is used:

Economic factors

- Society has to consider the **cost** of implementing changes based on scientific conclusions — e.g. the cost of reducing the UK's carbon emissions to limit the human contribution to **global warming**.
- Scientific research is often **expensive**. E.g. in areas such as astronomy, the Government has to **justify** spending money on a new telescope rather than pumping money into, say, the **NHS** or **schools**.

Social factors

- **Decisions** affect **people's lives** — e.g. when looking for a site to build a **nuclear power station**, you need to consider how it would affect the lives of the people in the **surrounding area**.

Environmental factors

- Many scientists suggest that building **wind farms** would be a **cheap** and **environmentally friendly** way to generate electricity in the future. But some people think that because **wind turbines** can **harm wildlife** such as birds and bats, other methods of generating electricity should be used.

So there you have it — how science works...

Hopefully these pages have given you a nice intro to how science works, e.g. what scientists do to provide you with 'facts'. You need to understand this, as you're expected to know how science works yourself — for the exam and for life.

Atomic Structure

*"So what did you do today Johnny?" "Particle Physics, Mum." "How nice dear — done with times tables then?" Yeah, well, it's not exactly the **easiest** topic in the world, but it's a darn sight more interesting than biology.*

Atoms are made up of Protons, Neutrons and Electrons

Inside **every atom**, there's a **nucleus** containing **protons** and **neutrons**.
Protons and **neutrons** are both known as **nucleons**. **Orbiting** this core are the **electrons**.

This is the **nuclear model** of the atom.

The diagram shows **neutral oxygen**, with **eight protons** and **eight electrons**.

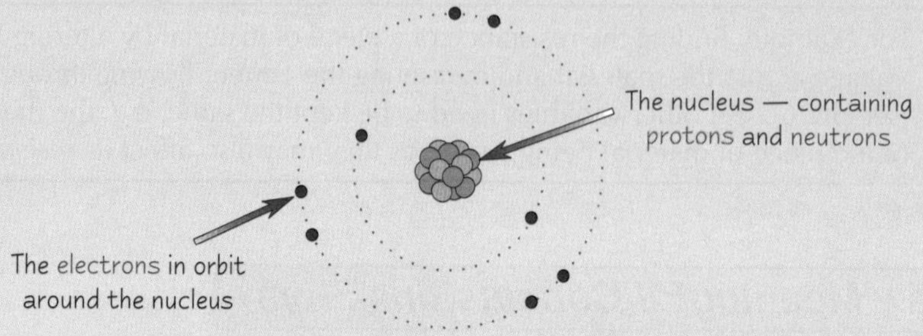

You have to know the **properties** of **electrons**, **protons** and **neutrons** for the exam — make sure you **learn this table**.

Particle	Charge	Relative Mass
Proton	Positive, +1	1
Neutron	Neutral, 0	1
Electron	Negative, –1	0.0005

Atoms are Really, Really Tiny

Each atom is about a **tenth of a nanometre** (**1×10^{-10} m**) in **diameter**. To give you that in context, you'd need to line up around **4 million iron atoms** side by side to give you a line **1 millimetre** long.

And if you think that's small, try the nucleus.

1) Although the **proton** and **neutron** are **2000 times** more **massive** than the **electron**, the nucleus only takes up a **tiny proportion** of the atom. The electrons orbit at relatively **vast distances**.
2) The nucleus is only one **10 000th the size** of the whole atom — most of the atom is **empty space**.
3) If we were to **shrink** the **Solar System** (including poor rejected pluto) so the **Sun** was the **size of a gold nucleus**, **Pluto** would only be **half as far** away as **gold's furthest electron**.

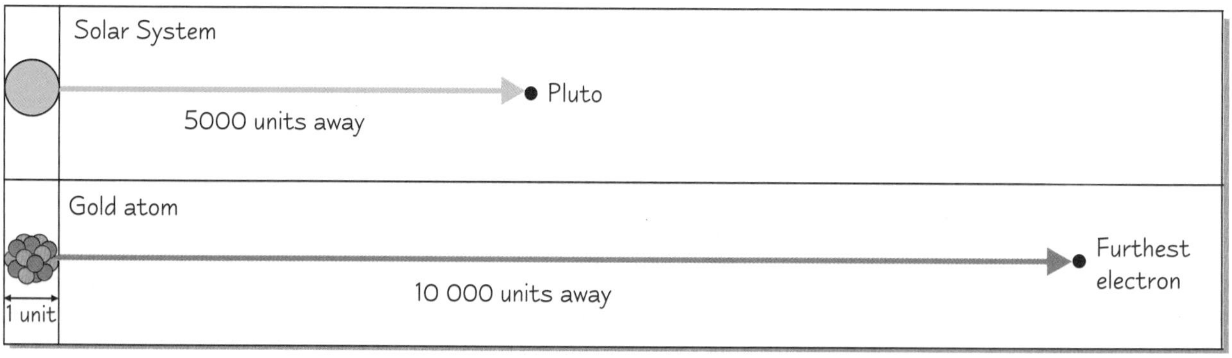

Mind-blowing isn't it. I try not to think too much about it...

Atomic Structure

The Proton Number is the Number of Protons in the Nucleus

No... really.

The **proton number** is sometimes called the **atomic number**, and has the **symbol Z** (I'm sure it makes sense to someone). **Z** is just the **number of protons** in the nucleus.

It's the **proton number** that **defines** the **element** — **no two elements** will have the **same** number of protons.

In a **neutral atom**, the number of **electrons equals** the number of **protons**.
The element's **reactions** and **chemical behaviour** depend on the number of **electrons**.
So the **proton number** tells you a lot about its **chemical properties**.

The Nucleon Number is the Total Number of Protons and Neutrons

The **nucleon number** is also called the **mass number**, and has the **symbol A** (*shrug*).
It tells you how many **protons** and **neutrons** are in the nucleus. Since each **proton or neutron** has an **atomic mass** of (approximately) **1** and the electrons weigh virtually nothing, the **number** of **nucleons** is the same as the **atom's mass**.

STANDARD NOTATION:

$$^{12}_{6}\mathrm{C}$$

- 12: The nucleon number — there are a total of 12 protons and neutrons in a carbon-12 atom.
- 6: The proton number — there are six protons in a carbon atom.
- C: The symbol for the element carbon

Isotopes have the Same Proton Number, but Different Nucleon Numbers

Atoms with the **same number of protons** but **different numbers of neutrons** are called **isotopes**.

Example: Hydrogen has three isotopes — hydrogen, deuterium and tritium

Hydrogen has 1 proton and 0 neutrons.
Deuterium has 1 proton and 1 neutron.
Tritium has 1 proton and 2 neutrons.

Changing the number of **neutrons doesn't affect** the atom's **chemical** properties.

The **number of neutrons** affects the **stability** of the nucleus though.

In **general**, the **greater** the number of **neutrons** compared with the number of **protons**, the **more unstable** the nucleus.

Unstable nuclei may be **radioactive** and **decay** to make themselves stable (see p7).

Practice Questions

Q1 List the particles that make up the atom and give their charges and relative masses.

Q2 Define the proton number and nucleon number.

Q3 What is an isotope?

Exam Questions

Q1 Describe the nuclear model of the atom. [3 marks]

Q2 How many protons, neutrons and electrons are there in a neutral atom of oxygen, $^{16}_{8}\mathrm{O}$? [2 marks]

Q3 Define the term 'isotope'. Describe the similarities and differences between the properties of two isotopes of the same element. [3 marks]

"Proton no. = no. of protons" — not exactly nuclear physics is it... oh wait...

I dunno, I just can't get my head round that size of a nucleus stuff — it's just mind-blowing. I have enough trouble imagining this huge Solar System, never mind something so small that, by rights, shouldn't even exist. It's like trying to imagine infinity or something. Still... in true Physicist fashion, I don't let that bother me — I just learn the powers of 10 and everything's OK.

Stable and Unstable Nuclei

Keeping the nucleus stable requires a lot of effort — a bit like AS Physics then...

*The **Strong Nuclear Force** Binds Nucleons Together*

There are several different **forces** acting on the nucleons in a nucleus. The two you already know about are **electrostatic** forces from the protons' electric charges, and **gravitational** forces due to the masses of the particles.

If you do the calculations (don't worry, *you* don't have to) you find the repulsion from the **electrostatic force** is much, much **bigger** than the **gravitational** attraction. If these were the only forces acting in the nucleus, the nucleons would **fly apart**. So there must be **another attractive force** that **holds the nucleus together** — called the **strong nuclear force**. (The gravitational force is so small, you can just ignore it.)

The **strong nuclear force** is quite **complicated**:

1) To **hold the nucleus together**, the strong nuclear force must be an **attractive force** that's **stronger** than the electrostatic force. So far so good.
2) Experiments have shown that the strong nuclear force has a **very short range**. It can only hold nucleons together when they're separated by up to **a few femtometres** (1 fm = 1×10^{-15} m) — the size of a nucleus.
3) The **strength** of the strong nuclear force **quickly falls** beyond this distance (see the graph below).
4) Experiments also show that the strong nuclear force **works equally between all nucleons**. This means that the size of the force is the same whether it's proton-proton, neutron-neutron or proton-neutron.
5) At **very small separations**, the strong nuclear force must be **repulsive** — otherwise there would be nothing to stop it **crushing** the nucleus to a **point**.

lime green, orange and day-glow pink — repulsive at small separations

*The **Size** of the Strong Nuclear Force **Varies with Nucleon Separation***

The **strong nuclear force** can be plotted on a **graph** to show how it changes with the **distance of separation** between **nucleons**. If the **electrostatic force** is also plotted, you can see the **relationship** between these **two forces**.

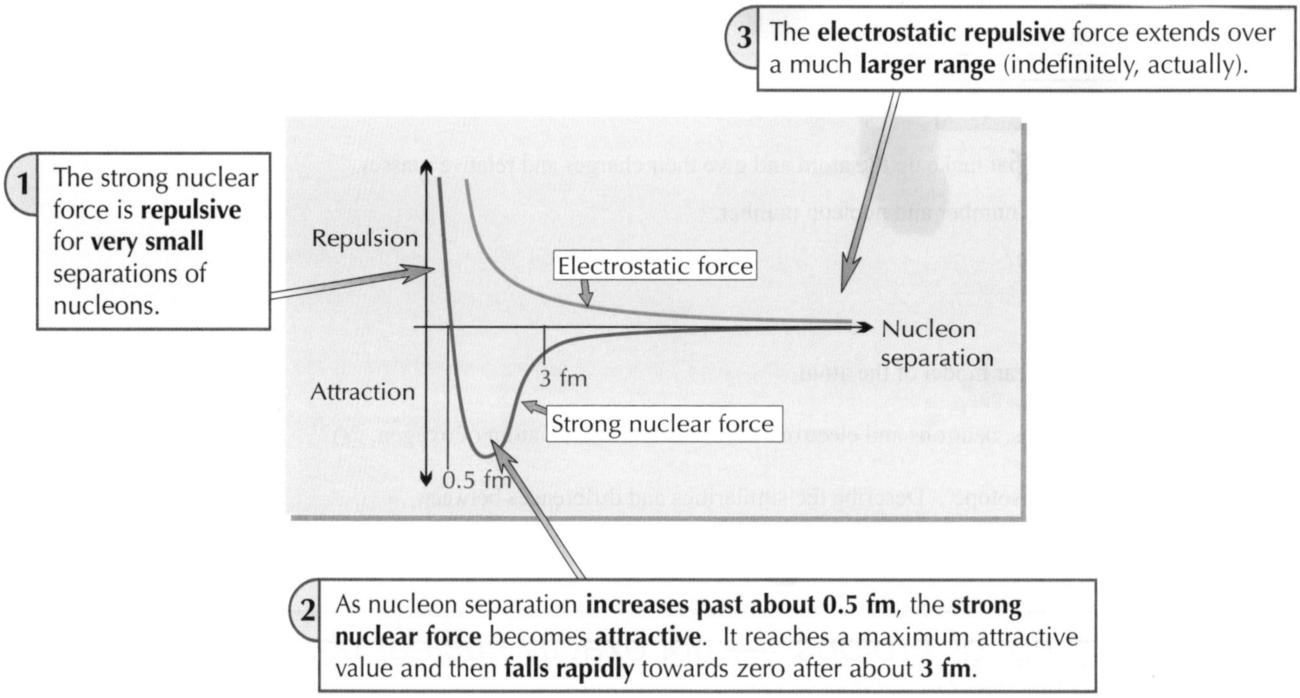

Stable and Unstable Nuclei

α Emission Happens in Very Big Nuclei

1) **Alpha emission** only happens in **very big** atoms (with more than 82 protons), like **uranium** and **radium**.
2) The **nuclei** of these atoms are just **too big** for the strong nuclear force to keep them stable.
3) When an alpha particle is **emitted**:

> The **proton number decreases** by **two**, and the **nucleon number decreases** by **four**.

nucleon number decreases by 4

$$^{238}_{92}U \longrightarrow\ ^{234}_{90}Th +\ ^{4}_{2}\alpha$$

proton number decreases by 2

β^- Emission Happens in Neutron-Rich Nuclei

1) **Beta-minus** (usually just called beta) decay is the emission of an **electron** from the **nucleus** along with an **antineutrino**.
2) Beta decay happens in isotopes that are **"neutron rich"** (i.e. have too many more **neutrons** than **protons** in their nucleus).
3) When a nucleus ejects a beta particle, one of the **neutrons** in the nucleus is **changed** into a **proton**.

> The **proton number increases** by **one**, and the **nucleon number stays the same**.

nucleon number stays the same

$$^{187}_{75}Re \longrightarrow\ ^{187}_{76}Os +\ ^{0}_{-1}\beta + \bar{\nu}$$

proton number increases by 1

> In beta decay, you get a **tiny neutral particle** called an **antineutrino** released. This antineutrino carries away some **energy** and **momentum**. There's more about neutrinos and antineutrinos on the next few pages.

Practice Questions

Q1 What causes an electrostatic force inside the nucleus?

Q2 What evidence suggests the existence of a strong nuclear force?

Q3 Is the strong interaction attractive or repulsive at a nucleon separation of 10 fm?

Q4 Describe the changes that happen in the nucleus during alpha and beta-minus decay.

Exam Questions

Q1 The strong nuclear force binds the nucleus together.

(a) Explain why the force must be repulsive at very short distances. [1 mark]

(b) How does the strong interaction limit the size of a stable nucleus? [2 marks]

Q2 Radium-226 and potassium-40 are both unstable isotopes.

(a) Radium-226 undergoes alpha decay to radon. Complete the balanced nuclear equation for this reaction: [3 marks]

$$^{226}_{88}Ra \longrightarrow \quad Rn +$$

(b) Potassium-40 (Z = 19, A = 40) undergoes beta decay to calcium.
Write a balanced nuclear equation for this reaction. [4 marks]

The strong interaction's like nuclear glue...

In every nuclear reaction energy, momentum, charge, and nucleon number (and several other things that you'll find out about over the next few pages) are conserved. That's why the antineutrino in beta decay ***has*** *to be there.*

Particles and Antiparticles

"I cannae do it Cap'n — their electron-antineutrino ray gun's interfering with my antineutron positron reading..."

Antiparticles *were* ***Predicted*** *Before they were* ***Discovered***

When **Paul Dirac** wrote down an equation obeyed by **electrons**, he found a kind of **mirror image** solution.

1) It predicted the existence of a particle like the **electron** but with **opposite electric charge**.
2) The **positron** turned up later in a cosmic ray experiment. Positrons have **identical mass** (and identical energy in MeV) to electrons but they carry a **positive** charge.

Nice one Paul. Now we've got twice as many things to worry about.

Every Particle *has an* ***Antiparticle***

Each particle type has a **corresponding antiparticle** with the **same mass** but with **opposite charge** (and in other ways too). For instance, an **antiproton** is a **negatively-charged** particle with the same mass as the **proton**.

Even the shadowy **neutrino** (see p12) has an antiparticle version called the **antineutrino** — it doesn't do much either.

Particle	Symbol	Charge	Antiparticle	Symbol	Charge
proton	p	+1	antiproton	$\bar{p}$	–1
neutron	n	0	antineutron	$\bar{n}$	0
electron	e	–1	positron	e^+	+1
electron-neutrino	ν_e	0	electron-antineutrino	$\bar{\nu}_e$	0

You can Create ***Matter*** *and* ***Antimatter*** *from* ***Energy***

You've probably heard about the **equivalence** of energy and mass. It all comes out of Einstein's Special Theory of Relativity. **Energy** can turn into **mass** and **mass** can turn into **energy** if you know how. You can work it all out using the formula $E = mc^2$ but you won't be expected to do the calculations for AS.

It's a good thing this doesn't randomly happen all the time or else you could end up with cute bunny rabbits popping up and exploding unexpectedly all over the place. Oh, the horror...

As you've probably guessed, there's a bit **more to it** than that:

When **energy** is converted into **mass** you get **equal amounts** of **matter** and **antimatter**.

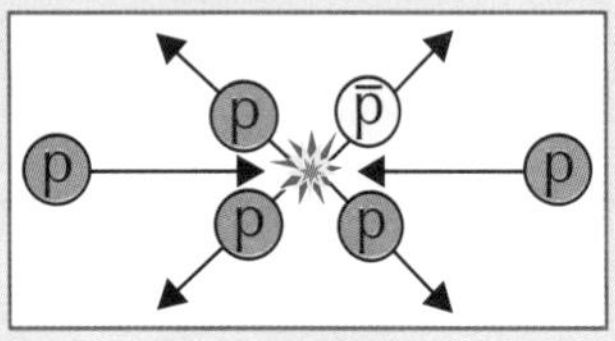

Fire **two protons** at each other at high speed and you'll end up with a lot of **energy** at the point of impact. This energy might be converted into **more particles**.

If an extra **proton** is formed then there will always be an **antiproton** to go with it. It's called **pair production**.

Particles and Antiparticles

Each Particle-Antiparticle Pair is Produced from a Single Photon

Pair production only happens if **one gamma ray photon** has enough energy to produce that much mass. It also tends to happen near a **nucleus**, which helps conserve momentum.

You usually get **electron-positron** pairs produced (rather than any other pair) — because they have a relatively **low mass**.

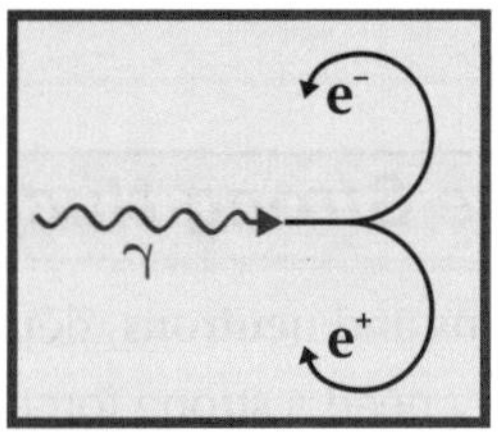

The particle tracks are curved because there's usually a magnetic field present in particle physics experiments. They curve in opposite directions because of the opposite charges on the electron and positron.

The Opposite of Pair-Production is Annihilation

When a **particle** meets its **antiparticle** the result is **annihilation**. All the **mass** of the particle and antiparticle gets converted back to **energy**. Antiparticles can only exist for a fraction of a second before this happens, so you don't get them in ordinary matter.

The electron and positron annihilate and their mass is converted into the energy of a pair of gamma ray photons.

OR

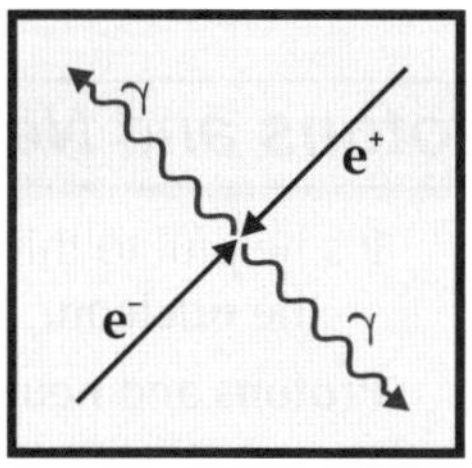

Mesons are Their Own Antiparticles

(Don't worry if you don't know what a meson is — look at page 11. This just seemed like a good place to put it.)

Just before you leave this bit it's worth mentioning that the π^- meson (a type of exotic particle) is just the **antiparticle** of the π^+ meson (another exotic particle), and the **antiparticle** of a π^0 meson is **itself**. You'll see why on p14.
So we don't need any more particles here... Phew.

Practice Questions

Q1 Describe the properties of an electron-antineutrino.

Q2 What is pair production?

Q3 What happens when a proton collides with an antiproton?

Exam Questions

Q1 Write down an equation for the reaction between a positron and an electron and give the name for this type of reaction. [2 marks]

Q2 According to Einstein, mass and energy are equivalent.
Explain why the mass of a block of iron cannot be converted directly into energy. [2 marks]

Q3 Give a reason why the reaction: $\mathbf{p} + \mathbf{p} \rightarrow \mathbf{p} + \mathbf{p} + \mathbf{n}$ is not possible. [1 mark]

This really is Physics at its ~~hardest~~ grooviest...

Inertial dampers are off-line Captain.........oops, no — it's just these false ears making me feel dizzy. Anyway — you'd need to carry an awful lot of antimatter to provide enough energy to run a spaceship. It's not easy to store, either. So it'd never work in real life.

Classification of Particles

There are loads of different types of particle apart from the ones you get in normal matter (protons, neutrons, etc.). They only appear in cosmic rays and in particle accelerators, and they often decay very quickly so they're difficult to get a handle on. Nonetheless, you need to learn about a load of them and their properties.

Stick with it — you'll get there.

Hadrons are *Particles* that Feel the *Strong Nuclear Force* (e.g. Protons and Neutrons)

1) The **nucleus** of an atom is made up of **protons** and **neutrons** (déjà vu).
2) Since the **protons** are **positively charged** they need a strong force to hold them together — the **strong nuclear force** or the **strong interaction** (who said physicists lack imagination...). See page 6 for details.
3) **Not all particles** can **feel** the **strong nuclear force** — the ones that **can** are called **hadrons**.
4) Hadrons aren't **fundamental** particles. They're made up of **smaller particles** called **quarks** (see pages 13–15).
5) There are **two** types of **hadrons** — **baryons** and **mesons**. They're classified according to the number of **quarks** that make them up, but don't worry about that for now.

(Leptons are an example of particles that can't. See page 12.)

Protons and *Neutrons* are *Baryons*

1) It's helpful to think of **protons** and **neutrons** as **two versions** of the **same particle** — the **nucleon**. They just have **different electric charges**.
2) **Protons** and **neutrons** are both **baryons**.
3) There are **other baryons** that you don't get in normal matter — like **sigmas** (Σ) — they're **short-lived** and you **don't** need to **know about them** for AS (woohoo!).

Baryon and Meson felt the strong interaction.

The *Proton* is the *Only Stable Baryon*

All **baryons** — except the proton — are **unstable**. This means that they **decay** to become other **particles**. The **particles** a baryon ends up as depends on what it started as, but it **always** includes a **proton**. **Protons** are the only **stable baryons** — they don't decay (as far as we know).

All baryons except protons decay to a **proton**.

Some theories predict that protons should decay with a very long half-life but there's no evidence for it at the moment.

Antiprotons and *Antineutrons* are *Antibaryons*

The **antiparticles** of protons and neutrons — **antiprotons** and **antineutrons** — are **antibaryons**. But, if you remember from the previous page, **antiparticles** are **annihilated** when they meet the corresponding **particle** — which means that you **don't** find **antibaryons** in ordinary matter.

The *Number of Baryons* in an interaction is called the *Baryon Number*

The **baryon number** is the number of baryons. (A bit like **nucleon number** but including unusual baryons like Σ too.) The **proton** and the **neutron** each have a baryon number **B = +1**. **Antibaryons** have a baryon number **B = –1**. **Other particles** (i.e. things that aren't baryons) are given a baryon number **B = 0**.

When an **interaction** happens, the **baryon number** on either side of the equation is the **same**. You can use this fact to **predict** whether an **interaction** will **happen** — if the numbers don't match, it can't happen.

The **total baryon number** in **any** particle interaction **never changes**.

Classification of Particles

Neutrons Decay into Protons

You saw on page 7 that **beta decay** involves a **neutron** changing into a **proton**. This happens when there are many **more neutrons** than **protons** in a nucleus. **Beta decay** is caused by the **weak interaction** (see page 14).

When a neutron decays, it forms a **proton**, an **electron** and an **antineutrino**:

$$n \rightarrow p + e^- + \bar{\nu}_e$$

Electrons and **antineutrinos** aren't baryons (they're **leptons**, as you'll see on the next page), so they have a baryon number **B = 0**. **Neutrons** and **protons** are baryons, so have a baryon number **B = 1**. This means that the **baryon numbers** on both sides are **equal** (to 1), so the interaction **can** happen.

The Mesons You Need to Know About are Pions and Kaons

The second type of hadron you need to know about is the **meson**.

1) **All mesons** are **unstable** and have **baryon number B = 0** (because they're not baryons).
2) **Pions** (π-mesons) are the **lightest mesons**. You get **three versions** with different **electric charges** — π^+, π^0 and π^-. Pions were **discovered** in **cosmic rays**. You get **loads** of them in **high-energy particle collisions** like those studied at the **CERN** particle accelerator.
3) **Kaons** (**K**-mesons) are **heavier** and more **unstable** than **pions**. You get different ones like K^+ and K^0.
4) Mesons **interact** with **baryons** via the **strong force**.

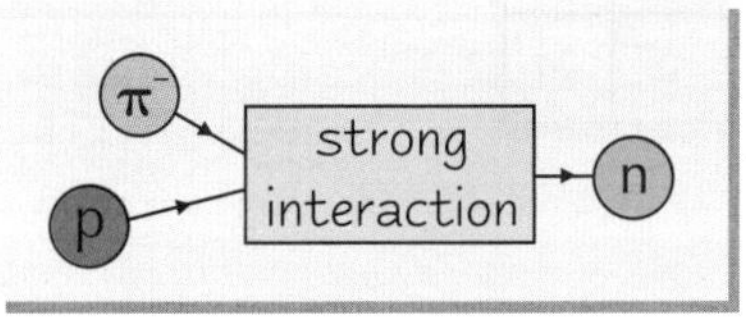

Pion interactions swap protons with neutrons and neutrons with protons, but leave the overall baryon number unchanged.

Summary of Hadron Properties

DON'T PANIC if you don't understand all this yet. For now, just **learn** these properties.

You'll need to work through to the end of page 15 to see how it **all fits together**.

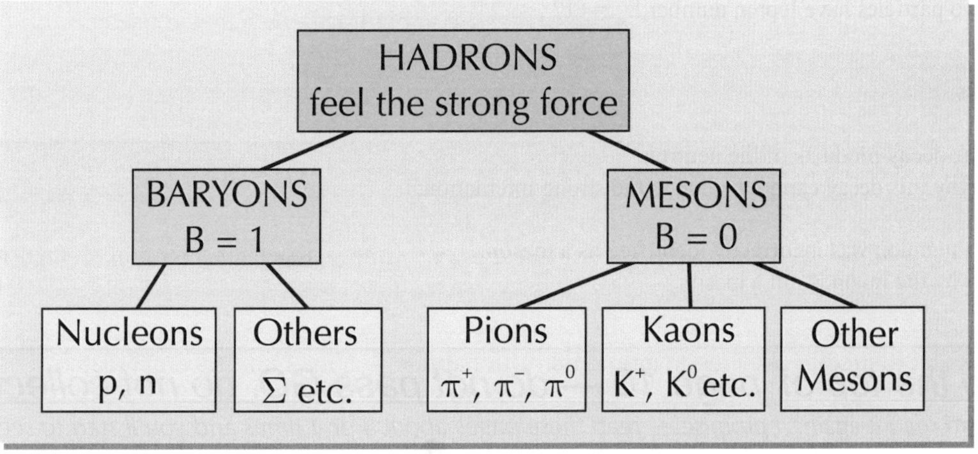

Classification of Particles

Leptons (e.g. Electrons and Neutrinos) Don't Feel the Strong Nuclear Force

1) **Leptons** are **fundamental particles** and they **don't** feel the **strong nuclear force**. They only really **interact** with other particles via the **weak interaction** (along with a bit of gravitational force and the electromagnetic force as well if they're charged).
2) **Electrons** (e^-) are **stable** and very **familiar**, but — you guessed it — there are also **two more leptons** called the **muon** (μ^-) and the **tau** (τ^-) that are just like **heavy electrons**.
3) **Muons** and **taus** are **unstable**, and **decay** eventually into **ordinary electrons**.
4) The **electron**, **muon** and **tau** leptons each come with their **own neutrino**, ν_e, ν_μ and ν_τ.
5) **Neutrinos** have **zero** or **almost zero mass**, and **zero electric charge** — so they don't do much. **Neutrinos** only take part in **weak interactions** (see p14). In fact, a neutrino can **pass right through the Earth** without **anything** happening to it.

You Have to Count the Three Types of Lepton Separately

Each lepton is given a **lepton number** of **+1**, but the **electron**, **muon** and **tau** types of lepton have to be **counted separately**.

You get **three different** lepton numbers, $\mathbf{L_e}$, $\mathbf{L_\mu}$ and $\mathbf{L_\tau}$.

Name	Symbol	Charge	L_e	L_μ	L_τ
electron	e^-	−1	+1	0	0
electron-neutrino	ν_e	0	+1	0	0
muon	μ^-	−1	0	+1	0
muon-neutrino	ν_μ	0	0	+1	0
tau	τ^-	−1	0	0	+1
tau-neutrino	ν_τ	0	0	0	+1

Like the baryon number, the lepton number is just the number of leptons.

Practice Questions

Q1 List the differences between a hadron and a lepton.

Q2 Which is the only stable baryon (probably)?

Q3 A particle collision at CERN produces 2 protons, 3 pions and 1 neutron.
What is the total baryon number of these particles?

Q4 Which two particles have lepton number $L_\tau = +1$?

Exam Questions

Q1 List all the decay products of the neutron.
Explain why this decay cannot be due to the strong interaction. [3 marks]

Q2 Initially, the muon was incorrectly identified as a meson.
Explain why the muon is not a meson. [3 marks]

Go back to the top of page 10 — do not pass GO, do not collect £200...

Do it. Go back and read it again. I promise — read these pages about 3 or 4 times and you'll start to see a pattern. There are hadrons that feel the force, leptons that don't. Hadrons are either baryons or mesons, and they're all weird except for those well-known baryons: protons and neutrons. There are loads of leptons, including good old electrons.

Quarks

*If you haven't read pages 8 to 12, do it now! For the rest of you — here are the **juicy bits** you've been waiting for. Particle physics makes **a lot more sense** when you look at quarks. More sense than it did before anyway.*

Quarks are **Fundamental Particles**

If that first sentence doesn't make much sense to you, read pages 8-12 — you have been warned... twice.

Quarks are the **building blocks** for **hadrons** (baryons and mesons).

1) To make **protons** and **neutrons** you only need two types of quark — the **up** quark (**u**) and the **down** quark (**d**).
2) An extra one called the **strange** quark (**s**) lets you make more particles with a property called **strangeness**.

Antiparticles of hadrons are made from **antiquarks**.

Quarks and **Antiquarks** have **Opposite Properties**

The **antiquarks** have **opposite properties** to the quarks — as you'd expect.

QUARKS

Name	Symbol	Charge	Baryon number	Strangeness
up	u	$+\frac{2}{3}$	$+\frac{1}{3}$	0
down	d	$-\frac{1}{3}$	$+\frac{1}{3}$	0
strange	s	$-\frac{1}{3}$	$+\frac{1}{3}$	–1

Particle physicists have found six different quarks altogether, but you only need to know about three of them.

ANTIQUARKS

Name	Symbol	Charge	Baryon number	Strangeness
anti-up	$\bar{u}$	$-\frac{2}{3}$	$-\frac{1}{3}$	0
anti-down	$\bar{d}$	$+\frac{1}{3}$	$-\frac{1}{3}$	0
anti-strange	$\bar{s}$	$+\frac{1}{3}$	$-\frac{1}{3}$	+1

Baryons are Made from **Three Quarks**

Evidence for quarks came from **hitting protons** with **high-energy electrons**.
The way the **electrons scattered** showed that there were **three concentrations of charge** (quarks) **inside** the proton.

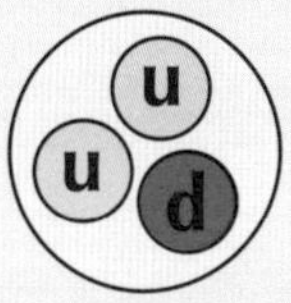

Proton = **uud**

Total charge
= 2/3 + 2/3 – 1/3 = 1

Baryon number
= 1/3 + 1/3 + 1/3 = 1

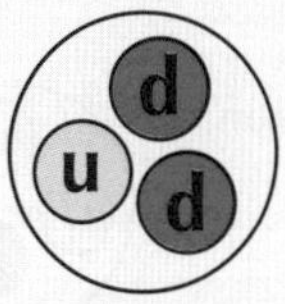

Neutron = **udd**

Total charge
= 2/3 – 1/3 – 1/3 = 0

Baryon number
= 1/3 + 1/3 + 1/3 = 1

Antiprotons are $\bar{u}\bar{u}\bar{d}$ and antineutrons are $\bar{u}\bar{d}\bar{d}$ — so no surprises there then.

Quarks

Mesons are a *Quark* and an *Antiquark*

Pions are just made from combinations of **up**, **down**, **anti-up** and **anti-down** quarks.
Kaons have **strangeness** so you need to put in **s** quarks as well (remember, the **s** quark has a strangeness of S = –1).

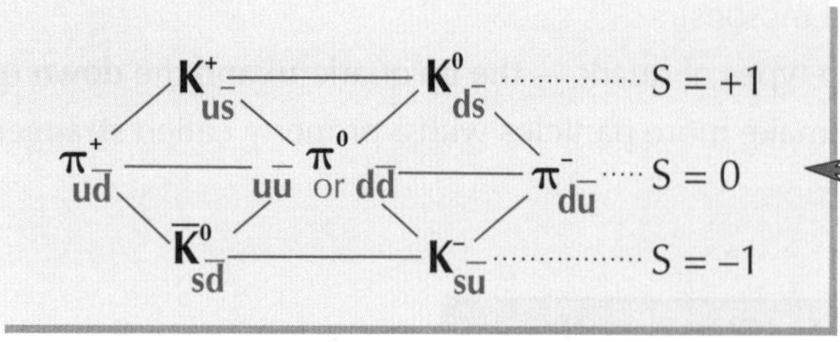

Physicists love patterns. Gaps in patterns like this predicted the existence of particles that were actually found later in experiments. Great stuff.

There's No Such Thing as a Free Quark

What if you **blasted** a **proton** with **enough energy** — could you **separate out** the quarks? Nope. Your energy just gets changed into more **quarks and antiquarks** — it's **pair production** again and you just make **mesons**. It's not possible to get a quark by itself — this is called **quark confinement**.

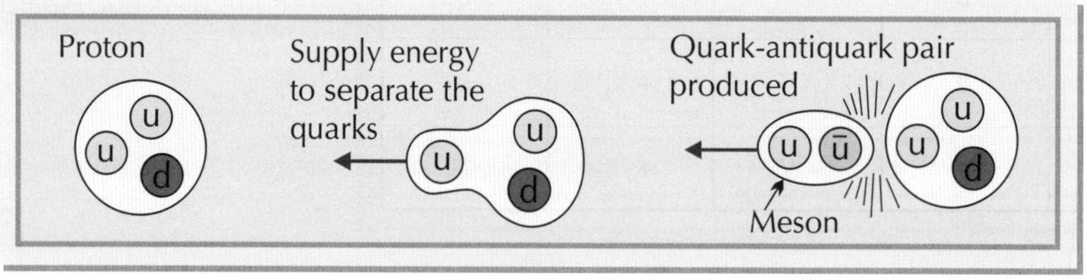

The Weak Interaction is something that Changes the Quark Type

In β⁻ decay a **neutron** is changed into a **proton** — in other words **udd** changes into **uud**. It means turning a **d** quark into a **u** quark. Only the weak interaction can do this.

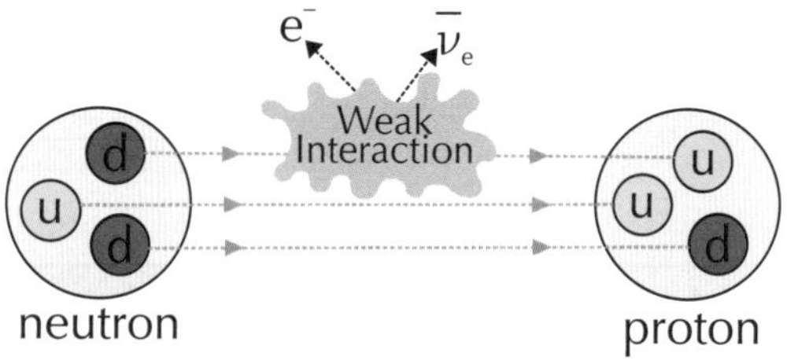

Some unstable isotopes like **carbon-11** decay by β⁺ emission. In this case a **proton** changes to a **neutron**, so a **u** quark changes to a **d** quark and we get:

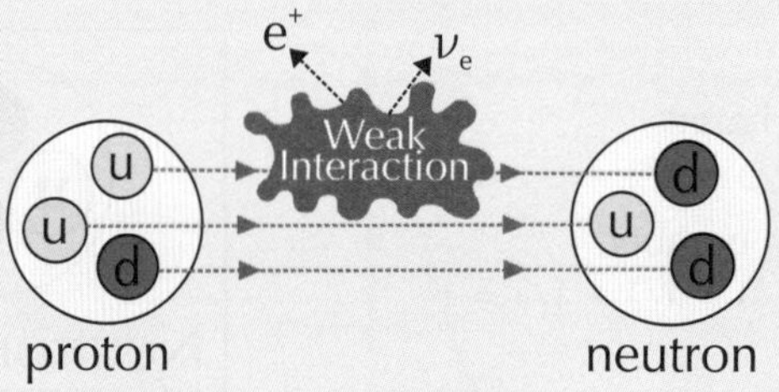

Quarks

Four Properties are Conserved in Particle Interactions

Charge is Always Conserved

In **any** particle interaction, the **total charge** after the interaction must equal the total charge before the interaction.

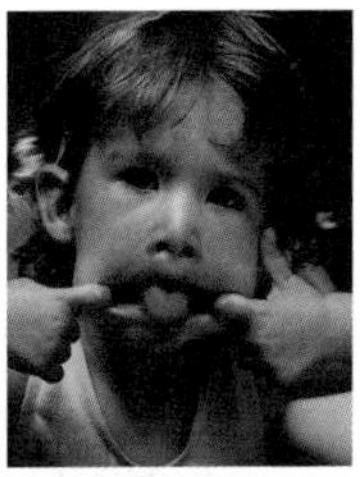

Dylan was committed to conserving strangeness.

Baryon Number is Always Conserved

Just like with charge, in **any** particle interaction, the **baryon number** after the interaction must equal the baryon number before the interaction.

Strangeness is Conserved in Strong Interactions

The **only** way to change the **type** of quark is with the **weak interaction**, so in strong interactions there has to be the same number of strange quarks at the beginning as at the end. The interaction $\mathbf{K^- + p \rightarrow n + \pi^0}$ is fine for **charge** and **baryon number** but not for **strangeness** — so it won't happen. The negative kaon has an **s** quark in it.

Conservation of Lepton Number is a Bit More Complicated

The **three types** of lepton number have to be conserved **separately**.

1) For example, the interaction $\pi^- \rightarrow \mu^- + \bar{\nu}_\mu$ has $L_\mu = 0$ at the start and $L_\mu = 1 - 1 = 0$ at the end, so it's OK.
2) On the other hand, the interaction $\nu_\mu + \mu^- \rightarrow e^- + \nu_e$ can't happen. At the start $L_\mu = 2$ and $L_e = 0$ but at the end $L_\mu = 0$ and $L_e = 2$.

Practice Questions

Q1 What is a quark?

Q2 Which type of particle is made from a quark and an antiquark?

Q3 Describe how a neutron is made up from quarks.

Q4 List four quantities that are conserved in particle interactions.

Exam Questions

Q1 Give the quark composition of the π^- and explain how the charges of the quarks give rise to its charge. [2 marks]

Q2 Explain how the quark composition is changed in the β^- decay of the neutron. [2 marks]

Q3 Give two reasons why the reaction $\mathbf{p + p \rightarrow p + K^+}$ does not happen. [2 marks]

A physical property called strangeness — how cool is that...

*True, there's a lot of information here, but this page really does **tie up** a lot of the stuff on the last few pages. Learn as much as you can from this double-page spread, then **go back** to page 8, and **work back** through to here. **Don't expect** to **understand** it all — but you will **definitely** find it **much easier to learn** when you can see how all the bits **fit in together**.*

Exchange Particles and Feynman Diagrams

*Having learnt about hadrons (baryons and mesons) and leptons, antiparticles and quarks, you now have the esteemed privilege of learning about yet another weirdy thing called a **gauge boson**. To the casual observer this might not seem **entirely fair**. And I have to say, I'd be with them.*

Forces are Caused by Particle Exchange

You can't have **instantaneous action at a distance** (according to Einstein, anyway). So, when two particles **interact**, something must **happen** to let one particle know that the other one's there. That's the idea behind **exchange particles**.

1) **Repulsion** — Each time the **ball** is **thrown or caught** the people get **pushed apart**. It happens because the ball carries **momentum**.

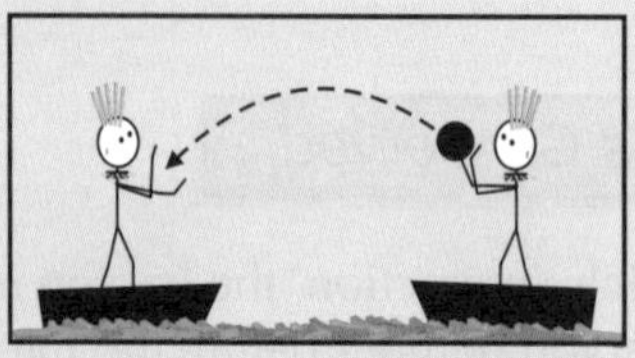

←REPULSION→

Particle exchange also explains **attraction**, but you need a bit more imagination.

2) **Attraction** — Each time the **boomerang** is **thrown or caught** the people **get pushed together**. (In real life, you'd probably fall in first.)

→ATTRACTION←

These exchange particles are called **gauge bosons**.

The **repulsion** between two **protons** is caused by the **exchange** of **virtual photons**, which are the gauge bosons of the **electromagnetic** force. Gauge bosons are **virtual** particles — they only exist for a **very short time**.

There are Four Fundamental Forces

All forces in nature are caused by four **fundamental** forces.
Each one has its **own gauge boson** and you have to learn their names:

Type of Interaction	Gauge Boson	Particles Affected
strong	gluon	hadrons only
electromagnetic	photon (symbol, γ)	charged particles only
weak	W^+, W^-, Z^0	all types
gravity	graviton?	all types

Particle physicists never **bother** about **gravity** because it's so incredibly **feeble** compared with the other types of interaction. Gravity only really **matters** when you've got **big masses** like **stars and planets**.

The graviton **may** exist but there's **no evidence** for it.

The Larger the Mass of the Gauge Boson, the Shorter the Range of the Force

1) The **W bosons** have a **mass** of about **100 times that of a proton**, which gives the weak force a **very short range**. Creating a **virtual W particle** uses **so much energy** that it can only exist for a **very short time** and it **can't travel far**.
2) On the other hand, the **photon** has **zero mass**, which gives you a force with **infinite range**.

Feynman Diagrams Show What's Going In and What's Coming Out

Richard Feynman was a brilliant physicist who was famous for explaining complicated ideas in a fun way that actually made sense. He worked out a really **neat way** of **solving problems** by **drawing pictures** rather than doing **calculations**.

1) **Gauge bosons** are represented by **wiggly lines** (technical term).
2) Other **particles** are represented by **straight lines**.

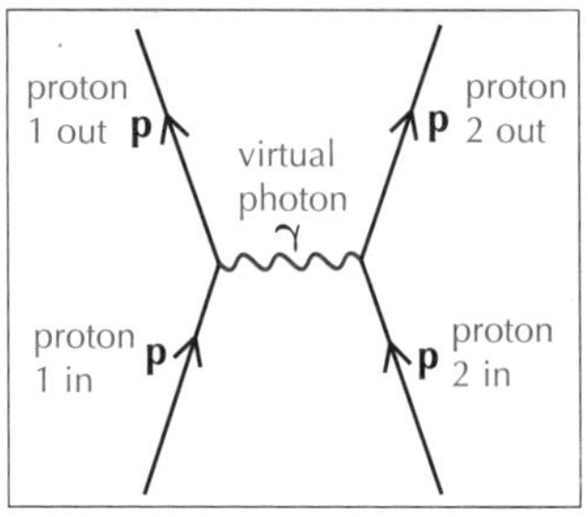

Exchange Particles and Feynman Diagrams

You can draw Feynman diagrams for **loads** of interactions but you **only** need to learn the ones on **this page** for your exam.

Beta-plus and *Beta-minus* Decay

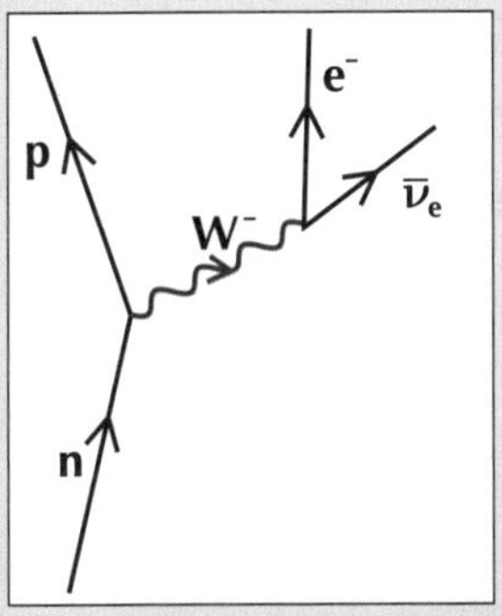

β^- decay

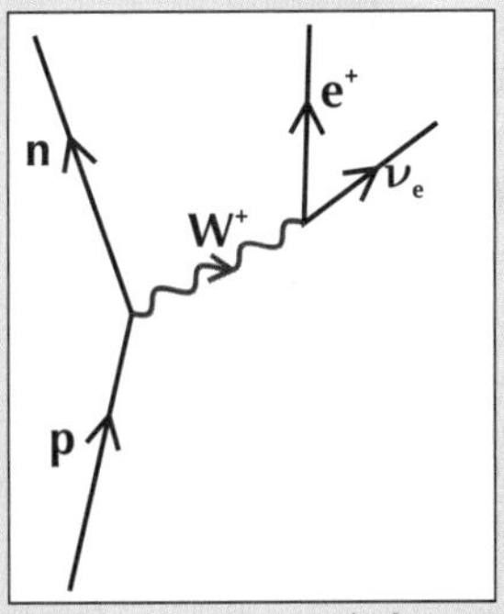

β^+ decay

$n \rightarrow p + e^- + \bar{\nu}_e$

$p \rightarrow n + e^+ + \nu_e$

You get an **antineutrino** in β^- decay and a **neutrino** in β^+ decay so that **lepton number** is conserved.

RULES FOR DRAWING FEYNMAN DIAGRAMS:

1) **Incoming** particles start at the bottom of the diagram and move upwards.
2) The **baryons** stay on one side of the diagram, and the **leptons** stay on the other side.
3) The **W** bosons carry **charge** from one side of the diagram to the other — make sure charges balance.
4) A W^- particle going to the **left** has the same effect as a W^+ particle going to the **right**.

A Proton Capturing an Electron

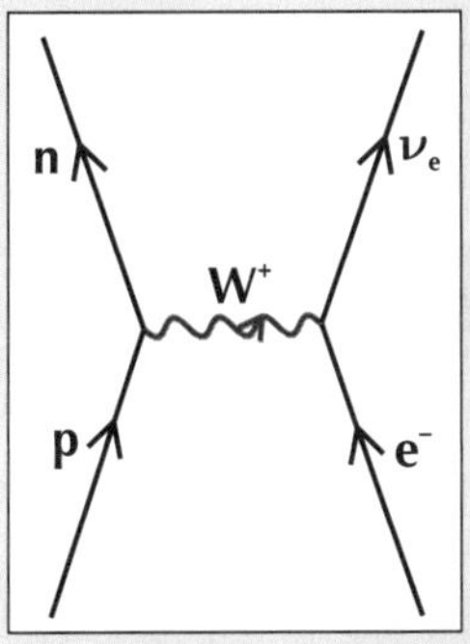

electron capture

Electrons and protons are of course attracted by the **electromagnetic interaction** between them, but if they **collide** the **weak interaction** can make this interaction happen.

$p + e^- \rightarrow n + \nu_e$

Neutrinos Interacting with Matter

There's a very **low probability** of a **neutrino interacting with matter**, but here's what happens when they do.

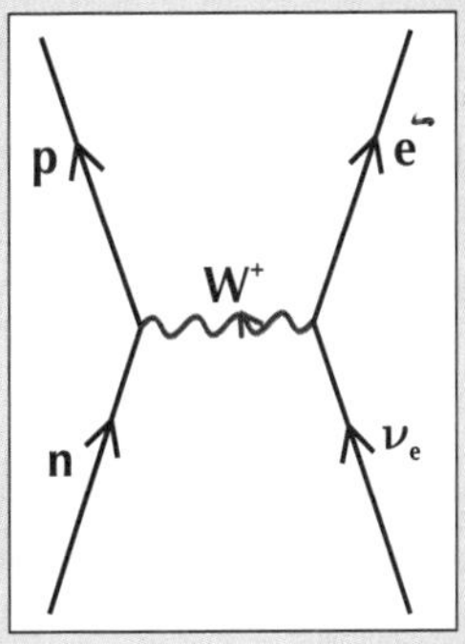

neutron-neutrino collision

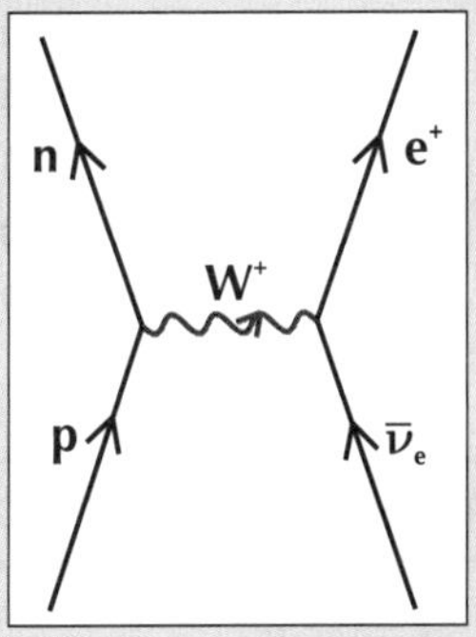

proton-antineutrino collision

$n + \nu_e \rightarrow p + e^-$

$p + \bar{\nu}_e \rightarrow n + e^+$

Practice Questions

Q1 List the four fundamental forces in nature.

Q2 Explain what a virtual particle is.

Q3 Draw the Feynman diagram for a neutrino-neutron interaction.

Q4 Which gauge bosons are exchanged in weak interactions?

Exam Questions

Q1 How is the force of electromagnetic repulsion between two protons explained by particle exchange? [2 marks]

Q2 Draw a Feynman diagram for the collision between an electron-antineutrino and a proton. Label the particles and state clearly which type of interaction is involved. [3 marks]

I need a drink...

Urrrgghhhh... eyes... glazed... brain... melting... ears... bleeding... help me... help me...

help me...

The Photoelectric Effect

I think they should rename 'the photoelectric effect' as 'the piece-of-cake effect' — it's not easy, I just like cake.

Shining Light on a Metal can Release Electrons

If you shine **light** of a **high enough frequency** onto the **surface of a metal**, it will **emit electrons**. For **most** metals, this **frequency** falls in the **U.V.** range.

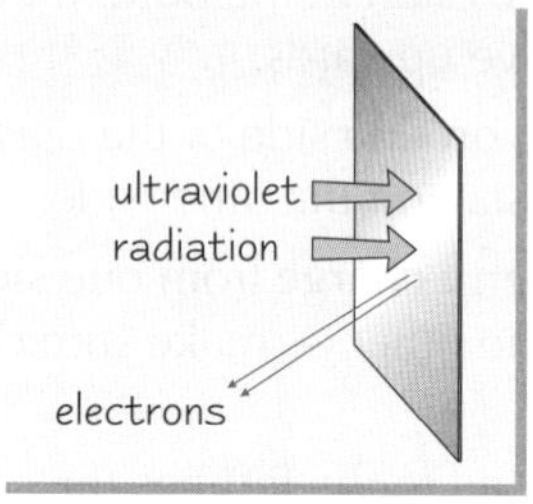

1) **Free electrons** on the **surface** of the metal **absorb energy** from the light, making them **vibrate**.
2) If an electron **absorbs enough** energy, the **bonds** holding it to the metal **break** and the electron is **released**.
3) This is called the **photoelectric effect** and the electrons emitted are called **photoelectrons**.

You don't need to know the details of any experiments on this, you just need to learn the three main conclusions:

Conclusion 1	For a given metal, **no photoelectrons are emitted** if the radiation has a frequency **below** a certain value — called the **threshold frequency**.
Conclusion 2	The photoelectrons are emitted with a variety of kinetic energies ranging from zero to some maximum value. This value of **maximum kinetic energy** increases with the **frequency** of the radiation, and is **unaffected** by the **intensity** of the radiation.
Conclusion 3	The **number** of photoelectrons emitted per second is **proportional** to the **intensity** of the radiation.

These are the two that had scientists puzzled. They can't be explained using wave theory.

The Photoelectric Effect Couldn't be Explained by Wave Theory

According to wave theory:

1) For a particular frequency of light, the **energy** carried is **proportional** to the **intensity** of the beam.
2) The energy carried by the light would be **spread evenly** over the wavefront.
3) **Each** free electron on the surface of the metal would gain a **bit of energy** from each incoming wave.
4) Gradually, each electron would gain **enough energy** to leave the metal.

SO... If the light had a **lower frequency** (i.e. was carrying less energy) it would take **longer** for the electrons to gain enough energy — but it would happen eventually. There is **no explanation** for the **threshold frequency**.
The **higher the intensity** of the wave, the **more energy** it should transfer to each electron — the kinetic energy should increase with **intensity**. There's **no explanation** for the **kinetic energy** depending only on the **frequency**.

Einstein came up with the Photon Model of Light

1) When Max Planck was investigating **black body radiation** (don't worry, you don't need to know about that just yet), he suggested that **EM waves** can **only** be **released** in **discrete packets**, or **quanta**.
2) The **energy carried** by one of these **wave-packets** had to be:

$$E = hf = \frac{hc}{\lambda}$$

where **h** = Planck's Constant = 6.63×10^{-34} Js
and **c** = speed of light in a vacuum = 3.00×10^{8} ms^{-1}

3) **Einstein** went **further** by suggesting that **EM waves** (and the energy they carry) can only **exist** in discrete packets. He called these wave-packets **photons**.
4) He saw these photons of light as having a **one-on-one**, **particle-like** interaction with **an electron** in a **metal surface**. It would **transfer all** its **energy** to that **one**, **specific electron**.

The Photoelectric Effect

The **Photon Model** Explained the **Photoelectric Effect** Nicely

According to the photon model:

1) When light hits its surface, the metal is **bombarded** by photons.
2) If one of these photons **collides** with a free electron, the electron will gain energy equal to **hf**.

Before an electron can **leave** the surface of the metal, it needs enough energy to **break the bonds holding it there**. This energy is called the **work function energy** (symbol ϕ (phi)) and its **value** depends on the **metal**.

*It Explains the **Threshold Frequency**...*

1) If the energy **gained** from the photon is **greater** than the **work function energy**, the electron is **emitted**.
2) If it **isn't**, the electron will just **shake about a bit**, then release the energy as another photon. The metal will heat up, but **no electrons** will be emitted.
3) Since, for **electrons** to be released, **hf** $\geq \phi$, the **threshold frequency** must be:

$$f = \frac{\phi}{h}$$

In theory, if a second photon hit an electron before it released the energy from the first, it could gain enough to leave the metal. This would have to happen very quickly though. An electron releases any excess energy after about 10^{-8} s. That's 0.000 000 01 s — safe to say, the chances of that happening are pretty slim.

*... and the **Maximum Kinetic Energy***

1) The **energy transferred** to an electron is **hf**.
2) The **kinetic energy** it will be carrying when it **leaves** the metal is hf **minus** any energy it's **lost** on the way out (there are loads of ways it can do that, which explains the **range** of energies).
3) The **minimum** amount of energy it can lose is the **work function energy**, so the **maximum kinetic energy** is given by the equation:

$$hf = \phi + \frac{1}{2}mv_{max}^2$$

4) The **kinetic energy** of the electrons is **independent of the intensity**, because they can **only absorb one photon** at a time.

Practice Questions

Q1 Explain what the photoelectric effect is.

Q2 What three conclusions were drawn from detailed experimentation?

Q3 What is meant by the work function energy of a metal?

Q4 How is the maximum kinetic energy of a photoelectron calculated?

Exam Questions

Q1 An isolated zinc plate with neutral charge is exposed to high-frequency ultraviolet light.
State and explain the effect of the ultraviolet light on the charge of the plate. [2 marks]

Q2 Explain why photoelectric emission from a metal surface only occurs when the frequency of the incident radiation exceeds a certain threshold value. [2 marks]

I'm so glad we got that all cleared up...

Well, that's about as hard as it gets at AS. The most important bits here are why wave theory doesn't explain the phenomenon, and why the photon theory does. A good way to learn conceptual stuff like this is to try and explain it to someone else. Make sure you know how to use those formulae as well — they're not the most straightforward.

Energy Levels and Photon Emission

Quantum Theory doesn't really make much sense — to anyone. It works though, so it's hard to argue with.

Electrons in Atoms Exist in Discrete Energy Levels

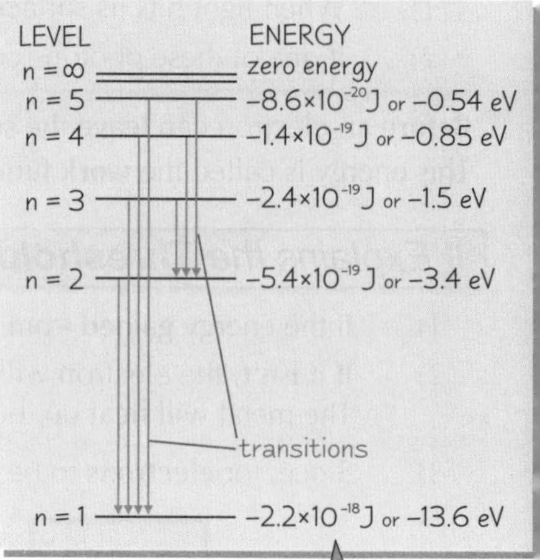

1) **Electrons** in an **atom** can **only exist** in certain **well-defined energy levels.** Each level is given a **number**, with **n = 1** representing the **ground state**.
2) Electrons can **move down** an energy level by **emitting** a **photon**.
3) Since these **transitions** are between **definite energy levels**, the **energy** of **each photon** emitted can **only** take a **certain allowed value**.
4) The diagram on the right shows the **energy levels** for **atomic hydrogen**.
5) The **energies involved** are **so tiny** that it makes sense to use a more **appropriate unit** than the **joule**. The **electronvolt (eV)** is defined as:

> The **kinetic energy carried** by an **electron** after it has been **accelerated** through a **potential difference** of **1 volt.**

energy gained by electron (eV) = accelerating voltage (V)

$1\text{ eV} = 1.6 \times 10^{-19}\text{ J}$

6) On the diagram, energies are labelled in **both units** for **comparison's** sake.
7) The **energy** carried by each **photon** is **equal** to the **difference in energies** between the **two levels**. The equation below shows a **transition** between levels **n = 2** and **n = 1**:

$$\Delta E = E_1 - E_2 = hf = \frac{hc}{\lambda}$$

The energies are only negative because of how "zero energy" is defined. Just one of those silly convention things — don't worry about it.

Fluorescent Tubes use Excited Electrons to Produce Light

1) **Fluorescent tubes** contain **mercury vapour**, across which a **high voltage** is applied.
2) When **electrons** in the **mercury collide** with **fast-moving free electrons** (accelerated by the **high voltage**), they're **excited** to a **higher energy level**.
3) When these **excited electrons** return to their **ground states**, they emit **photons** in the **UV** range.
4) A **phosphorus coating** on the **inside** of the tube **absorbs** these **photons**, exciting its **electrons** to **much higher orbits**. These electrons then **cascade** down the **energy levels**, **emitting** many **lower energy photons** in the form of **visible light**.

Fluorescent Tubes Produce Line Emission Spectra

1) If you **split** the light from a **fluorescent tube** with a **prism** or a **diffraction grating** (see pages 88-89), you get a **line spectrum**.
2) A line spectrum is seen as a **series** of **bright lines** against a **black background**.
3) Each **line** corresponds to a **particular wavelength** of light **emitted** by the source.
4) Since only **certain photon energies** are **allowed,** you only see the **corresponding wavelengths**.

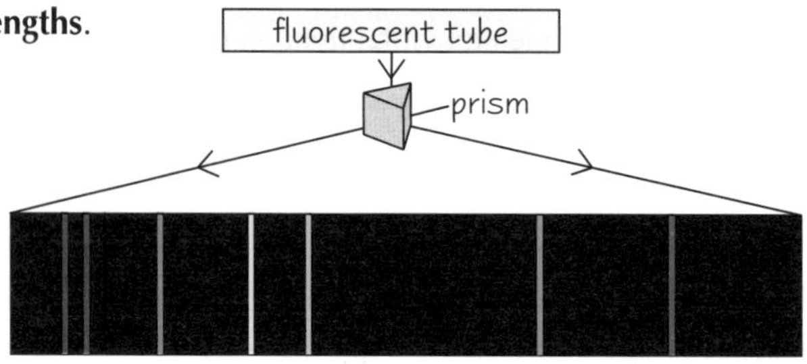

Energy Levels and Photon Emission

Shining **White Light** *through a* **Cool Gas** *gives an* **Absorption Spectrum**

Continuous Spectra Contain **All** Possible **Wavelengths**

1) The **spectrum** of **white light** is **continuous**.
2) If you **split** the **light** up with a **prism**, the **colours** all **merge** into each other — there **aren't** any **gaps** in the spectrum.
3) **Hot things** emit a **continuous spectrum** in the visible and infrared.

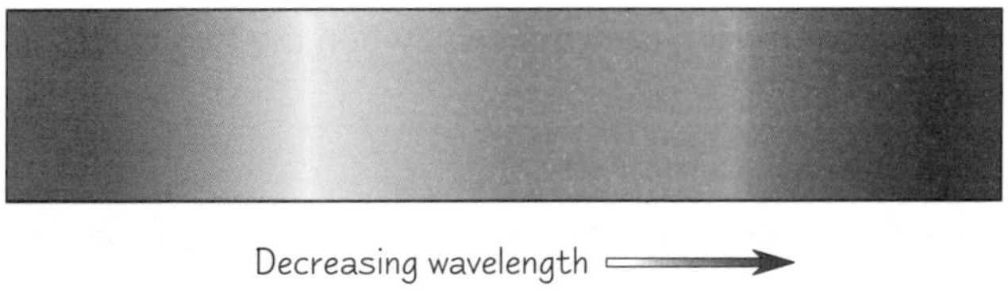

Cool Gases **Remove** Certain **Wavelengths** *from the Continuous Spectrum*

1) You get a **line absorption spectrum** when **light** with a **continuous spectrum** of **energy** (white light) passes through a cool gas.
2) At **low temperatures**, **most** of the **electrons** in the **gas atoms** will be in their **ground states**.
3) **Photons** of the **correct wavelength** are **absorbed** by the **electrons** to **excite** them to **higher energy levels**.
4) These **wavelengths** are then **missing** from the **continuous spectrum** when it **comes out** the other side of the gas.
5) You see a **continuous spectrum** with **black lines** in it corresponding to the **absorbed wavelengths**.
6) If you **compare** the **absorption** and **emission spectra** of a **particular gas**, the **black lines** in the **absorption spectrum match up** to the **bright lines** in the **emission spectrum**.

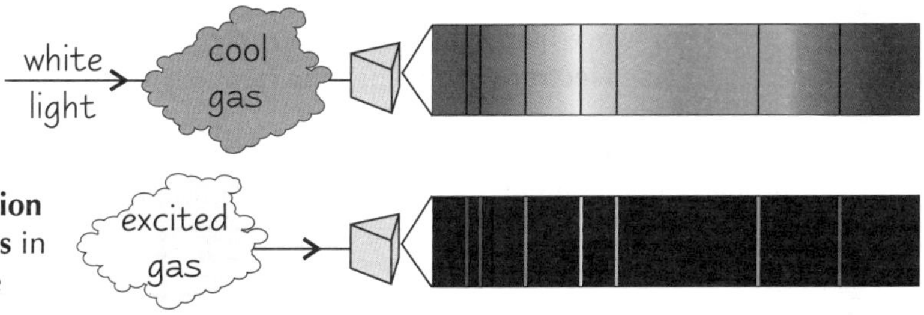

Practice Questions

Q1 Describe line absorption and line emission spectra. How are these two types of spectra produced?

Q2 Use the size of the energy level transitions involved to explain how the coating on a fluorescent tube converts UV into visible light.

Exam Question

Q1 An electron is accelerated through a potential difference of 12.1 V.
($e = -1.6 \times 10^{-9}$ C)

(a) How much kinetic energy has it gained in (i) eV and (ii) joules? [1 mark]

(b) This electron hits a hydrogen atom and excites it.
  (i) Explain what is meant by excitation. [1 mark]
  (ii) Using the energy values on the right, work out to which energy level the electron from the hydrogen atom is excited. [1 mark]
  (iii) Calculate the energies of the three photons that might be emitted as the electron returns to its ground state. [3 marks]

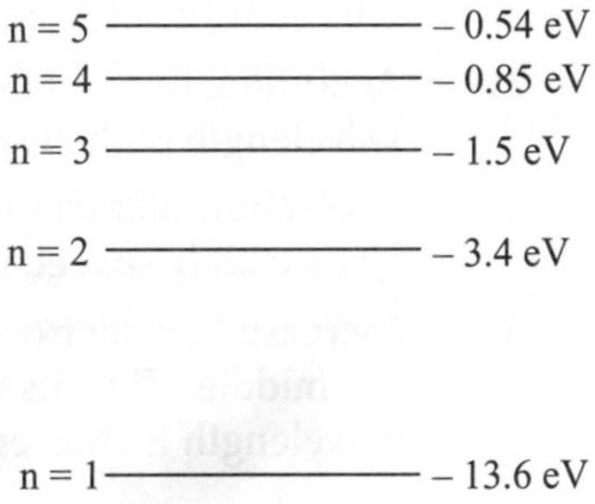

I can honestly say I've never got so excited that I've produced light...

This is heavy stuff, it really is. Quite interesting though, as I was just saying to Dom a moment ago. He's doing a psychology book. Psychology's probably quite interesting too — and easier. But it won't help you become an astrophysicist.

Wave-Particle Duality

Is it a wave? Is it a particle? No, it's a wave. No, it's a particle. No it's not, it's a wave. No don't be daft, it's a particle. (etc.

Interference *and* ***Diffraction*** *show* ***Light*** *as a* ***Wave***

1) Light produces **interference** and **diffraction** patterns — **alternating bands** of **dark** and **light**.
2) These can **only** be explained using **waves interfering constructively** (when two waves overlap in phase) or **interfering destructively** (when the two waves are out of phase). (See p.81.)

The ***Photoelectric Effect*** *Shows* ***Light*** *Behaving as a* ***Particle***

1) **Einstein** explained the results of **photoelectricity experiments** (see p.18) by thinking of the **beam of light** as a series of **particle-like photons**.
2) If a **photon** of light is a **discrete** bundle of energy, then it can **interact** with an **electron** in a **one-to-one way**.
3) **All** the **energy** in the **photon** is **given** to one **electron**.

De Broglie *Came up With the* ***Wave-Particle Duality Theory***

1) Louis de Broglie made a **bold suggestion** in his **PhD thesis**:

> If **'wave-like' light** showed **particle properties** (photons), **'particles'** like **electrons** should be expected to show **wave-like properties**.

2) The **de Broglie equation** relates a **wave property** (**wavelength**, λ) to a **moving particle property** (**momentum**, ***mv***). ***h*** = Planck's constant = 6.63×10^{-34} Js.

$$\lambda = \frac{h}{mv}$$

3) The **de Broglie wave** of a particle can be interpreted as a '**probability wave**'.
(The probability of finding a particle at a point is directly proportional to the square of the amplitude of the wave at that point — but you don't need to know that for your exam.)
4) Many physicists at the time **weren't very impressed** — his ideas were just **speculation**. But later experiments **confirmed** the wave nature of electrons.

I'm not impressed — this is just speculation. What do you think Dad?

Electron Diffraction *shows the* ***Wave Nature*** *of* ***Electrons***

1) **Diffraction patterns** are observed when **accelerated electrons** in a vacuum tube **interact** with the **spaces** in a graphite **crystal**.
2) This **confirms** that electrons show **wave-like** properties.
3) According to wave theory, the **spread** of the **lines** in the diffraction pattern **increases** if the **wavelength** of the wave is **greater**.
4) In electron diffraction experiments, a **smaller accelerating voltage**, i.e. **slower** electrons, gives **widely spaced** rings.
5) **Increase** the **electron speed** and the diffraction pattern circles **squash together** towards the **middle**. This fits in with the **de Broglie** equation above — if the **velocity** is **higher**, the **wavelength** is **shorter** and the **spread** of lines is **smaller**.

> In general, λ for **electrons** accelerated in a **vacuum tube** is about the **same size** as **electromagnetic waves** in the **X-ray** part of the spectrum.

Wave-Particle Duality

Particles Don't show Wave-Like Properties All the Time

You **only** get **diffraction** if a particle interacts with an object of about the **same size** as its **de Broglie wavelength**. A **tennis ball**, for example, with **mass 0.058 kg** and **speed 100 ms^{-1}** has a **de Broglie wavelength** of **10^{-34} m**. That's **10^{19} times smaller** than the **nucleus** of an **atom**! There's nothing that small for it to interact with.

Example An electron of mass 9.11×10^{-31} kg is fired from an electron gun at 7×10^{6} ms^{-1}. What size object will the electron need to interact with in order to diffract?

Momentum of electron = $mv = 6.38 \times 10^{-24}$ kg ms^{-1}

$\lambda = h/mv = 6.63 \times 10^{-34} / 6.38 \times 10^{-24} = 1 \times 10^{-10}$ m

Only crystals with atom layer spacing around this size are likely to cause the diffraction of this electron.

A **shorter wavelength** gives **less diffraction effects**. This fact is used in the **electron microscope**.

Diffraction effects **blur detail** on an image. If you want to **resolve tiny detail** in an **image**, you need a **shorter wavelength**. **Light** blurs out detail more than '**electron-waves**' do, so an **electron microscope** can resolve **finer detail** than a **light microscope**. They can let you look at things as tiny as a single strand of DNA... which is nice.

Practice Questions

Q1 Which observations show light to have a 'wave-like' character?

Q2 Which observations show light to have a 'particle' character?

Q3 What happens to the de Broglie wavelength of a particle if its velocity increases?

Q4 Which observations show electrons to have a 'wave-like' character?

Exam Questions

$h = 6.63 \times 10^{-34}$ Js ; $c = 3.00 \times 10^{8}$ ms^{-1}; electron mass = 9.11×10^{-31} kg ; proton mass = $1840 \times$ electron mass

Q1 (a) State what is meant by the wave-particle duality of electromagnetic radiation. [1 mark]

(b) (i) Calculate the energy in joules and in electronvolts of a photon of wavelength 590 nm. [3 marks]

(ii) Calculate the speed of an electron which will have the same wavelength as the photon in (b)(i). [2 marks]

Q2 Electrons travelling at a speed of 3.5×10^{6} ms^{-1} exhibit wave properties.

(a) Calculate the wavelength of these electrons. [2 marks]

(b) Calculate the speed of protons which would have the same wavelength as these electrons. [2 marks]

(c) Both electrons and protons were accelerated from rest by the same potential difference. Explain why they will have different wavelengths. (Hint: if they're accelerated by the same p.d., they have the same K.E.) [3 marks]

Q3 An electron is accelerated through a potential difference of 6.0 kV.

(a) Calculate its kinetic energy in joules, assuming no energy is lost in the process. [2 marks]

(b) Using the data above, calculate the speed of the electron. [2 marks]

(c) Calculate the de Broglie wavelength of the electron. [2 marks]

Don't hide your wave-particles under a bushel...

*Right — I think we'll all agree that quantum physics is a wee bit strange when you come to think about it. What it's saying is that electrons and photons aren't really waves, and they aren't really particles — they're **both**... at the **same time**. It's what quantum physicists like to call a 'juxtaposition of states'. Well they would, wouldn't they...*

Current, Potential Difference and Resistance

You wouldn't reckon there was that much to know about electricity... just plug something in, and bosh — electricity. Ah well, never mind the age of innocence — here are all the gory details...

Current is the Rate of Flow of Charge

1) The **current** in a **wire** is like **water** flowing in a **pipe**. The **amount** of water that flows depends on the **flow rate** and the **time**. It's the same with electricity — **current is the rate of flow of charge**.

$$\Delta Q = I\Delta t \quad \text{or} \quad I = \frac{\Delta Q}{\Delta t}$$

Where ΔQ is the charge in coulombs, I is the current and Δt is the time taken.

Remember that conventional current flows from + to -, the opposite way from electron flow.

2) The **coulomb** is the unit of **charge**.

One **coulomb** (**C**) is defined as the **amount of charge** that passes in **1 second** when the **current** is **1 ampere**.

3) You can measure the current flowing through a part of a circuit using an **ammeter**. Remember — you always need to attach an ammeter in **series** (so that the current through the ammeter is the same as the current through the component — see page 34).

Potential Difference is the Energy per Unit Charge

1) To make electric charge flow through a conductor, you need to do work on it.
2) **Potential difference** (p.d.), or **voltage**, is defined as the **energy converted per unit charge moved**.

$$V = \frac{W}{Q}$$

W is the energy in joules. It's the work you do moving the charge.

Back to the 'water analogy' again. The p.d. is like the pressure that's forcing water along the pipe.

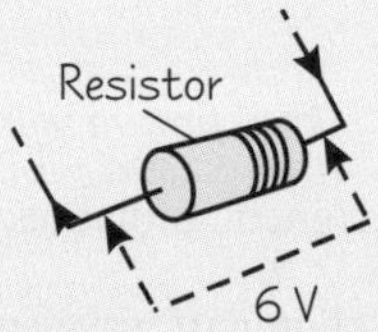

Here you do 6 J of work moving each coulomb of charge through the resistor, so the p.d. across it is 6 V. The energy gets converted to heat.

Definition of the Volt

The **potential difference** across a component is **1 volt** when you convert **1 joule** of energy moving **1 coulomb** of charge through the component.

$$1\ \text{V} = 1\ \text{J}\text{C}^{-1}$$

Everything has Resistance

1) If you put a **potential difference** (p.d.) across an **electrical component**, a **current** will flow.
2) **How much** current you get for a particular **p.d.** depends on the **resistance** of the component.
3) You can think of a component's **resistance** as a **measure** of how **difficult** it is to get a **current** to **flow** through it.

Mathematically, **resistance** is: $R = \frac{V}{I}$

This equation **defines** resistance.

4) **Resistance** is measured in **ohms** (**Ω**)

A component has a resistance of **1 Ω** if a **potential difference** of **1 V** makes a **current** of **1 A** flow through it.

Current, Potential Difference and Resistance

For an *Ohmic Conductor*, *R* is a *Constant*

1) A chap called **Ohm** did most of the early work on resistance. He developed a rule to **predict** how the **current** would **change** as the applied **potential difference increased**, for **certain types** of conductor.
2) The rule is now called **Ohm's law** and the conductors that **obey** it (mostly metals) are called **ohmic conductors**.

Provided the **temperature** is **constant**, the **current** through an ohmic conductor is **directly proportional** to the **potential difference** across it.

$I \propto V$

1) This graph shows what happens if you plot voltage against current for an ohmic conductor.
2) As you can see it's a **straight-line** graph — **doubling** the **p.d. doubles** the **current**.
3) What this means is that the **resistance** is **constant** — the gradient is always a fixed value.
4) Often **factors** such as **light level** or **temperature** will have a **significant effect** on resistance (the resistivity changes), so you need to remember that Ohm's law is **only** true for **ohmic conductors** at **constant temperature**.
5) Ohm's law is a **special case** — lots of components aren't ohmic conductors and have characteristic current-voltage (I–V) graphs of their very own (see page 26).

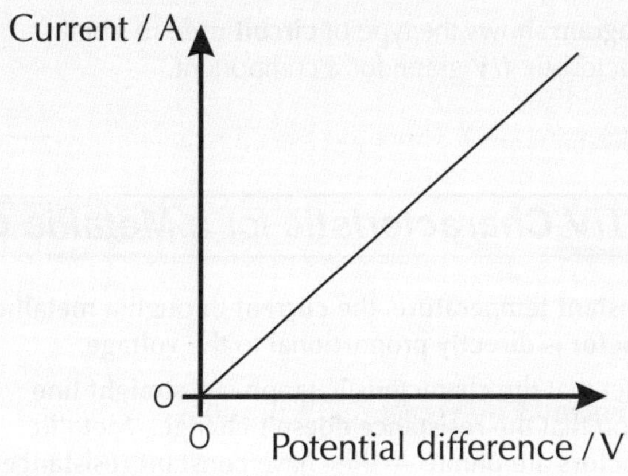

Practice Questions

Q1 Describe in words how current and charge are related.

Q2 Define the coulomb.

Q3 Define potential difference.

Q4 Name one environmental factor likely to alter the resistance of a component.

Q5 What is special about an ohmic conductor?

Exam Questions

Q1 A battery delivers 4500 C of electric charge to a circuit in 10 minutes. Calculate the average current. [2 marks]

Q2 An electric motor runs off a 12 V d.c. supply and has an overall efficiency of 75%.
Calculate how much electric charge will pass through the motor when it does 90 J of work. [3 marks]

Q3 A current of 60 amps flows through an ohmic conductor when a potential difference of 10 V is applied across it.
Assume the temperature of the conductor remains constant.

(a) What is the resistance of the conductor? [1 mark]

(b) Find the current flowing through the conductor when the potential difference across it is 35 V. [1 mark]

(c) Sketch the I-V graph for the conductor when potential differences of up to 35 V are applied to it. [1 mark]

I can't even be bothered to make the current joke...

Talking of currant jokes, I saw this bottle of wine the other day called 'raisin d'être' — 'raison d'être' of course meaning 'reason for living', but spelled slightly different to make 'raisin', meaning 'grape'. Ho ho. Chuckled all the way out of Tesco.

I/V Characteristics

Woohoo — real physics. This stuff's actually kind of interesting.

I/V Graphs Show how Resistance Varies

The term '***I/V* characteristic**' refers to a **graph** which shows how the **current** (***I***) flowing through a **component changes** as the **potential difference** (***V***) across it is increased.

The **shallower** the **gradient** of a characteristic ***I/V*** graph, the **greater** the **resistance** of the component.

A **curve** shows that the resistance is **changing**.

The **diagram** shows the type of **circuit** used to obtain a characteristic ***I/V*** graph for a component.

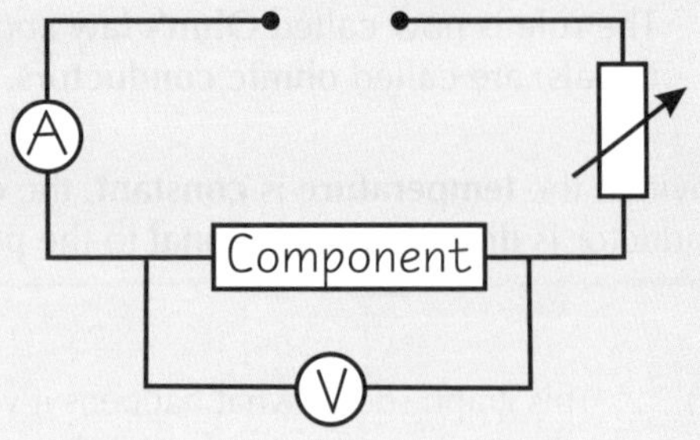

The *I/V* Characteristic for a Metallic Conductor is a Straight Line

At **constant temperature**, the **current** through a **metallic conductor** is **directly proportional** to the **voltage**.

The fact that the characteristic graph is a **straight line** tells you that the **resistance doesn't change**. **Metallic conductors** are **ohmic** — they have **constant resistance provided** their temperature doesn't change.

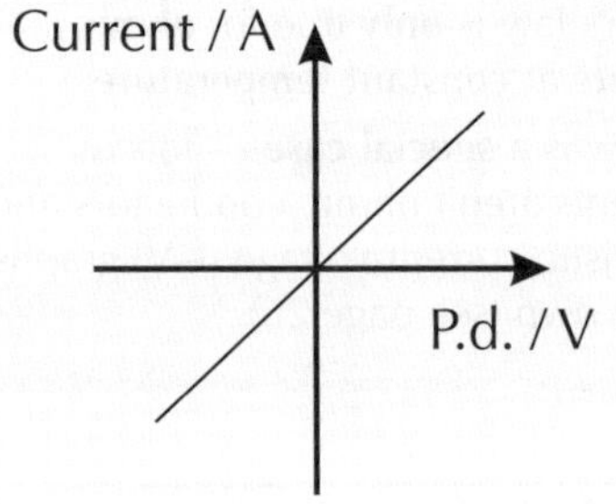

The *I/V* Characteristic for a Filament Lamp is a Curve

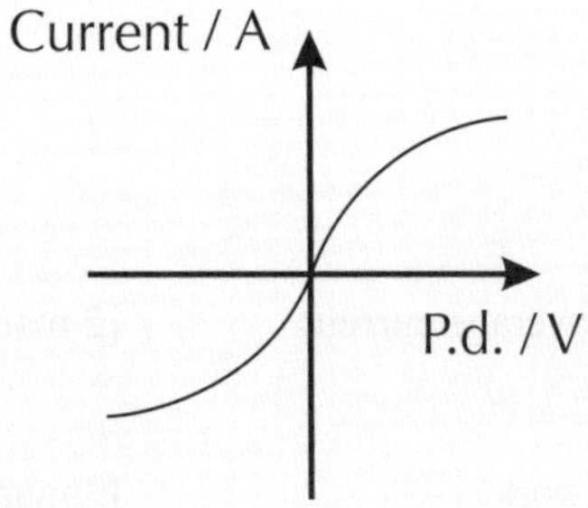

The characteristic graph for a **filament lamp** is a **curve** that starts **steep** but gets **shallower** as the **voltage rises**.

The **filament** in a lamp is just a **coiled-up** length of **metal wire**, so you might think it should have the **same characteristic graph** as a **metallic conductor**. It doesn't because it **gets hot**. **Current** flowing through the lamp **increases** its **temperature**.

The **resistance** of a **metal increases** as the **temperature increases**.

Semiconductors are Used in Sensors

Semiconductors are **nowhere near** as good at **conducting** electricity as **metals**. This is because there are far, far **fewer charge carriers** available. However, if **energy** is supplied to the semiconductor, **more charge carriers** can be **released**. This means that they make **excellent sensors** for detecting **changes** in their **environment**.

You need to know about the **three** semiconductor components on the next page — **thermistors**, **LDRs** and **diodes**.

I/V Characteristics

The Resistance of a Thermistor Depends on Temperature

Thermistor circuit symbol:

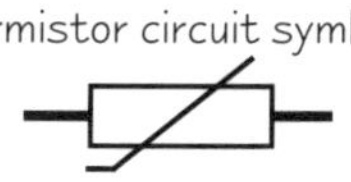

A **thermistor** is a **resistor** with a **resistance** that depends on its **temperature**. You only need to know a bout **NTC** thermistors — NTC stands for 'Negative Temperature Coefficient'. This means that the **resistance decreases** as the **temperature goes up**. The characteristic graph for an NTC thermistor curves upwards.

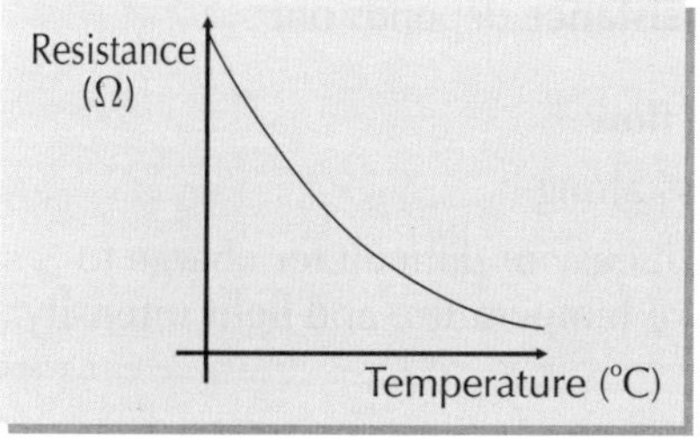

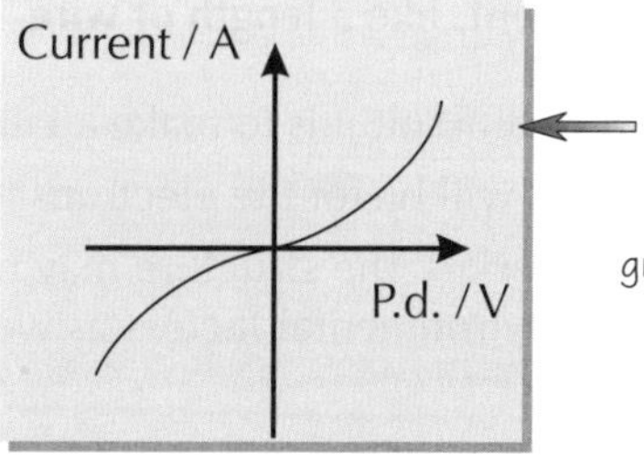

Increasing the current through the thermistor increases its temperature. The increasing gradient of this characteristic graph tells you that the resistance is decreasing.

Warming the thermistor gives more **electrons** enough **energy** to **escape** from their atoms. This means that there are **more charge carriers** available, so the resistance is lower.

The Resistance of an LDR depends on Light Intensity

LDR circuit symbol:

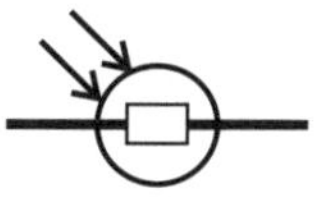

LDR stands for **Light-Dependent Resistor**. The **greater** the intensity of **light** shining on an LDR, the **lower** its **resistance**.

The explanation for this is similar to that for the thermistor. In this case, **light** provides the **energy** that releases more electrons. More charge carriers means a lower resistance.

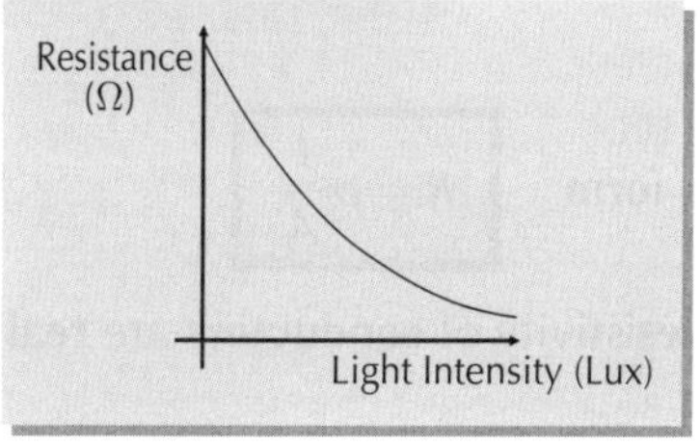

Large Dayglow Rabbit

Diodes Only Let Current Flow in One Direction

Diode and LED circuit symbols:

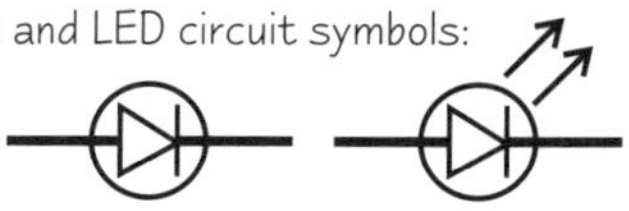

diode LED

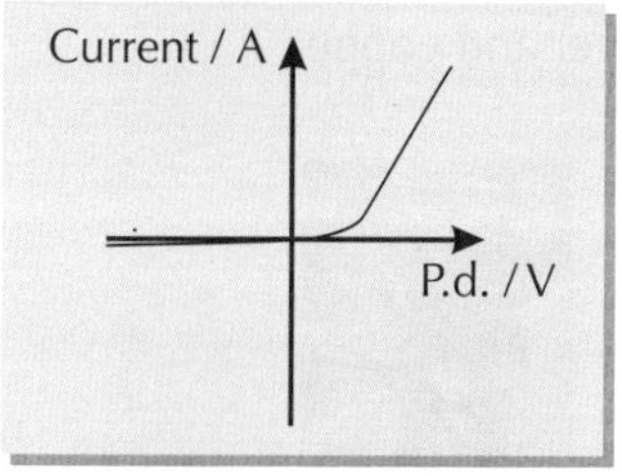

Diodes (including light emitting diodes (LEDs)) are designed to let **current flow** in **one direction** only. You don't need to be able to explain how they work, just what they do.

1) **Forward bias** is the **direction** in which the **current** is **allowed to flow**.
2) **Most** diodes require a **threshold voltage** of about **0.6 V** in the **forward direction** before they will conduct.
3) In **reverse bias**, the **resistance** of the diode is **very high** and the current that flows is **very tiny**.

Practice Questions

Q1 Sketch the circuit used to determine the ***I/V*** characteristics of a component.

Q2 Draw an ***I/V*** characteristic graph for a diode.

Q3 What is an LDR?

Q4 If an ***I/V*** graph is curved, what does this tell you about the resistance?

Exam Question

Q1 (a) Sketch a characteristic ***I/V*** graph for a filament lamp. [1 mark]

(b) State how the resistance of the filament changes as the temperature increases. [1 mark]

Thermistor man — temperature-dependent Mr Man...

Learn the graphs on this page, and make sure you can explain them. Whether it's a light-dependent resistor or a thermistor, the same principle applies. More energy releases more charge carriers, and more charge carriers means a lower resistance.

Resistivity and Superconductivity

Resistance is all well and good, but it depends on the size of the thing doing the resisting. If you want to compare two materials, you need a quantity that doesn't depend on size. Enter 'resistivity'...

Three Things Determine Resistance

If you think about a nice, **simple electrical component**, like a **length of wire**, its **resistance** depends on:

1) **Length** (l). The **longer** the wire, the **more difficult** it is to make a **current flow**.
2) **Area** (A). The **wider** the wire, the **easier** it will be for the electrons to pass along it.
3) **Resistivity** (ρ), which **depends** on the **material**. The **structure** may make it easy or difficult for charge to flow. In general, resistivity depends on **environmental factors** as well, like **temperature** and **light intensity**.

The **resistivity** of a material is defined as the **resistance** of a **1 m length** with a **1 m² cross-sectional area**. It is measured in **ohm-metres** (Ωm).

This is the Greek letter rho, the symbol for resistivity.

$$\rho = \frac{RA}{l}$$

where A = cross-sectional area in m², l = length in m

You'll more **usually** see the equation in the **form**:

$$R = \rho \frac{l}{A}$$

Typical values for the **resistivity** of **conductors** are **really small**, e.g., for **copper** (at 25 °C) $\rho = 1.72 \times 10^{-8}$ Ωm.

Superconductors Have Zero Resistivity

1) Normally, all materials have **some resistivity** — even really good conductors like silver and copper.
2) That resistance means that whenever electricity flows through them, they **heat up**, and some of the electrical energy is **wasted** as heat.
3) You can **lower** the resistivity of many materials like metals by **cooling them down**.
4) If you **cool** some materials down to below a '**transition temperature**', their **resistivity disappears entirely** and they become a **superconductor**.
5) Without any resistance, **none** of the electrical energy is turned into heat, so **none** of it's wasted. That means you can start a current flowing in a circuit using a magnetic field, take away the magnet and the current would carry on flowing **forever**. Neat.
6) There's a catch, though. Most 'normal' conductors, e.g. metals, have transition temperatures below **10 kelvin** (**–263 °C**). Getting things that cold is **hard**, and **really expensive**.
7) Solid-state physicists all over the world are trying to develop **room-temperature superconductors**. So far, they've managed to get some weird **metal oxide** things to superconduct at about **140 K** (**–133 °C**), which is a much easier temperature to get down to. They've still got a long way to go though.

I couldn't find a conductor, so you'll have to make do with this instead.

Uses of Superconductors

Using superconducting wires you could make:

1) **Power cables** that transmit electricity without any **loss** of power.
2) Really **strong electromagnets** that **don't** need a constant power source (for use in medical applications and Maglev trains).
3) **Electronic circuits** that work really **fast**, because there's no resistance to slow them down.

Resistivity and Superconductivity

Get used to *Using the Resistivity Formula*

Jess tested a piece of Mr T's bling to see if it was made of real gold. A cross-section of the medallion is shown below. Jess found its length to be 0.5 cm and its resistance to be $1.22 \times 10^{-8}\,\Omega$. The resistivity of gold is $2.44 \times 10^{-8}\,\Omega$m. Is the medallion made of gold?

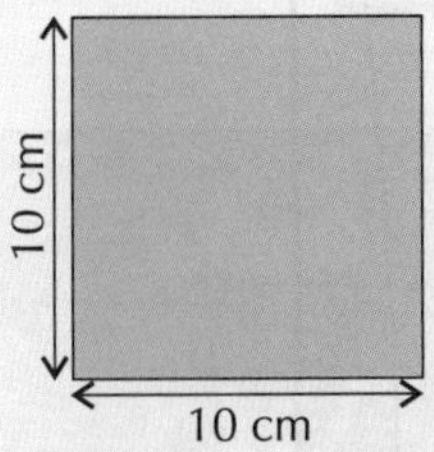

Convert all lengths into metres.

The cross-sectional area of the medallion = 0.1 m × 0.1 m = 0.01 m^2

Length = l = 0.005 m

The resistivity of the medallion $= \rho = \frac{RA}{l} = \frac{(1.22 \times 10^{-8} \times 0.01)}{0.005} = \mathbf{2.44 \times 10^{-8}\ \Omega m}$

So Mr T's bling is the real deal.

The heating element in a toaster is a bare nichrome wire with a radius of 0.2 mm and a length of 0.8 m. Find the resistance of the wire. (Resistivity of nichrome = $1.10 \times 10^{-6}\,\Omega$m.)

First of all, you need to find the cross-sectional area of the wire, A, in m^2.
Assuming the wire is cylindrical, its cross-sectional area will be the area of a circle, πr^2.

Wire radius = 0.2 mm = 2.0×10^{-4} m, so $A = \pi(2.0 \times 10^{-4})^2 = 1.26 \times 10^{-7}$ m^2 (3 s.f.)

Rearrange the resistivity equation to get $R = \rho\frac{l}{A}$,

and plug in the numbers: $R = 1.10 \times 10^{-6}\left(\frac{0.8}{1.26 \times 10^{-7}}\right) = \underline{7.0\ \Omega}$ (2 s.f.)

Practice Questions

Q1 What happens to a superconductor when it's cooled to its transition temperature?

Q2 What three factors does the resistance of a length of wire depend on?

Q3 What are the units for resistivity?

Exam Questions

Q1 Aluminium has a resistivity of $2.8 \times 10^{-8}\,\Omega$m at 20 °C and a transition temperature of 1.2 K.

(a) Calculate the resistance of a pure aluminium wire of length 4 m and diameter 1 mm, at 20 °C. [3 marks]

(b) The wire is cooled to a temperature of 1 K. What is its resistance now? Explain your answer. [2 marks]

Q2 The table below shows some measurements taken by a student during an experiment investigating an unknown electrical component.

Potential Difference (V)	Current (mA)
2.0	2.67
7.0	9.33
11.0	14.67

(a) Use the first row of the table to calculate the resistance of the component when a p.d. of 2 V is applied. [2 marks]

(b) By means of further calculation, or otherwise, decide whether the component is an ohmic conductor. [3 marks]

Superconductors and Johnny Depp — just too cool to resist...

Superconducting electromagnets are used in magnetic resonance imaging (MRI) scanners in hospitals. That way, the huge magnetic fields they need can be generated without using up a load of electricity. Great stuff...

Electrical Energy and Power

Power and energy are pretty familiar concepts — and here they are again. Same principles, just different equations.

Power is the Rate of Transfer of Energy

Power (*P*) is **defined** as the **rate** of **transfer** of **energy**.
It's measured in **watts** (*W*), where **1 watt** is equivalent to **1 joule per second**.

or $P = \frac{E}{t}$

There's a really simple formula for **power** in **electrical circuits**:

$P = VI$

This makes sense, since:

1) **Potential difference** (*V*) is defined as the **energy transferred** per **coulomb**.
2) **Current** (*I*) is defined as the **number** of **coulombs** transferred per **second**.
3) So **p.d.** × **current** is **energy transferred per second**, i.e. **power**.

He didn't know when, he didn't know where... but one day this PEt would get his revenge.

You know from the definition of **resistance** that: $V = IR$

Combining the **two equations** gives you loads of **different ways** to **calculate power**.

$P = VI$ $\quad P = \frac{V^2}{R}$ $\quad P = I^2R$

Obviously, which equation you should use depends on what **quantities** you're given in the **question**.

Phew... that's quite a few equations to learn and love. And as if they're not exciting enough, here are some examples to get your teeth into...

Example 1

A 24 W car headlamp is connected to a 12 V car battery.
Assume the wires connecting the lamp to the battery have negligible resistance.
(a) How much energy will the lamp convert into light and heat energy in 2 hours?
(b) Find the total resistance of the lamp.

(a) Number of seconds in 2 hours = 120 × 60 = 7200 s
$E = P \times t = 24 \times 7200 = 172\,800$ J = **172.8 kJ**

(b) Rearrange the equation $P = \frac{V^2}{R}$, $R = \frac{V^2}{P} = \frac{12^2}{24} = \frac{144}{24} = \underline{6\,\Omega}$

Example 2

A robotic mutant Santa from the future converts 750 J of electrical energy into heat every second.
(a) What is the power rating of the robotic mutant Santa?
(b) All of the robotic mutant Santa's components are connected in series, with a total resistance of 30 Ω. What current flows through his wire veins?

(a) Power = $E \div t = 750 \div 1 = \underline{750\text{ W}}$

(b) Rearrange the equation $P = I^2R$, $I = \sqrt{\frac{P}{R}} = \sqrt{\frac{750}{30}} = \sqrt{25} = \underline{5\text{ A}}$

Electrical Energy and Power

Energy is Easy to *Calculate* if you Know the *Power*

Sometimes it's the **total energy** transferred that you're interested in. In this case you simply need to **multiply** the **power** by the **time**. So:

$E = VIt$ (or $E = \frac{V^2}{R}t$ or $E = I^2Rt$)

You've got to make sure that the time is in seconds.

Example

Betty pops the kettle on to make a brew.
It takes 4.5 minutes for the kettle to boil the water inside it.
A current of 4 A flows through the kettle's heating element once it is connected to the mains (230 V).

(a) What is the power rating of the kettle?

(b) How much energy does the kettle's heating element transfer to the water in the time it takes to boil?

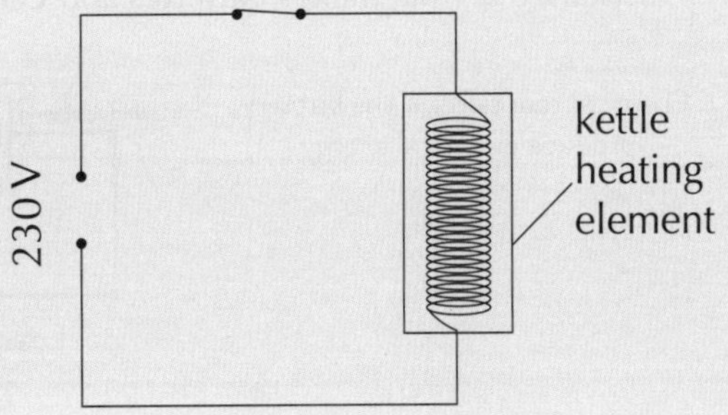

(a) Use $P = V \times I = 230 \times 4 =$ **920 W**

(b) Time the kettle takes to boil in seconds = 4.5 × 60 = 270 s.
Use the equation $E = Pt$ and your answer to part (a): $E = 920 \times 270 = 248\,400$ J = **248.4 kJ**

Practice Questions

Q1 Write down the equation linking power, current and resistance.

Q2 What equation links power, voltage and resistance?

Q3 Power is measured in watts. What is 1 watt equivalent to?

Exam Questions

Q1 This question concerns a mains-powered hairdryer, the circuit diagram for which is given below.

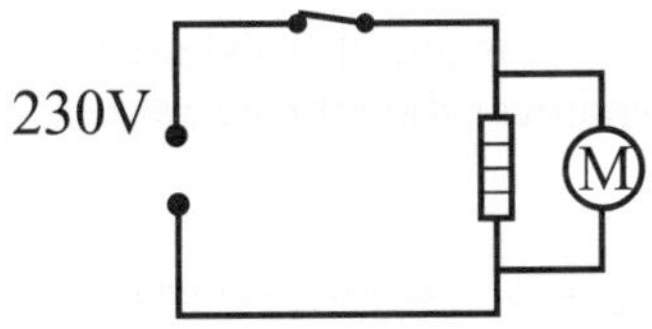

(a) The heater has a power of 920 W in normal operation. Calculate the current in the heater. [2 marks]

(b) The motor has a resistance of 190 Ω.
What current will flow through the motor when the hairdryer is in use? [2 marks]

(c) Show that the total power of the hairdryer in normal operation is just under 1.2 kW. [3 marks]

Q2 A 12 V car battery supplies a current of 48 A for 2 seconds to the car's starter motor.
The total resistance of the connecting wires is 0.01 Ω.

(a) Calculate the energy transferred from the battery. [2 marks]

(b) Calculate the energy wasted as heat in the wires. [2 marks]

Ultimate cosmic powers...

Whenever you get equations in this book, you know you're gonna have to learn them. Fact of life. I used to find it helped to stick big lists of equations all over my walls in the run-up to the exams. But as that's possibly the least cool wallpaper imaginable, I don't advise inviting your friends round till after the exams...

E.m.f. and Internal Resistance

There's resistance everywhere — inside batteries, in all the wires and in the components themselves. No one's for giving current an easy ride.

From now on, I'm assuming that the resistance of the wires in the circuit is zero. In practice, they do have a small resistance.

Batteries have **Resistance**

Resistance comes from **electrons colliding** with **atoms** and **losing energy**.

In a **battery**, **chemical energy** is used to make **electrons move**. As they move, they collide with atoms inside the battery — so batteries **must** have resistance. This is called **internal resistance**.

Internal resistance is what makes **batteries** and **cells warm up** when they're used.

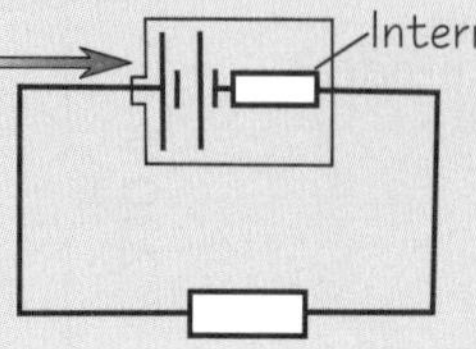

1) The amount of **electrical energy** the battery produces for each **coulomb** of charge is called its **electromotive force** or **e.m.f.** (ε). Be careful — e.m.f. **isn't** actually a force. It's measured in **volts**.
2) The **potential difference** across the **load resistance** (***R***) is the **energy transferred** when **one coulomb** of charge flows through the **load resistance**. This potential difference is called the **terminal p.d.** (***V***).
3) If there was **no internal resistance**, the **terminal p.d**. would be the **same** as the **e.m.f.** However, in **real** power supplies, there's **always some energy lost** overcoming the internal resistance.
4) The **energy wasted per coulomb** overcoming the internal resistance is called the **lost volts** (***v***).

Conservation of energy tells us:

energy per coulomb supplied by the source = **energy per coulomb used in load resistance** + **energy per coulomb wasted in internal resistance**

There are Loads of **Calculations** with **E.m.f.** and **Internal Resistance**

Examiners can ask you to do **calculations** with **e.m.f.** and **internal resistance** in loads of **different** ways. You've got to be ready for whatever they throw at you.

$$\varepsilon = V + v \qquad \varepsilon = IR + Ir$$

$$V = \varepsilon - v \qquad V = \varepsilon - Ir$$

Learn these equations for the exam. Only this one ($\varepsilon = IR + Ir$) will be on your formula sheet.

These are all basically the **same equation**, just written differently. If you're given enough information you can calculate the e.m.f. (ε), terminal p.d. (***V***), lost volts (***v***), current (***I***), load resistance (***R***) or internal resistance (***r***). Which equation you should use depends on what information you've got, and what you need to calculate.

Most Power Supplies Need **Low Internal Resistance**

A **car battery** has to deliver a **really high current** — so it needs to have a **low internal resistance**. The cells used to power a **torch** or a **personal stereo** are the **same**. **Generally**, **batteries** have an **internal resistance** of **less than 1 Ω**.

Since **internal resistance** causes **energy loss**, you'd think **all** power supplies should have a **low internal resistance**.

High voltage power supplies are the **exception**. **HT** (high tension) and **EHT** (extremely high tension) **supplies** are designed with **very high** internal resistances. This means that if they're **accidentally short-circuited** only a **very small current** can flow. Much **safer**.

E.m.f. and Internal Resistance

Use this **Circuit** to **Measure Internal Resistance** and **E.m.f.**

By **changing** the value of R (**load resistance**) in this circuit and **measuring** the **current** (I) and **p.d.** (V), you can work out the **internal resistance** of the source.

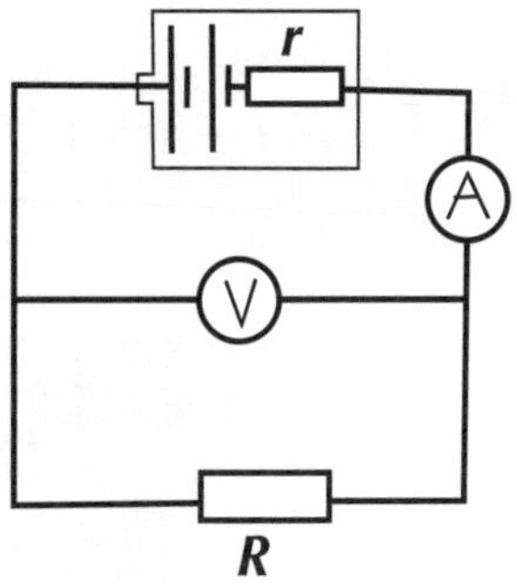

Start with the equation:

$$V = \varepsilon - Ir$$

Plot a graph of V against I.

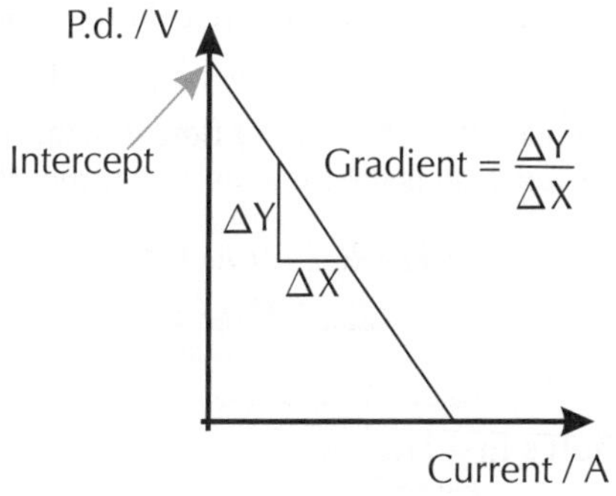

1) Rearrange the equation: $V = -rI + \varepsilon$
2) Since ε and r are constants, that's just the equation of a **straight line** (in the form: $y = mx + c$).
3) So the intercept on the vertical axis is ε.
4) And the gradient is $-r$.

Equation of a straight line
$y = mx + c$
gradient (m), y-intercept (c)

An **easier** way to **measure** the **e.m.f.** of a **power source** is by connecting a high-resistance **voltmeter** across its **terminals**. A **small current flows** through the **voltmeter**, so there must be some **lost volts** — this means you measure a value **very slightly less** than the **e.m.f.** In **practice** the difference **isn't** usually **significant**.

Practice Questions

Q1 What causes internal resistance?

Q2 What is meant by 'lost volts'?

Q3 What is the difference between e.m.f. and terminal p.d.?

Q4 Write the equation used to calculate the terminal p.d. of a power supply.

Exam Questions

Q1 A large battery with an internal resistance of 0.8 Ω and e.m.f. 24 V is used to power a dentist's drill with resistance 4 Ω.

(a) Calculate the current in the circuit when the drill is connected to the power supply. [2 marks]

(b) Calculate the voltage across the drill while it is being used. [1 mark]

Q2 A student mistakenly connects a 10 Ω ray box to an HT power supply of 500 V. The ray box does not light, and the student measures the current flowing to be only 50 mA.

(a) Calculate the internal resistance of the HT power supply. [2 marks]

(b) Explain why this is a sensible internal resistance for an HT power supply. [2 marks]

You're UNBELIEVABLE... [Frantic air guitar]... Ueuuurrrrghhh... Yeah...

Wanting power supplies to have a low internal resistance makes sense, you wouldn't want your MP3 player battery melting if you listened to music for more than half an hour. Make sure you know your e.m.f. equations, they're an exam fave. A good way to get them learnt is to keep trying to getting from one equation to another... dull, but it can help.

Conservation of Energy & Charge in Circuits

There are some things in Physics that are so fundamental that you just have to accept them. Like the fact that there's loads of Maths in it. And that energy is conserved. And that Physicists get more homework than everyone else.

Charge Doesn't 'Leak Away' Anywhere — it's Conserved

1) As **charge flows** through a circuit, it **doesn't** get **used up** or **lost**.
2) This means that whatever **charge flows into** a junction will **flow out** again.
3) Since **current** is **rate of flow of charge**, it follows that whatever **current flows into** a junction is the same as the current **flowing out** of it.

e.g.

CHARGE FLOWING IN 1 SECOND

$Q_1 = 6\text{ C} \Rightarrow I_1 = 6\text{ A}$ → splits into $Q_2 = 2\text{ C} \Rightarrow I_2 = 2\text{ A}$ and $Q_3 = 4\text{ C} \Rightarrow I_3 = 4\text{ A}$

$I_1 = I_2 + I_3$

Kirchhoff's first law says:

The total **current entering a junction** = the total **current leaving it.**

Energy conservation is vital.

Energy is Conserved too

1) **Energy is conserved**. You already know that. In **electrical circuits**, **energy** is **transferred round** the circuit. Energy **transferred to** a charge is **e.m.f.**, and energy **transferred from** a charge is **potential difference**.
2) In a **closed loop**, these two quantities must be **equal** if energy is conserved (which it is).

Kirchhoff's second law says:

The **total e.m.f.** around a **series circuit** = the **sum** of the **p.d.s** across each component. (or $\varepsilon = \Sigma IR$ in symbols)

Exam Questions get you to Apply Kirchhoff's Laws to Combinations of Resistors

A **typical exam question** will give you a **circuit** with bits of information missing, leaving you to fill in the gaps. Not the most fun... but on the plus side you get to ignore any internal resistance stuff (unless the question tells you otherwise)... hurrah. You need to remember the **following rules**:

SERIES Circuits

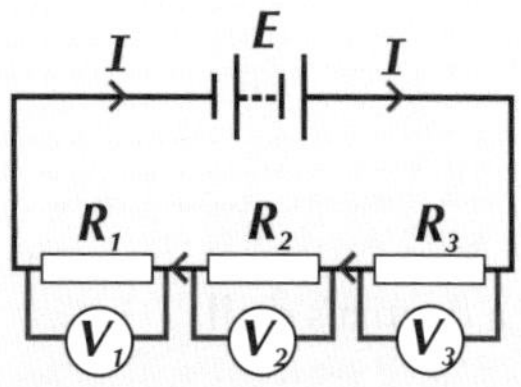

1) **same current** at **all points** of the circuit (since there are no junctions)
2) **e.m.f. split** between **components** (by Kirchhoff's 2nd law), so: $E = V_1 + V_2 + V_3$
3) $V = IR$, so if I is constant: $IR_{total} = IR_1 + IR_2 + IR_3$
4) cancelling the Is gives:

$$R_{total} = R_1 + R_2 + R_3$$

PARALLEL Circuits

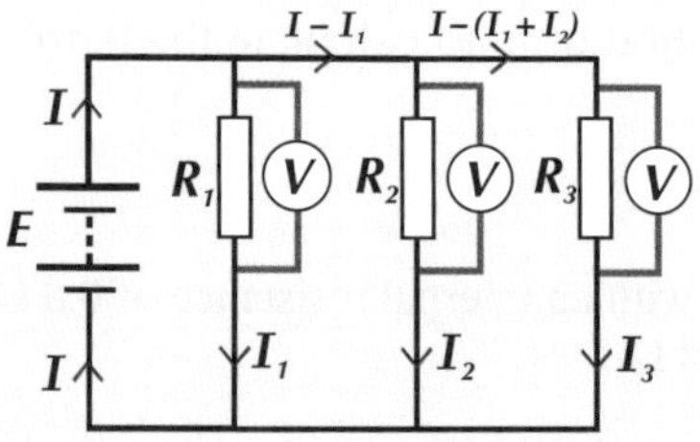

1) **current** is **split** at each **junction**, so: $I = I_1 + I_2 + I_3$
2) **same p.d.** across **all components** (three separate loops — within each loop the e.m.f. equals sum of individual p.d.s)
3) so, $V/R_{total} = V/R_1 + V/R_2 + V/R_3$
4) cancelling the Vs gives:

$$1/R_{total} = 1/R_1 + 1/R_2 + 1/R_3$$

...and there's an example on the next page to make sure you know what to do with all that...

Conservation of Energy & Charge in Circuits

Worked Exam Question

A battery of e.m.f. 16 V and negligible internal resistance is connected in a circuit as shown:

a) Show that the group of resistors between X and Y could be replaced by a single resistor of resistance 15 Ω.

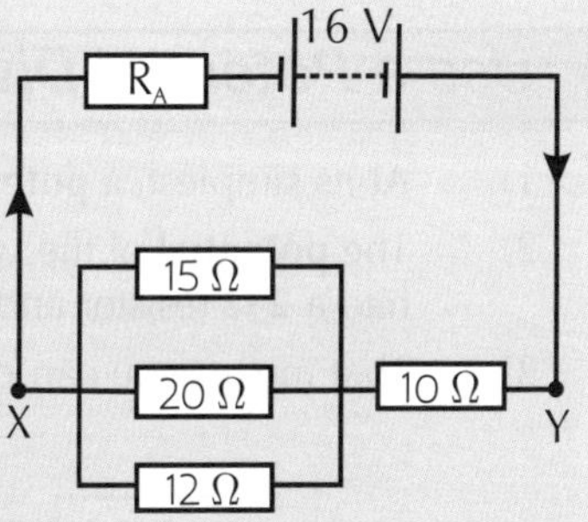

You can find the **combined resistance** of the 15 Ω, 20 Ω and 12 Ω resistors using:

$1/R = 1/R_1 + 1/R_2 + 1/R_3 = 1/15 + 1/20 + 1/12 = 1/5 \quad \Rightarrow R = 5\ \Omega$

So **overall resistance** between **X** and **Y** can be found by $\mathbf{R = R_1 + R_2 = 5 + 10 = 15\ \Omega}$

b) If $R_A = 20\ \Omega$:
(i) calculate the potential difference across R_A,

Careful — there are a few steps here. You need the p.d. across R_A, but you don't know the current through it. So start there:
total resistance in circuit = 20 + 15 = 35 Ω, **so** current through R_A can be found using $I = V_{total}/R_{total}$:

$I = 16/35$ A

then you can use $V = IR_A$ to find the p.d. across R_A: $V = 16/35 \times 20 = \mathbf{9.1\ V}$

(ii) calculate the current in the 15 Ω resistor.

You know the **current flowing** into the group of three resistors and out of it, but not through the individual branches. But you know that their **combined resistance** is **5** Ω (from part a) so you can work out the p.d. across the group:

$V = IR = 16/35 \times 5 = 16/7$ V

The p.d. across the **whole group** is the same as the p.d. across each **individual resistor**, so you can use this to find the current through the 15 Ω resistor:

$I = V/R = (16/7) / 15 = \mathbf{0.15\ A}$

Practice Questions

Q1 State Kirchhoff's laws.

Q2 Find the current through and potential difference across each of two 5 Ω resistors when they are placed in a circuit containing a 5 V battery, and are wired: a) in series, b) in parallel.

Exam Question

Q1 For the circuit on the right:

(a) Calculate the total effective resistance of the three resistors in this combination. [2 marks]

(b) Calculate the main current, I_3. [2 marks]

(c) Calculate the potential difference across the 4 Ω resistor. [1 mark]

(d) Calculate the potential difference across the parallel pair of resistors. [1 mark]

(e) Using your answer from 1 (d), calculate the currents I_1 and I_2. [2 marks]

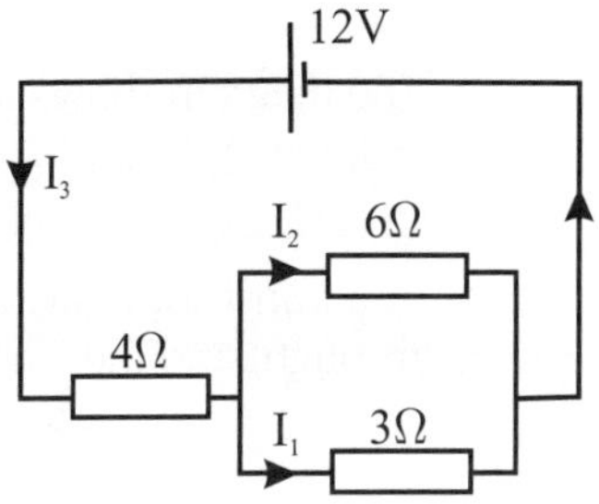

This is a very purple page — needs a bit of yellow I think...

V = IR is the formula you'll use most often in these questions. Make sure you know whether you're using it on the overall circuit, or just one specific component. It's amazingly easy to get muddled up — you've been warned.

The Potential Divider

I remember the days when potential dividers were pretty much the hardest thing they could throw at you. Then along came AS Physics. Hey ho.

Anyway, in context this doesn't seem too hard now, so get stuck in.

Use a Potential Divider to get a Fraction of a Source Voltage

1) At its simplest, a **potential divider** is a circuit with a **voltage source** and a couple of **resistors** in series.
2) The **potential** of the voltage source (e.g. a power supply) is **divided** in the **ratio** of the **resistances**. So, if you had a **2 Ω** resistor and a **3 Ω** resistor, you'd get **2/5** of the p.d. across the **2 Ω** resistor and **3/5** across the **3 Ω**.
3) That means you can **choose** the **resistances** to get the **voltage** you **want** across one of them.

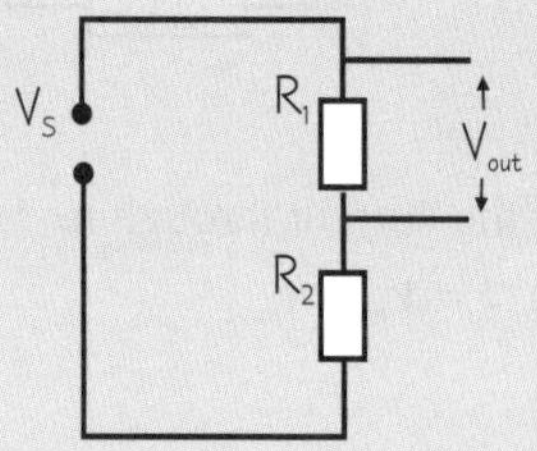

In the circuit shown, R_1 has $\frac{R_1}{R_1 + R_2}$ of the total resistance.

So: $$V_{out} = \frac{R_1}{R_1 + R_2} V_S$$

E.g. if V_s = **9 V** and you want V_{out} to be **6 V**,

then you need: $$\frac{R_1}{R_1 + R_2} = \frac{2}{3} \quad \textit{which gives } R_1 = 2R_2$$

So you could have, say, R_1 = **200 Ω**, R_2 = **100 Ω**

4) This circuit is mainly used for **calibrating voltmeters**, which have a **very high resistance**.
5) If you put something with a **relatively low resistance** across R_1 though, you start to run into **problems**. You've **effectively** got **two resistors** in **parallel**, which will **always** have a **total** resistance **less** than R_1. That means that V_{out} will be **less** than you've calculated, and will depend on what's connected across R_1. Hrrumph.

Add an LDR or Thermistor for a Light or Temperature Switch

1) A **light-dependent resistor** (LDR) has a very **high resistance** in the **dark**, but a **lower resistance** in the **light**.
2) An **NTC thermistor** has a **high resistance** at **low temperatures**, but a much **lower resistance** at **high temperatures** (it varies in the opposite way to a normal resistor, only much more so).
3) Either of these can be used as one of the **resistors** in a **potential divider**, giving an **output voltage** that **varies** with the **light level** or **temperature**.
4) Add a **transistor** and you've got yourself a **switch**, e.g. to turn on a light or a heating system.

The diagram shows a type of **burglar alarm**.
When light shines on the LDR its **resistance decreases**, so V_{out} increases.
The transistor is switched on, current flows through, and the **alarm sounds**.

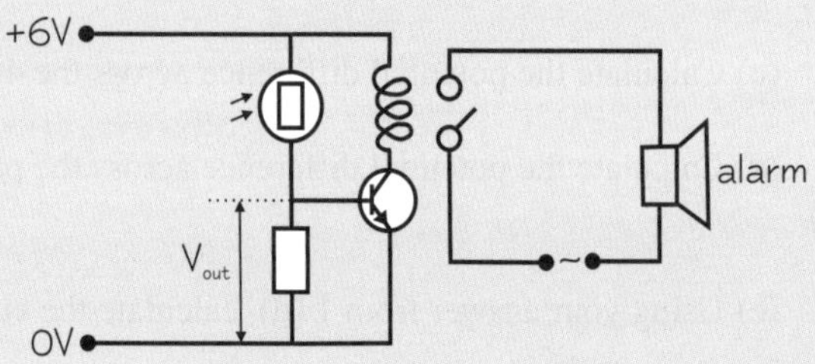

You can think of a transistor as a kind of switch. It's off when the voltage across it is low, and on when the voltage is high.

The Potential Divider

A *Potentiometer* uses a *Variable Resistor* to give a *Variable Voltage*

1) A **potentiometer** has a variable resistor replacing R_1 and R_2 of the potential divider, but it uses the **same idea** (it's even sometimes **called** a potential divider just to confuse things).
2) You move a **slider** or turn a knob to **adjust** the **relative sizes** of R_1 and R_2. That way you can vary V_{out} from **0 V** up to the source voltage.
3) This is dead handy when you want to be able to **change** a **voltage continuously**, like in the **volume control** of a stereo.

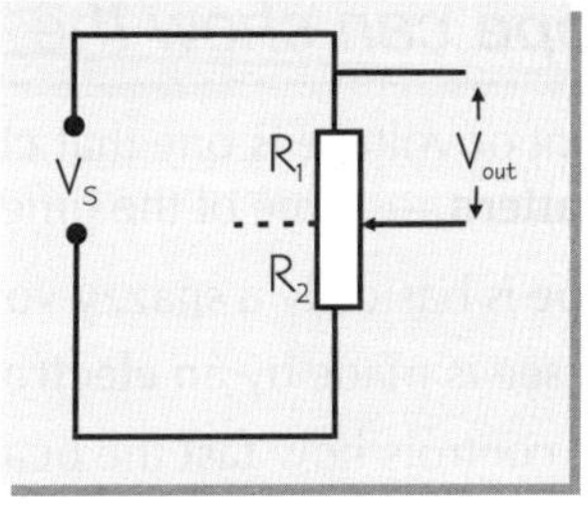

Here, V_S is replaced by the input signal (e.g. from a CD player) and V_{out} is the output to the amplifier and loudspeaker.

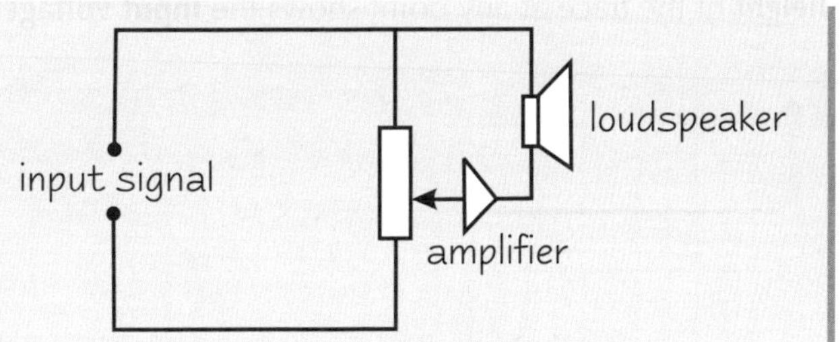

Practice Questions

Q1 Look at the burglar alarm circuit on page 36. How could you change the circuit so that the alarm sounds when the light level decreases?

Q2 The LDR in the burglar alarm circuit has a resistance of 300 Ω when light and 900 Ω when dark. The fixed resistor has a value of 100 Ω. Show that V_{out} (light) = 1.5 V and V_{out} (dark) = 0.6 V.

Exam Questions

Q1 In the circuit on the right, all the resistors have the same value. Calculate the p.d. between:

(i) A and B. [1 mark]

(ii) A and C. [1 mark]

(iii) B and C. [1 mark]

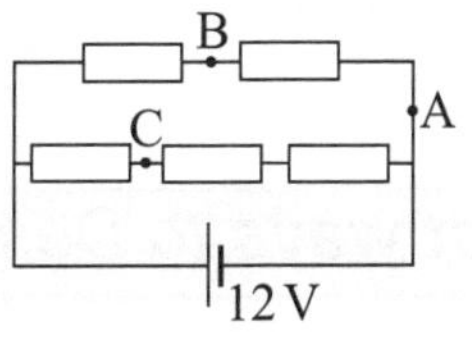

Q2 Look at the circuit on the right.

(a) Calculate the p.d. between A and B as shown by a high resistance voltmeter placed between the two points. [1 mark]

(b) A 40 Ω resistor is now placed between points A and B. Calculate the p.d. across AB and the current flowing through the 40 Ω resistor. [4 marks]

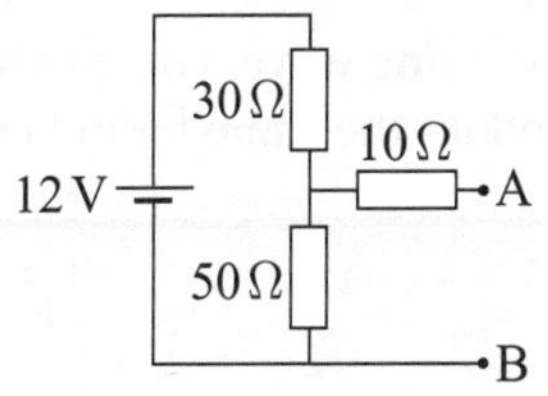

OI...YOU... [bang bang bang]... turn that potentiometer down...

You'll probably have to use a potentiometer in every experiment you do with electricity from now on in, so you'd better get used to them. I can't stand the things myself, but then lab and me don't mix — far too technical.

Alternating Current

Just when you think you've got the hang of this electricity lark, they spring alternating current on you. Here's where it all gets way more complicated. You can't use a normal voltmeter and ammeter any more — enter 'the oscilloscope'.

An **Oscilloscope** can show the **Waveform** of an **Alternating Current**

An **alternating** current or voltage is one that **changes with time**. The voltage goes up and down in a **regular pattern** — some of the time it's **positive** and some of the time it's **negative**.

1) An **oscilloscope** is basically a snazzy **voltmeter**.
2) The **trace** you see is made by an **electron beam** moving across a screen.
3) The **time base** controls how **fast** the beam is moved across the screen. You can **set** this using a **dial** on the **front** of the oscilloscope.
4) The **vertical height** of the trace at any point shows the **input voltage** at that point.

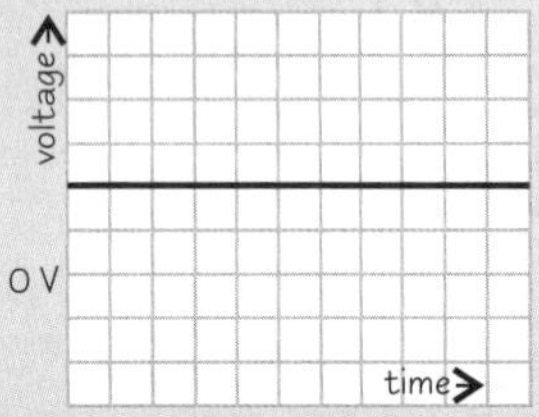

A **direct current** (d.c.) source is always at the same voltage, so you get a **horizontal line**.

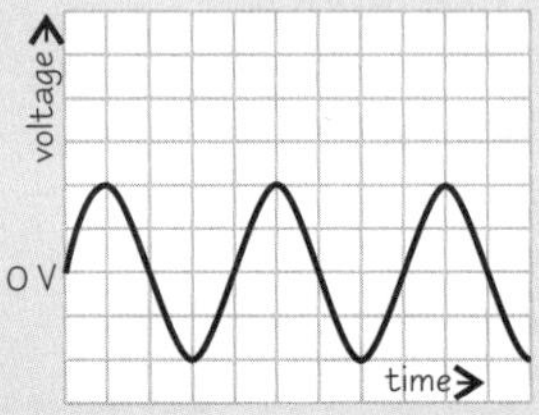

An **alternating current** (a.c.) source gives a regularly **repeating waveform**.

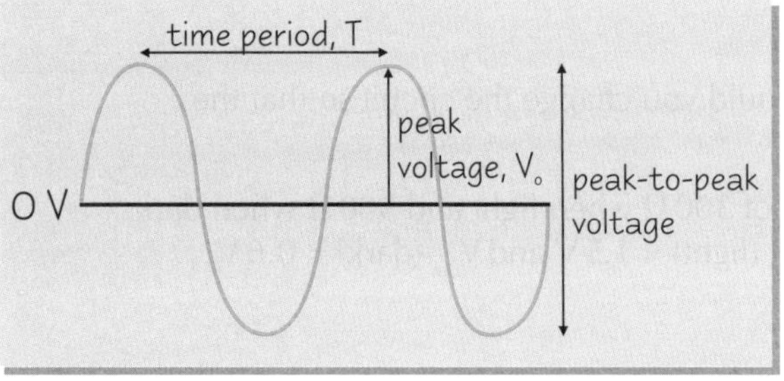

Measuring the **distance** between successive **peaks** along the time axis gives you the **time period** (as long as you know the time base setting).

You can use this to calculate the **frequency**:

$$\textit{frequency} = \frac{1}{\textit{time period}} \qquad f = \frac{1}{T}$$

This is the same for any type of wave.

The **peak voltage** is useful to know, though it's often **easier** to measure the **peak-to-peak voltage** and **halve** it.

V_{rms} and I_{rms} are often the **best ways** to Describe an **Alternating Current**

1) An **a.c. supply** with a **peak voltage** of **2 V** will be **below 2 V** most of the time. That means it won't have as high a power output as a **2 V d.c. supply**.
2) To **compare** them properly, you need to **average** the **a.c. voltage** somehow.
3) A **normal average won't work**, because the **positive** and **negative** bits **cancel out**. It turns out that something called the **root mean square (r.m.s.) voltage** does the trick.
4) For a **sine wave**, you get this by **dividing** the **peak voltage**, V_0, by $\sqrt{2}$. And it's the **same** for the **current**:

$$V_{rms} = \frac{V_0}{\sqrt{2}} \qquad I_{rms} = \frac{I_0}{\sqrt{2}}$$

Even though this is only strictly true if the a.c. signal is a sine wave, that's what you get from a generator. It's also the only one on your syllabus, so I wouldn't worry about it.

5) If you want to work out the **power**, just replace I and V in the power formula with the **r.m.s. values**.

$$\textbf{Power} = V_{rms} \times I_{rms}$$

Alternating Current

For the **resistance** it doesn't matter, as the **peak values** will give you the same answer:

$$\textbf{Resistance, } R = \frac{V_{rms}}{I_{rms}} = \frac{V_o}{I_o}$$

It's usually the **r.m.s. value** that's stated on a **power supply**.

For example, the value of **230 V** stated for the **UK mains electricity supply** is the **r.m.s. value**.
Just use the equation above to get the **peak** value, or double that to get the **peak-to-peak** value:

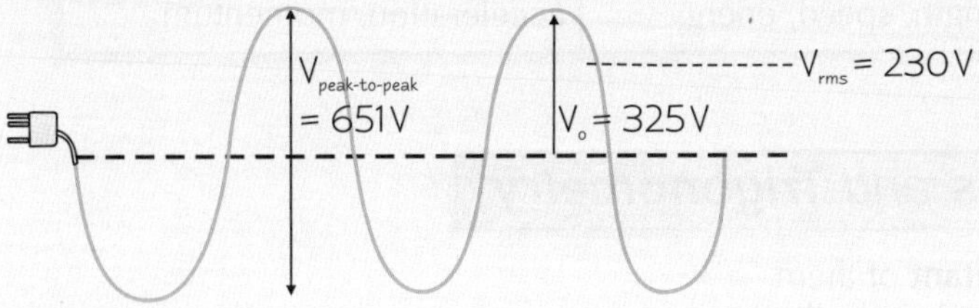

$$V_{peak} = \sqrt{2} \times V_{rms} = 325 \text{ V}$$

$$V_{peak\text{-}to\text{-}peak} = 2 \times V_{peak} = 651 \text{ V}$$

Practice Questions

Q1 Define the terms 'time period' and 'frequency' for an alternating signal.

Q2 Define the r.m.s. voltage of an alternating sinusoidal signal in terms of the peak voltage.

Q3 Why are the r.m.s. values for current and voltage used to work out the power of a.c. supplies rather than the peak values?

Exam Questions

Q1 A CRO is used to study the waveform of a sinusoidal alternating voltage of frequency 100 Hz and peak voltage of 2.0 V. The time base is set to 2.0 ms cm^{-1} horizontally and the voltage sensitivity vertically is 0.5 Vcm^{-1}.

Draw the trace you expect on a 10 cm square grid and show all your working. [4 marks]

Q2 An oscilloscope is set up initially with a horizontal line at zero volts, and the time base set to 5.0 ms per division. The Y-input voltage sensitivity is set at 1.0 volt per division.

Sketch the traces you would expect if each of the following were connected to the oscilloscope.

(a) A 2.0 V cell. [2 marks]

(b) An a.c. signal of 1.5 V peak value and frequency 50 Hz. [3 marks]

Q3 Five cycles of a.c. are observed on a CRO screen marked with 10 divisions horizontally and vertically. The frequency of the signal is 2500 Hz.

Work out:

(a) The time period of the signal. [1 mark]

(b) The horizontal time-base in both seconds per division and ms per division. [2 marks]

Yay, oscilloscopes — cool...

This stuff on a.c. electricity isn't too complicated — you just need to make sure you know all the equations on these two pages. Remember, any current or voltage you read off an a.c. power supply will be r.m.s.

Scalars and Vectors

Mechanics is one of those things that you either love or hate. I won't tell you which side of the fence I'm on.

Scalars Only Have Size, but Vectors Have Size and Direction

1) A **scalar** has **no direction** — it's **just an amount** of something, like the **mass** of a **sack of meaty dog food**.
2) A **vector** has magnitude (**size**) and **direction** — like the **speed and direction** of next door's **cat** running away.
3) **Force** and **velocity** are both **vectors** — you need to know **which way** they're going as well as **how big** they are.
4) Here are a few examples to get you started:

Scalars	Vectors
mass, temperature, time, length, speed, energy	displacement, force, velocity, acceleration, momentum

Adding Vectors Involves Pythagoras and Trigonometry

Adding two or more vectors is called finding the **resultant** of them.
You find the resultant of two vectors by drawing them '**tip-to-tail**'.

Example Jemima goes for a walk. She walks 3 m North and 4 m East. She has walked 7 m but she isn't 7 m from her starting point. Find the magnitude and direction of her displacement.

Jemima's 'displacement' gives her position relative to her starting point.

First, draw the vectors **tip-to-tail**. Then draw a line from the **tail** of the first vector to the **tip** of the last vector to give the **resultant**:

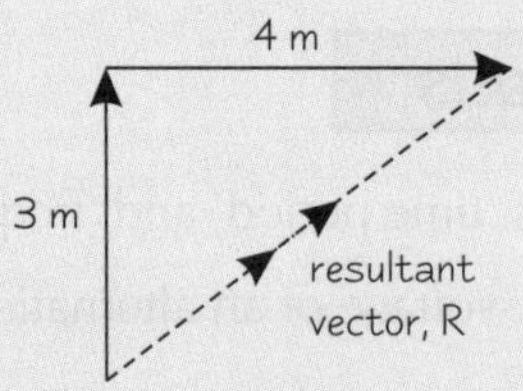

Because the vectors are at right angles, you get the **magnitude** of the resultant using Pythagoras:

$R^2 = 3^2 + 4^2 = 25$

So $R = 5$ m

Now find the **bearing** of Jemima's new position from her original position.

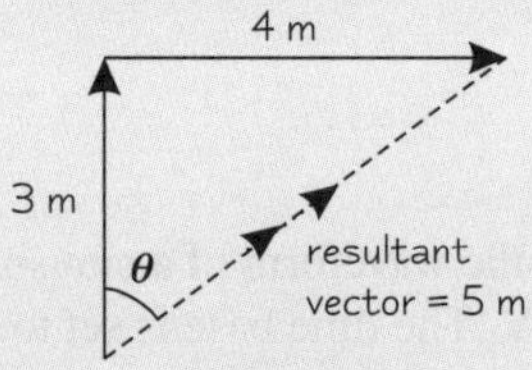

Jemima

You use the triangle again, but this time you need to use trigonometry. You know the opposite and the adjacent sides, so you need to use:

$\tan\theta = 4/3$

$\theta = 53.1°$ Trig's really useful in mechanics — so make sure you're completely okay with it. Remember SOH CAH TOA.

Use the Same Method for Resultant Forces or Velocities

If the vectors aren't at right angles, you'll need to do a scale drawing.

Always start by drawing a diagram.

Example

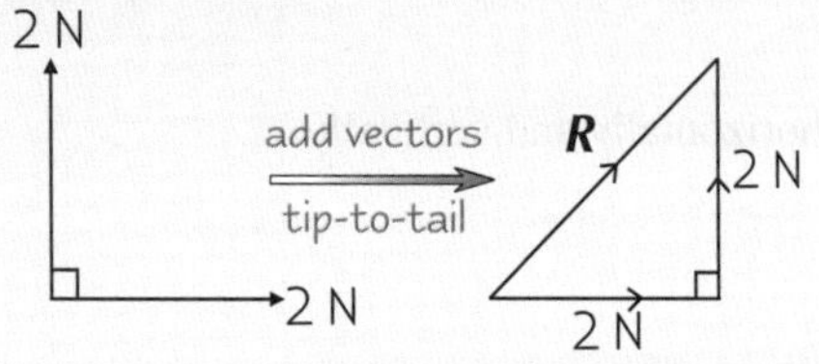

You know the resultant force is at 45° to the horizontal (since both forces are the same size).

So all you need to do is use Pythagoras:

$R^2 = 2^2 + 2^2 = 8$

which gives $R = 2.83$ N at 45° to the horizontal.

Don't forget to take the square root.

Example

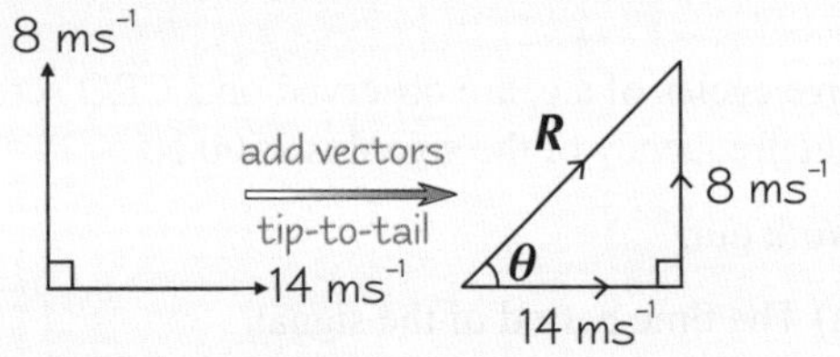

Start with: $R^2 = 14^2 + 8^2 = 260$

so you get: $R = 16.1\ \text{ms}^{-1}$.

Then: $\tan\theta = 8/14 = 0.5714$

$\theta = 29.7°$

Scalars and Vectors

Sometimes you have to do it backwards.

It's Useful to Split a ***Vector*** *into* ***Horizontal*** *and* ***Vertical Components***

This is the opposite of finding the resultant — you start from the resultant vector and split it into two **components** at right angles to each other. You're basically **working backwards** from the examples on the other page.

Resolving a vector v into horizontal and vertical components

You get the **horizontal** component v_x like this:

$$\cos\theta = v_x / v$$

$$v_x = v\cos\theta$$

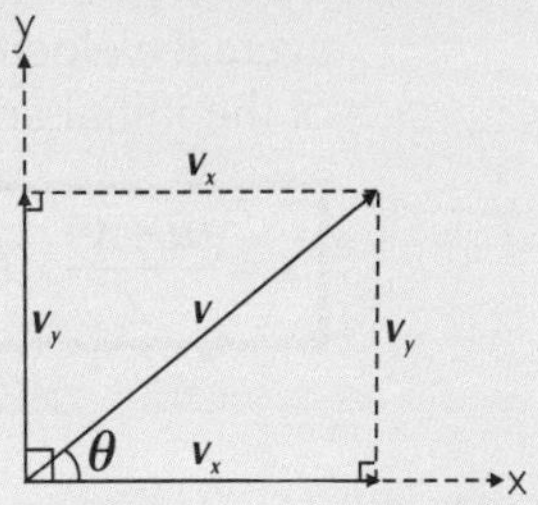

...and the **vertical** component v_y like this:

$$\sin\theta = v_y / v$$

$$v_y = v\sin\theta$$

See pages 52 and 53 for more on resolving.

θ is measured anticlockwise from the horizontal.

Example

Charley's amazing floating home is travelling at a speed of 5 ms^{-1} at an angle of 60° up from the horizontal. Find the vertical and horizontal components.

Charley's mobile home was the envy of all his friends.

The **horizontal** component v_x is:

$v_x = v\cos\theta = \mathbf{5\cos 60°} = 2.5\ ms^{-1}$

The vertical component v_y is:

$v_y = v\sin\theta = \mathbf{5\sin 60°} = 4.33\ ms^{-1}$

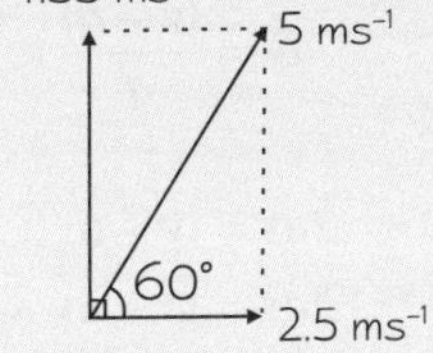

Resolving is dead useful because the two components of a vector **don't affect each other**. This means you can deal with the two directions **completely separately**.

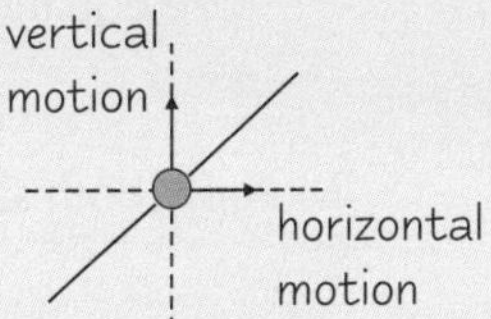

Only the vertical component is affected by gravity.

Practice Questions

Q1 Explain the difference between a scalar quantity and a vector quantity.

Q2 Jemima has gone for a swim in a river which is flowing at 0.35 ms^{-1}. She swims at 0.18 ms^{-1} at right angles to the current. Show that her resultant velocity is 0.39 ms^{-1} at an angle of 27.2° to the current.

Q3 Jemima is pulling on her lead with a force of 40 N at an angle of 26° below the horizontal. Show that the horizontal component of this force is about 36 N.

Exam Questions

Q1 The wind is creating a horizontal force of 20 N on a falling rock of weight 75 N. Calculate the magnitude and direction of the resultant force. [2 marks]

Q2 A glider is travelling at a velocity of 20.0 ms^{-1} at an angle of 15° below the horizontal. Find the horizontal and vertical components of the glider's velocity. [2 marks]

His Dark Vectors Trilogy — displacement, velocity and acceleration...

I always think that the hardest part of vector questions is sketching the initial diagram. Once you've worked out what's going on and done your sketch, they're all the same — they're a piece of cake taking a walk in the park. Easy as pie.

Motion with Constant Acceleration

Uniform Acceleration is *Constant* Acceleration

Acceleration could mean a change in speed or direction or both.

Uniform means **constant** here. It's nothing to do with what you wear.

There are **four main equations** that you use to solve problems involving **uniform acceleration**. You need to be able to **use them**, but you don't have to know how they're **derived** — we've just put it in to help you learn them.

1) **Acceleration is the rate of change of velocity.**
From this definition you get:

$$a = \frac{(v-u)}{t} \quad \text{so} \quad \boxed{v = u + at}$$

where:
u = initial velocity
v = final velocity
a = acceleration
t = time taken

2) **s = average velocity × time**
If acceleration is constant, the average velocity is just the average of the initial and final velocities, so:

$$\boxed{s = \frac{(u+v)}{2} \times t}$$

s = displacement

3) Substitute the expression for v from equation 1 into equation 2 to give:

$$s = \frac{(u+u+at) \times t}{2} = \frac{2ut + at^2}{2}$$

$$\boxed{s = ut + \tfrac{1}{2}at^2}$$

4) You can **derive** the fourth equation from equations **1** and **2**:

Use equation **1** in the form: $a = \frac{v-u}{t}$

Multiply both sides by s, where: $s = \frac{(u+v)}{2} \times t$

This gives us: $as = \frac{(v-u)}{t} \times \frac{(u+v)t}{2}$

The t's on the right cancel, so:

$2as = (v - u)(v + u)$

$2as = v^2 - uv + uv - u^2$

so: $\boxed{v^2 = u^2 + 2as}$

Example

A tile falls from a roof 25 m high. Calculate its speed when it hits the ground and how long it takes to fall. Take $g = 9.8\ ms^{-2}$.

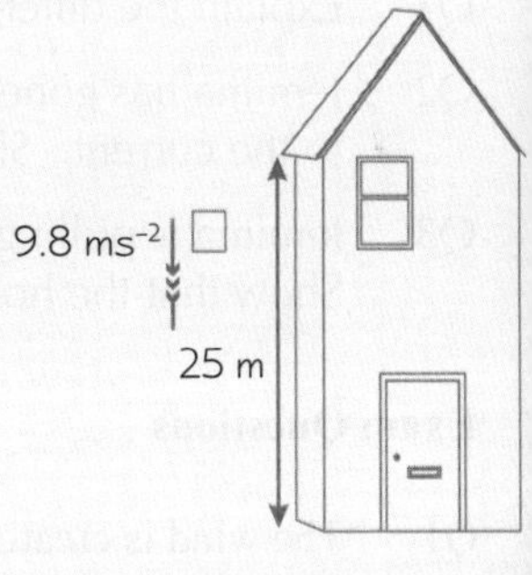

First of all, write out what you know:

s = 25 m

$u = 0\ ms^{-1}$ since the tile's stationary to start with

$a = 9.8\ ms^{-2}$ due to gravity

v = ? t = ?

Usually you take upwards as the positive direction. In this question it's probably easier to take downwards as positive, so you get $g = +9.8\ ms^{-2}$ instead of $g = -9.8\ ms^{-2}$.

Then, choose an equation with only **one unknown quantity**.

So start with $v^2 = u^2 + 2as$

$v^2 = 0 + 2 \times 9.8 \times 25$

$v^2 = 490$

$v = 22.1\ ms^{-1}$

Now, find t using:

$s = ut + \tfrac{1}{2}at^2$

$25 = 0 + \tfrac{1}{2} \times 9.8 \times t^2$

$t^2 = \frac{25}{4.9}$

Final answers:

$t = 2.3$ s

$v = 22.1\ ms^{-1}$

Motion with Constant Acceleration

Example

A car accelerates steadily from rest at a rate of 4.2 ms^{-2} for 6 seconds.

a) Calculate the final speed.

b) Calculate the distance travelled in 6 seconds.

Remember — always start by writing down what you know.

a) $a = 4.2\ ms^{-2}$ choose the right equation... $v = u + at$

$u = 0\ ms^{-1}$ $v = 0 + 4.2 \times 6$

$t = 6\ s$ *Final answer:* $v = 25.2\ ms^{-1}$

$v = ?$

b) $s = ?$

$t = 6\ s$

$u = 0\ ms^{-1}$

$a = 4.2\ ms^{-2}$

$v = 25.2\ ms^{-1}$

you can use: $s = \frac{(u+v)t}{2}$ → $s = \frac{(0+25.2)\times 6}{2}$ → *Final answer:* $s = 75.6$ m

or: $s = ut + \frac{1}{2}at^2$ → $s = 0 + \frac{1}{2} \times 4.2 \times (6)^2$ → $s = 75.6$ m

You Have to **Learn** the Constant Acceleration **Equations**

Make sure you learn the equations. There are only four of them and these questions are always dead easy marks in the exam, so you'd be dafter than a hedgehog in a helicopter not to learn them...

Practice Questions

Q1 Write out the four constant acceleration equations.

Q2 Show how the equation $s = ut + \frac{1}{2}at^2$ can be derived.

Exam Questions

Q1 A skydiver jumps from an aeroplane when it is flying horizontally. She accelerates due to gravity for 5 s.
(a) Calculate her maximum vertical velocity. (Assume no air resistance.) [2 marks]
(b) How far does she fall in this time? [2 marks]

Q2 A motorcyclist slows down uniformly as he approaches a red light. He takes 3.2 seconds to come to a halt and travels 40 m in this time.
(a) How fast was he travelling initially? [2 marks]
(b) Calculate his acceleration. (N.B. a negative value shows a deceleration.) [2 marks]

Q3 A stream provides a constant acceleration of 6 ms^{-2}. A toy boat is pushed directly against the current and then released from a point 1.2 m upstream from a small waterfall. Just before it reaches the waterfall, it is travelling at a speed of 5 ms^{-1}.
(a) Find the initial velocity of the boat. [2 marks]
(b) What is the maximum distance upstream from the waterfall the boat reaches? [2 marks]

Constant acceleration — it'll end in tears...

If a question talks about "uniform" or "constant" acceleration, it's a dead giveaway they want you to use one of these equations. The tricky bit is working out which one to use — start every question by writing out what you know and what you need to know. That makes it much easier to see which equation you need. To be sure. Arrr.

Free Fall and Projectile Motion

Here's a double-page spread on how to calculate the air-speed velocity of an unladen swallow. And stuff.

Free Fall is when there's Only Gravity and Nothing Else

Free fall is defined as the motion of an object undergoing an acceleration of '**g**'. You need to remember:

1) Acceleration is a **vector quantity** — and '**g**' acts **vertically downwards**.
2) The magnitude of '**g**' is usually taken as **9.81 ms^{-2}**, though it varies slightly at different points on the Earth's surface
3) The **only force** acting on an object in free fall is its **weight**.
4) Objects can have an initial velocity in any direction and still undergo **free fall** as long as the **force** providing the initial velocity is **no longer acting**.

Newton — All Objects in Free Fall Accelerate at the Same Rate

Another gravity experiment.

1) It sounds simple enough... but it took a loooong time to understand that all free-falling objects accelerate towards the ground at the **same** rate.
2) For over 1000 years the generally accepted theory was that heavier objects would fall towards the ground quicker than lighter objects. It was challenged a few times, but it was finally overturned when **Galileo** came on the scene.
3) The difference with Galileo was that he set up **systematic** and **rigorous experiments** to **test** his theories — just like in modern science. These experiments could be repeated and the results described **mathematically** and compared.
4) Galileo believed that all objects fall at the same rate. The problem in trying to prove it was that free-falling objects **fell too quickly** for him to be able to take any accurate measurements (he only had a water clock), and **air resistance** affects the rate at which objects fall.

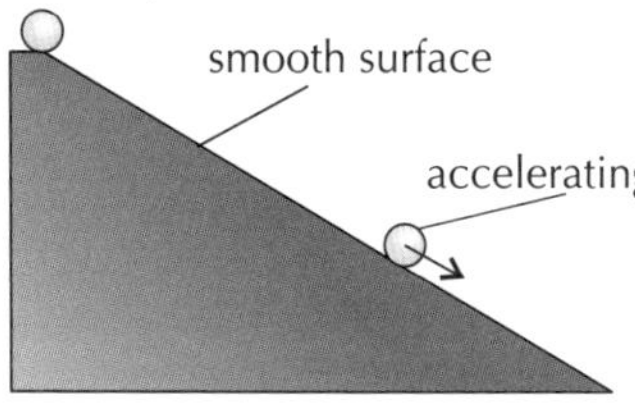

5) Galileo measured the time a ball took to roll down a **smooth** groove in an inclined plane. He killed two birds with one stone by rolling it down a plane, which **slows** the ball's fall as well as reducing the effect of **air resistanc**
6) By rolling the ball along different fractions of the total length of the slope, he found that the distance the ball travelled was proportional to the square of the time taken. The ball was **accelerating** at a **constant rate**.
7) In the end it took **Newton** to bring it all together to show and explain why **all** free falling objects have the same acceleration. He showed **mathematically** that all objects are attracted towards the Earth due to a force he called **gravity**. Ah, good ol' Newton...

You can Just Replace a with g in the Equations of Motion

You need to be able to work out **speeds**, **distances** and **times** for objects moving vertically with an **acceleration of g**. As **g** is a **constant acceleration** you can use the **equations of motion**. But because **g** acts downwards, you need to be careful about directions. To make it clear, there's a sign convention: **upwards is positive**, **downwards is negative**.

Case 1: No initial velocity (it's just falling)
Initial velocity $u = 0$
Acceleration $a = g = -9.81$ ms^{-2}. Hence the equations of motion become:

$$v = gt \qquad v^2 = 2gs$$
$$s = \frac{1}{2}gt^2 \qquad s = \frac{vt}{2}$$

Case 2: An initial velocity upwards (it's thrown up into the air)
The equations of motion are just as normal, but with $a = g = -9.81$ ms^{-2}

Sign Conventions — Learn Them:
g is always downwards so it's usually negative
t is always positive
u and **v** can be either positive or negative
s can be either positive or negative

Case 3: An initial velocity downwards (it's thrown down)

Example: Alex throws a stone downwards from the top of a cliff. She throws it with a downwards velocity of 2 ms^{-1}. It takes 3 s to reach the water below. How high is the cliff?

1) You know $u = -2$ ms^{-1}, $a = g = -9.81$ ms^{-2} and $t = 3$ s. You need to find **s**.
2) Use $s = ut + \frac{1}{2}gt^2 = (-2 \times 3) + \left(\frac{1}{2} \times -9.81 \times 3^2\right) = -50.15$ m. The cliff is **50.15 m** high.

s is negative because the stone ends up further down than it started. Height is a scalar quantity, so is always positive.

Free Fall and Projectile Motion

Any object given an initial velocity and then left to move freely under gravity is a projectile.
If you're doing AS Maths, you've got all this to look forward to in M1 as well, quite likely. Fun for all the family.

You have to think of Horizontal and Vertical Motion Separately

Example Sharon fires a scale model of a TV talent show presenter horizontally with a velocity of 100 ms^{-1} from 1.5 m above the ground. How long does it take to hit the ground, and how far does it travel? Assume the model acts as a particle, the ground is horizontal and there is no air resistance.

Think about the vertical motion first:

1) It's **constant acceleration** under gravity...
2) You know $u = 0$ (no vertical velocity at first), $s = -1.5$ m and $a = g = -9.81$ ms^{-2}. You need to find t.
3) Use $s = \frac{1}{2}gt^2 \Rightarrow t = \sqrt{\frac{2s}{g}} = \sqrt{\frac{2 \times -1.5}{-9.81}} = 0.55$ s
4) So the model hits the ground after **0.55** seconds.

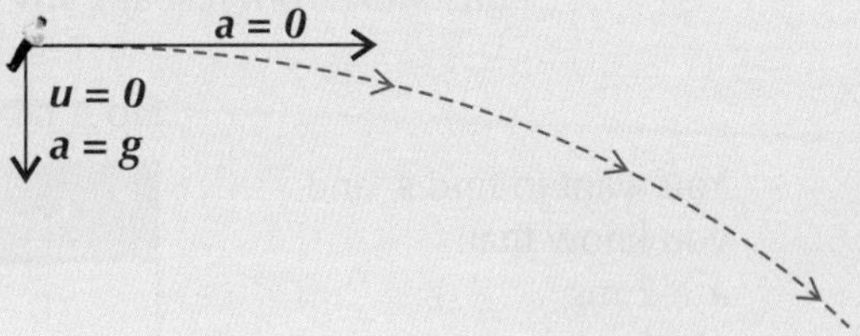

Then do the horizontal motion:

1) The horizontal motion isn't affected by gravity or any other force, so it moves at a **constant speed**.
2) That means you can just use good old **speed = distance / time**.
3) Now $v_h = 100$ ms^{-1}, $t = 0.55$ s and $a = 0$. You need to find s_h.
4) $s_h = v_h t = 100 \times 0.55 = \underline{55\text{ m}}$

Where v_h is the horizontal velocity, and s_h is the horizontal distance travelled (rather than the height fallen).

It's Slightly Trickier if it Starts Off at an Angle

If something's projected at an angle (like, say, a javelin) you start off with both horizontal and vertical velocity:

Method:
1) Resolve the initial velocity into horizontal and vertical components.
2) Use the vertical component to work out how long it's in the air and/or how high it goes.
3) Use the horizontal component to work out how far it goes while it's in the air.

Practice Questions

Q1 What is the initial vertical velocity for an object projected horizontally with a velocity of 5 ms^{-1}?

Q2 How does the horizontal velocity of a free-falling object change with time?

Q3 What is the main reason Galileo's ideas became generally accepted in place of the old theory?

Exam Questions

Q1 Jason stands on a vertical cliff edge throwing stones into the sea below.
He throws a stone horizontally with a velocity of 20 ms^{-1}, 560 m above sea level.
(a) How long does it take for the stone to hit the water from leaving Jason's hand?
Use $g = 9.81$ ms^{-2} and ignore air resistance. [2 marks]
(b) Find the distance of the stone from the base of the cliff when it hits the water. [2 marks]

Q2 Robin fires an arrow into the air with a vertical velocity of 30 ms^{-1}, and a horizontal velocity of 20 ms^{-1}, from 1 m above the ground. Find the maximum height from the ground reached by his arrow.
Use $g = 9.81$ ms^{-2} and ignore air resistance. [3 marks]

So it's this "Galileo" geezer who's to blame for my practicals...

Ah, the ups and downs and er... acrosses of life. Make sure you're happy splitting an object's motion into horizontal and vertical bits — it comes up all over mechanics. Hmmm... I wonder what Galileo would be proudest of — insisting on the systematic, rigorous experimental method on which modern science hangs... or getting in a Queen song? Magnificoooooo...

Displacement-Time Graphs

Drawing graphs by hand — oh joy. You'd think examiners had never heard of the graphical calculator. Ah well, until they manage to drag themselves out of the dark ages, you'll just have to grit your teeth and get on with it.

Acceleration* Means a *Curved Displacement-Time Graph

A graph of displacement against time for an **accelerating object** always produces a **curve**.
If the object is accelerating at a **uniform rate**, then the **rate of change** of the **gradient** will be constant.

Example

Plot a displacement-time graph for a lion who accelerates constantly from rest at 2 ms^{-2} for 5 seconds.

You want to find **s**, and you know that:
$\boldsymbol{a} = 2\ ms^{-2}$
$\boldsymbol{u} = 0\ ms^{-1}$

Use $\boldsymbol{s} = \boldsymbol{ut} + \frac{1}{2}\boldsymbol{at}^2$
If you substitute in **u** and **a**, this simplifies to:
$\boldsymbol{s} = 0 \times \boldsymbol{t} + \frac{1}{2} \times 2\boldsymbol{t}^2$
$\boldsymbol{s} = \boldsymbol{t}^2$

Do a **table of values:**

t (s)	s (m)
0	**0**
1	**1**
2	**4**
3	**9**
4	**16**
5	**25**

...then plot the **graph:**

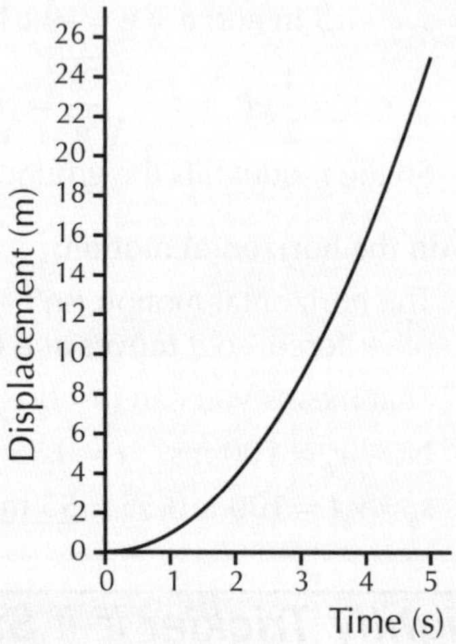

*Different Accelerations Have **Different Gradients***

In the example above, if the lion has a **different acceleration** it'll change the **gradient** of the curve like this:

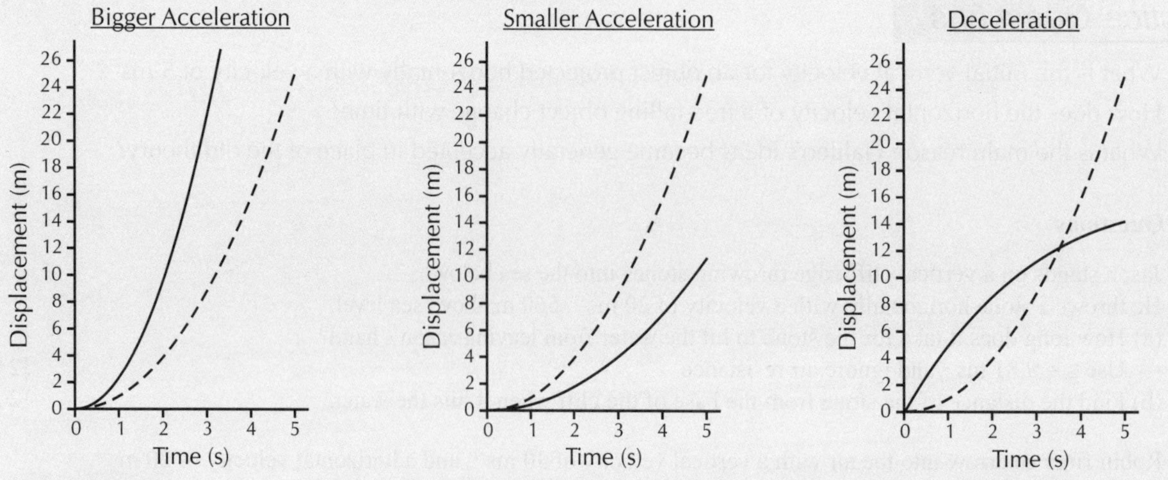

deceleration — the line has a decreasing gradient and curves the other way.

Gerald (the lion).
Ooo, he's mean...

Displacement-Time Graphs

The **Gradient** of a **Displacement-Time Graph** Tells You the Velocity

When the velocity is constant, the graph's a **straight line**.
Velocity is defined as...

$$\text{velocity} = \frac{\text{change in displacement}}{\text{time taken}}$$

On the graph, this is $\frac{\text{change in } y\ (\Delta y)}{\text{change in } x\ (\Delta x)}$, i.e. the gradient.

So to get the velocity from a displacement-time graph, just find the gradient.

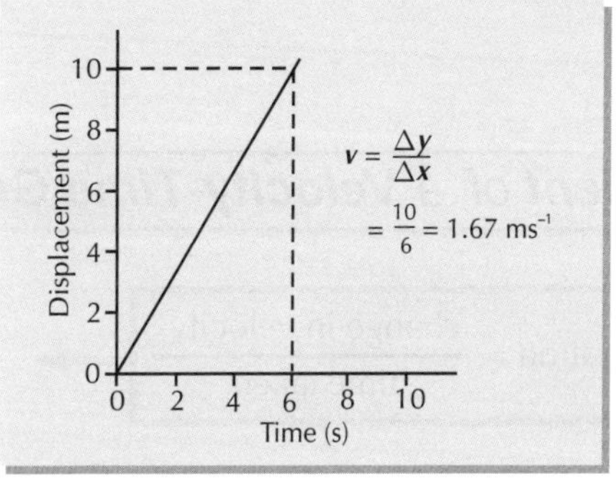

It's the Same with **Curved Graphs**

If the gradient **isn't constant** (i.e. if it's a curved line), it means the object is **accelerating**.

To find the **velocity** at a certain point you need to draw a **tangent** to the curve at that point and find its gradient.

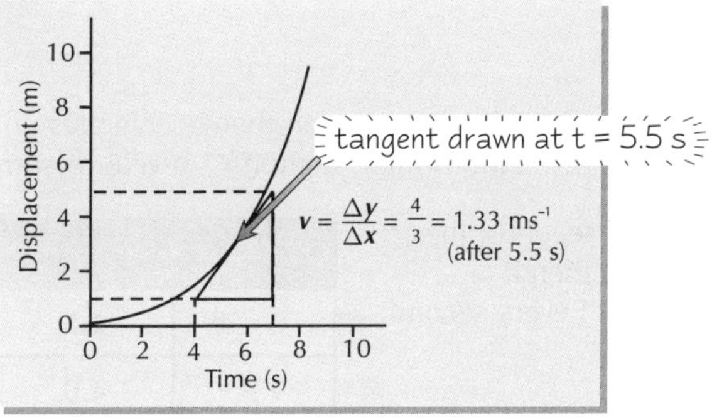

Practice Questions

Q1 What is given by the slope of a displacement-time graph?

Q2 Sketch a displacement-time graph to show: a) constant velocity, b) acceleration, c) deceleration

Exam Questions

Q1 Describe the motion of the cyclist as shown by the graph below. [4 marks]

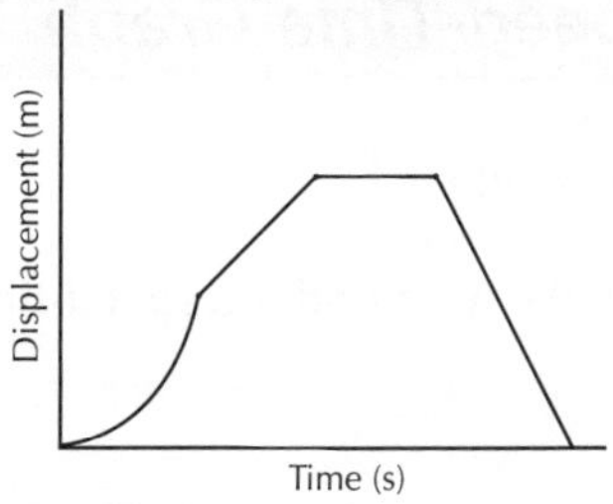

Q2 A baby crawls 5 m in 8 seconds at a constant velocity. She then rests for 5 seconds before crawling a further 3 m in 5 seconds. Finally, she makes her way back to her starting point in 10 seconds, travelling at a constant speed all the way.

(a) Draw a displacement-time graph to show the baby's journey. [4 marks]

(b) Calculate her velocity at all the different stages of her journey. [2 marks]

Some curves are bigger than others...

Whether it's a straight line or a curve, the steeper it is, the greater the velocity. There's nothing difficult about these graphs — the main problem is that it's easy to get them muddled up with velocity-time graphs (next page). If in doubt, think about the gradient — is it velocity or acceleration, is it changing (curve), is it constant (straight line), is it 0 (horizontal line)...

Velocity-Time Graphs

Speed-time graphs and velocity-time graphs are pretty similar. The big difference is that velocity-time graphs can have a negative part to show something travelling in the opposite direction:

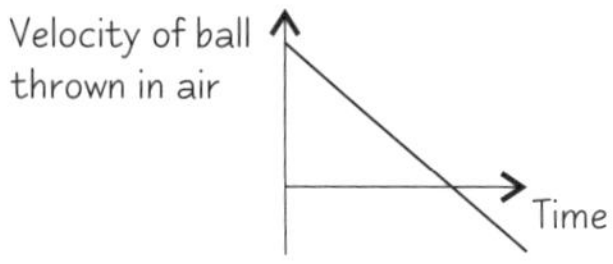

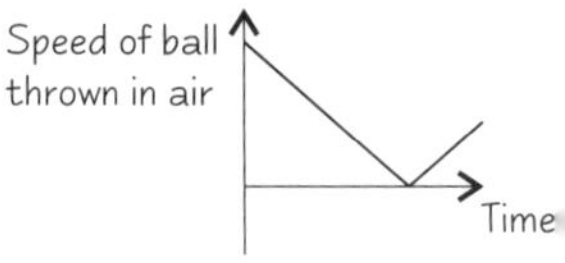

*The **Gradient** of a **Velocity-Time Graph** tells you the **Acceleration***

$$\text{acceleration} = \frac{\text{change in velocity}}{\text{time taken}}$$

likewise for a speed-time graph

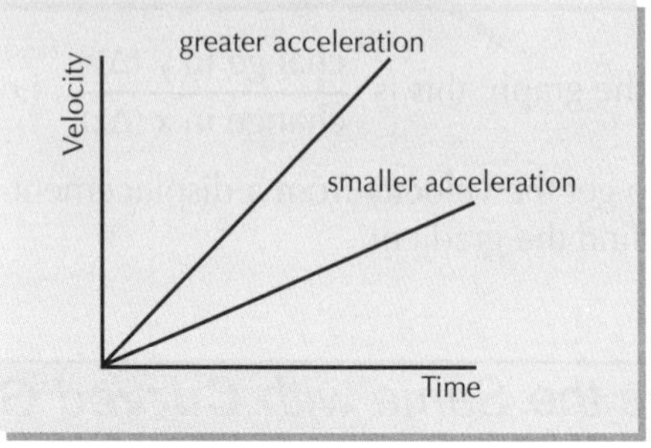

So the acceleration is just the **gradient** of a **velocity-time graph**.

1) **Uniform** acceleration is always a **straight line**.
2) The **steeper** the **gradient**, the **greater** the **acceleration**.

Example

A lion strolls along at 1.5 ms^{-1} for 4 s and then accelerates uniformly at a rate of 2.5 ms^{-2} for 4 s. Plot this information on a velocity-time graph.

Norman (the lion)...

So, for the first four seconds, the velocity is 1.5 ms^{-1}, then it increases by **2.5 ms^{-1} every second**:

t (s)	v (ms^{-1})
0 – 4	**1.5**
5	**4.0**
6	**6.5**
7	**9.0**
8	**11.5**

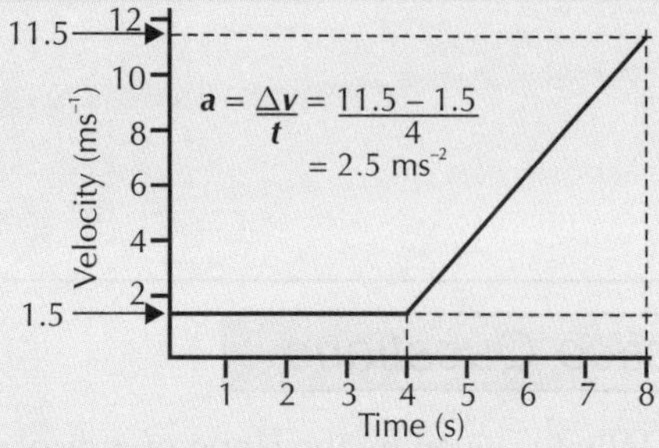

You can see that the **gradient of the line** is **constant** between 4 s and 8 s and has a value of 2.5 ms^{-2}, representing the **acceleration of the lion**.

Distance** Travelled = **Area** under **Speed-Time Graph

You know that:

distance travelled = average speed × time

So you can find the distance travelled by working out the **area under a speed-time graph**.

Example

A racing car accelerates uniformly from rest to 40 ms^{-1} in 10 s. It maintains this speed for a further 20 s before coming to rest by decelerating at a constant rate over the next 15 s. Draw a velocity-time graph for this journey and use it to calculate the total distance travelled by the racing car.

Split the **graph** up into **sections**: A, B and C

Calculate the **area** of each and **add** the three results together.

A: Area = ½ base × height = ½ × 10 × 40 = 200 m

B: Area = b × h = 20 × 40 = 800 m

C: Area = ½ b × h = ½ × 15 × 40 = 300 m

Total distance travelled = 1300 m

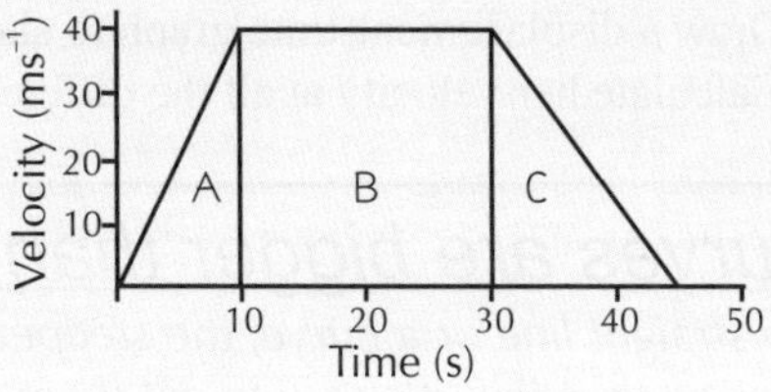

Velocity-Time Graphs

Non-Uniform Acceleration is a Curve on a V-T Graph

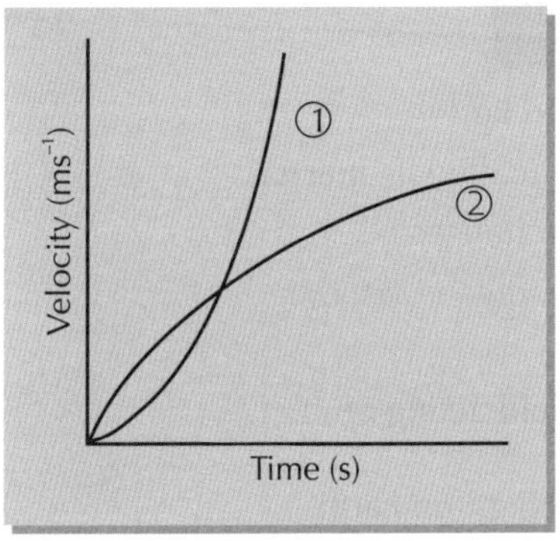

1) If the acceleration is changing, the gradient of the velocity-time graph will also be changing — so you **won't** get a **straight line**.
2) **Increasing acceleration** is shown by an **increasing gradient** — like in curve ①.
3) **Decreasing acceleration** is shown by a **decreasing gradient** — like in curve ②.

Simple enough...

You Can Draw Displacement-Time and Velocity-Time Graphs Using ICT

Instead of gathering distance and time data using **traditional methods**, e.g. a stopwatch and ruler, you can be a bit more **high-tech**.

A fairly **standard** piece of kit you can use for motion experiments is an **ultrasound position detector**. This is a type of **data-logger** that automatically records the **distance** of an object from the sensor several times a second.

If you attach one of these detectors to a computer with **graph-drawing software**, you can get **real-time** displacement-time and velocity-time graphs.

The main **advantages** of data-loggers over traditional methods are:

1) The data is more **accurate** — you don't have to allow for human reaction times.
2) Automatic systems have a much higher **sampling** rate than humans — most ultrasound position detectors can take a reading ten times every second.
3) You can see the data displayed in **real time**.

Practice Questions

Q1 How do you calculate acceleration from a velocity-time graph?

Q2 How do you calculate the distance travelled from a speed-time graph?

Q3 Sketch velocity-time graphs for constant velocity and constant acceleration.

Q4 Describe the main advantages of ICT over traditional methods for the collection and display of motion data.

Exam Question

Q1 A skier accelerates uniformly from rest at 2 ms^{-2} down a straight slope.

(a) Sketch a velocity-time graph for the first 5 s of his journey. [2 marks]

(b) Use a constant acceleration equation to calculate his displacement at t = 1, 2, 3, 4 and 5 s, and plot this information onto a displacement-time graph. [5 marks]

(c) Suggest another method of calculating the skier's distance travelled after each second and use this to check your answers to part (b). [2 marks]

Still awake — I'll give you five more minutes...

There's a really nice sunset outside my window. It's one of those ones that makes the whole landscape go pinky-yellowish. And that's about as much interest as I can muster on this topic. Normal service will be resumed on page 51, I hope.

Mass, Weight and Centre of Gravity

I'm sure you know all this 'mass', 'weight' and 'density' stuff from GCSE. But let's just make sure...

The Mass** of a **Body** makes it **Resist Changes** in **Motion

1) The **mass** of an object is the **amount of 'stuff'** (or **matter**) in it. It's measured in **kg**.
2) The greater an object's mass, the greater its **resistance** to a **change in velocity** (called its **inertia**).
3) The **mass** of an object **doesn't change** if the strength of the **gravitational field** changes.
4) Weight is a **force**. It's measured in **newtons** (N), like all forces.
5) Weight is the **force experienced by a mass** due to a **gravitational field**.
6) The weight of an object **does vary** according to the size of the **gravitational field** acting on it.

weight = mass × gravitational field strength (W = mg) where $g = 9.81\ \text{Nkg}^{-1}$ on Earth.

This table shows Derek (the lion*)'s mass and weight on the Earth and the Moon.

Name	Quantity	Earth ($g = 9.81\ \text{Nkg}^{-1}$)	Moon ($g = 1.6\ \text{Nkg}^{-1}$)
Mass	Mass (scalar)	150 kg	150 kg
Weight	Force (vector)	1471.5 N	240 N

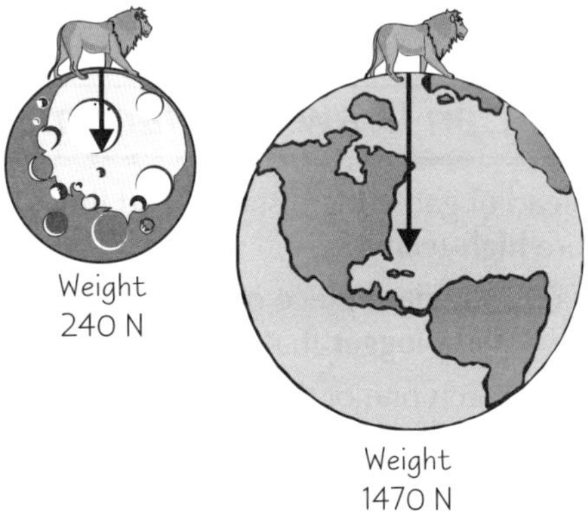

Density** is **Mass** per **Unit Volume

Density is a measure of the 'compactness' (for want of a better word) of a substance.
It relates the mass of a substance to how much space it takes up.

$$\textit{density} = \frac{\textit{mass}}{\textit{volume}} \qquad \rho = \frac{m}{V}$$

The symbol for density is a Greek letter rho (ρ) — it looks like a p but it isn't.

The **units** of **density** are **g cm^{-3}** or **kg m^{-3}**
N.B. $1\ \text{g cm}^{-3} = 1000\ \text{kg m}^{-3}$

1) The density of an object depends on what it's made of.
   Density **doesn't vary** with **size or shape**.
2) The **average density** of an object determines whether it **floats** or **sinks**.
3) A solid object will **float** on a fluid if it has a **lower density** than the **fluid**.

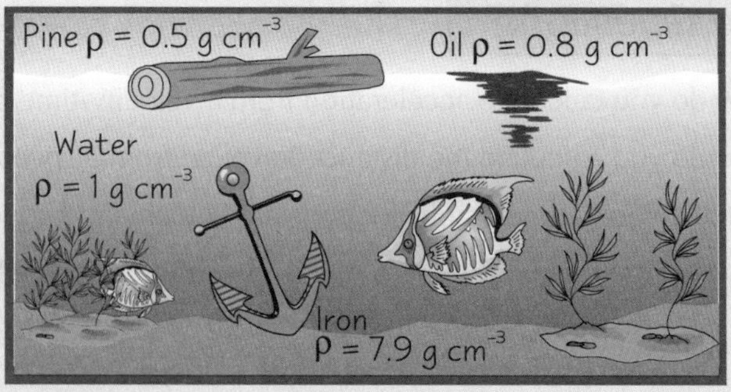

Centre of Gravity** — Assume **All the Mass** is in **One Place

1) The **centre of gravity** (or centre of mass) of an object is the **single point** that you can consider its **whole weight** to **act through** (whatever its orientation).
2) The object will always **balance** around this **point**, although in some cases the **centre of gravity** will **fall outside** the object.

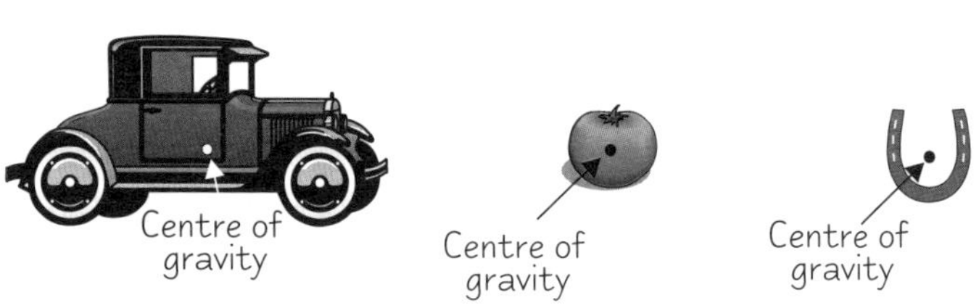

*Yes, I know — I just like lions, O

Mass, Weight and Centre of Gravity

Find the *Centre of Gravity* either by *Symmetry* or *Experiment*

Experiment to find the Centre of Gravity of an Irregular Object

1) Hang the object freely from a point (e.g. one corner).
2) Draw a vertical line downwards from the point of suspension — use a plumb bob to get your line exactly vertical.
3) Hang the object from a different point.
4) Draw another vertical line down.
5) The centre of gravity is where the two lines cross.

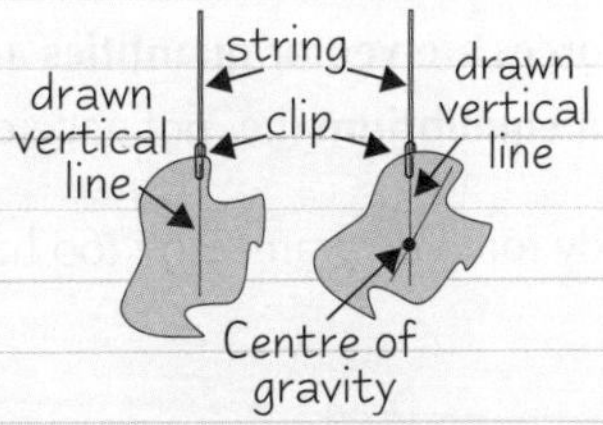

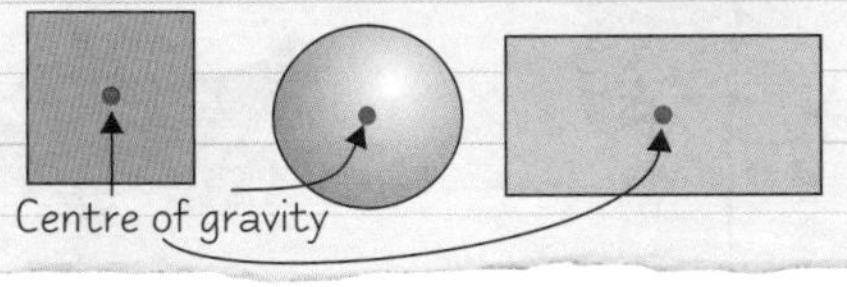

For a regular object you can just use symmetry. The centre of gravity of any regular shape is at its centre.

How High the *Centre of Gravity* is tells you *How Stable* the Object is

1) An object will be nice and **stable** if it has a **low centre** of **gravity** and a **wide base area**. This idea is used a lot in design, e.g. Formula 1 racing cars.

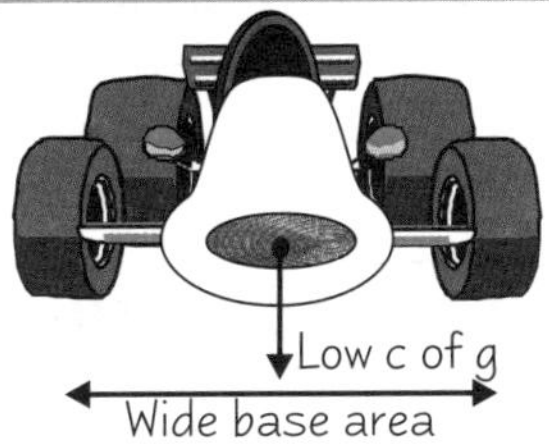

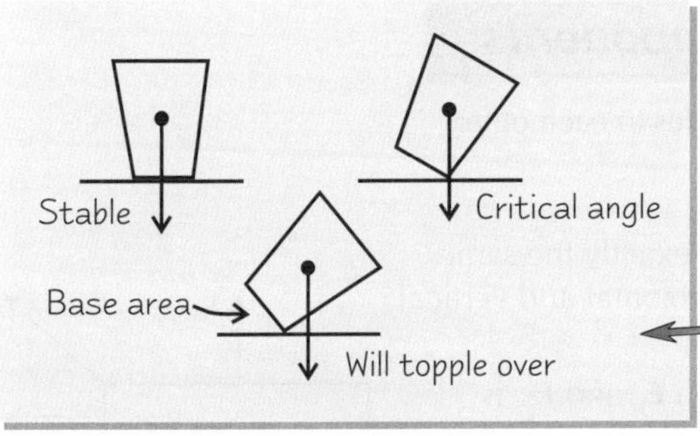

2) The **higher** the **centre of gravity**, and the **smaller** the **base area**, the **less stable** the object will be. Think of unicyclists...
3) An object will topple over if a **vertical line** drawn **downwards** from its **centre of gravity** falls **outside** its **base area**.

Practice Questions

Q1 A lioness has a mass of 200 kg. What would be her mass and weight on the Earth (where $g = 9.8\ \text{Nkg}^{-1}$) and on the Moon (where $g = 1.6\ \text{Nkg}^{-1}$)?

Q2 What is meant by the centre of gravity of an object?

Exam Questions

Q1 (a) Define **density**. [1 mark]

(b) A cylinder of aluminium, radius 4 cm and height 6 cm, has a mass of 820 g. Calculate its density. [3 marks]

(c) Use the information from part (b) to calculate the mass of a cube of aluminium of side 5 cm. [1 mark]

Q2 Describe an experiment to find the centre of gravity of an object of uniform density with a constant thickness and irregular cross-section. Identify one major source of uncertainty and suggest a way to reduce its effect on the accuracy of your result. [5 marks]

The centre of gravity of this book should be round about page 108...

This is a really useful area of physics. To would-be nuclear physicists it might seem a little dull, but if you want to be an engineer — something a bit more useful (no offence Einstein) — then things like centre of gravity and density are dead important things to understand. You know, for designing things like cars and submarines... yep, pretty useful I'd say.

Forces

Remember the vector stuff from the beginning of the section... good, you're going to need it...

Free-Body Force Diagrams show All Forces on a Single Body

1) **Free-body force** diagrams show a **single body** on its own.
2) The diagram should include all the **forces** that **act on** the body, but **not** the **forces it exerts** on the rest of the world.
3) Remember **forces** are **vector quantities** and so the **arrow labels** should show the **size** and **direction** of the forces.
4) If a body is in **equilibrium** (i.e. not accelerating) the **forces** acting on it will be **balanced**.

Drawing free-body force diagrams isn't too hard — you just need practice. Here are a few **examples**:

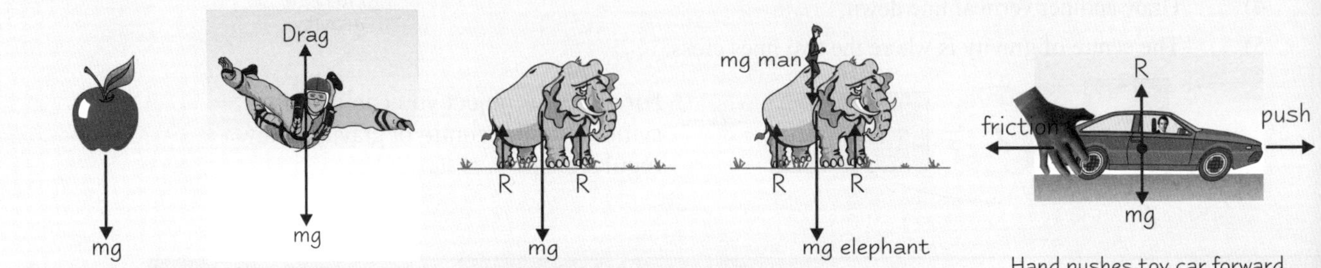

Gravity pulls apple downwards (weight = m × g).

Gravity pulls man down. Air resistance pushes man up.

Gravity pulls elephant down. Earth pushes elephant up (reaction).

Gravity pulls elephant down. Man pushes elephant down. Earth pushes elephant up.

Hand pushes toy car forward. Friction pulls car backwards. Gravity pulls car down. Earth pushes car up.

Resolving a Force means Splitting it into Components

1) Forces can be in **any direction**, so they're not always at right angles to each other. This is sometimes a bit **awkward** for **calculations**.
2) To make an 'awkward' force easier to deal with, you can think of it as **two separate forces**, acting at **right angles** to **each other**.

The force ***F*** has exactly the same effect as the horizontal and vertical forces, F_H and F_V.
Replacing ***F*** with F_H and F_V is called **resolving the force *F***.

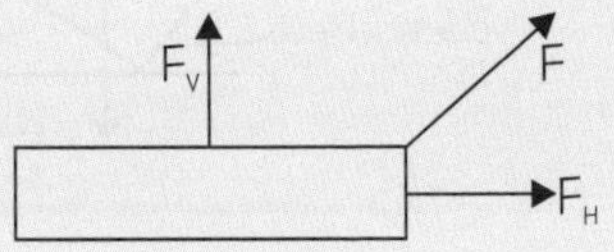

3) To find the size of a component force in a particular direction, you need to use trigonometry (see page 41). Forces are vectors, so you treat them in the same way as velocities — put them end to end.

So this...

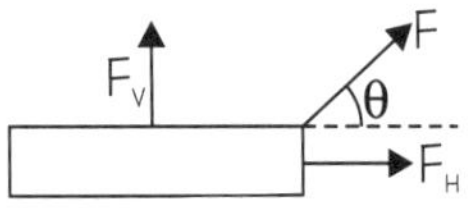

...could be drawn like this:

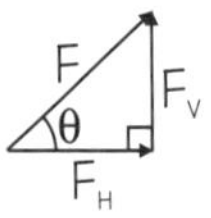

Using trig. you get:

$$\frac{F_H}{F} = \cos\theta \quad \textbf{or} \quad F_H = F\cos\theta$$

And:

$$\frac{F_V}{F} = \sin\theta \quad \textbf{or} \quad F_V = F\sin\theta$$

Example

A tree trunk is pulled along the ground by an elephant exerting a force of 1200 N at an angle of 25° to the horizontal. Calculate the components of this force in the horizontal and vertical directions.

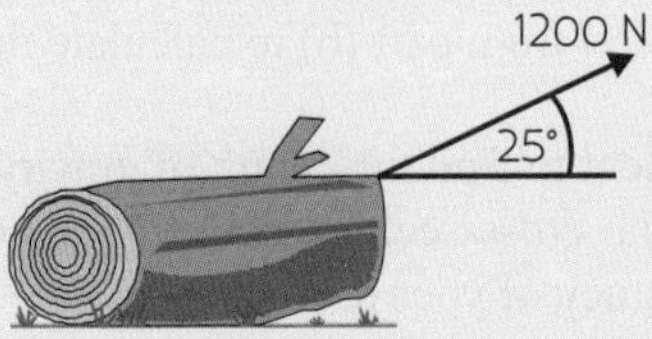

Horizontal force = 1200 × cos 25° = **1088 N**

Vertical force = 1200 × sin 25° = **507 N**

Forces

You Add the Components Back Together to get the Resultant Force

1) If **two forces** act on an object, you find the **resultant** (total) **force** by adding the **vectors** together and creating a **closed triangle**, with the resultant force represented by the **third side**.
2) Forces are vectors (as you know), so you use **vector addition** — draw the forces as vector arrows put 'tail to top'.
3) Then it's yet more trigonometry to find the **angle** and the **length** of the third side.

Example

Two dung beetles roll a dung ball along the ground at constant velocity. Beetle A applies a force of 0.5 N northwards while beetle B exerts a force of only 0.2 N eastwards. What is the resultant force on the dung ball?

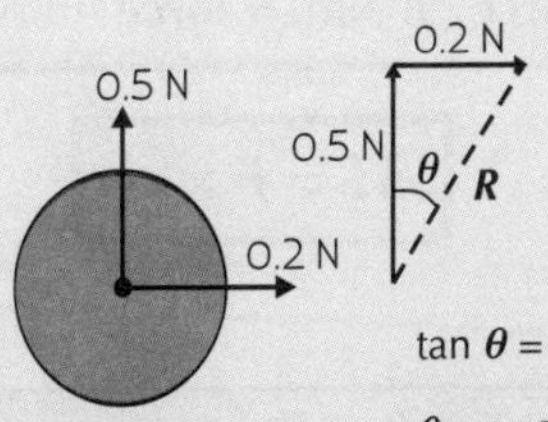

By Pythagoras

$R^2 = 0.5^2 + 0.2^2$
$R = \sqrt{0.29}$
$= 0.54$ N

$\tan\theta = \frac{0.2}{0.5}$
$\theta = \tan^{-1}0.4$
$\theta = 21.8°$

The resultant force is **0.54 N** at an angle of **21.8°** from North.

Choose sensible Axes for Resolving

Use directions that **make sense** for the situation you're dealing with. If you've got an object on a slope, choose your directions **along the slope** and **at right angles to it**. You can turn the paper to an angle if that helps.

Always choose sensible axes

Examiners like to call a slope an "inclined plane".

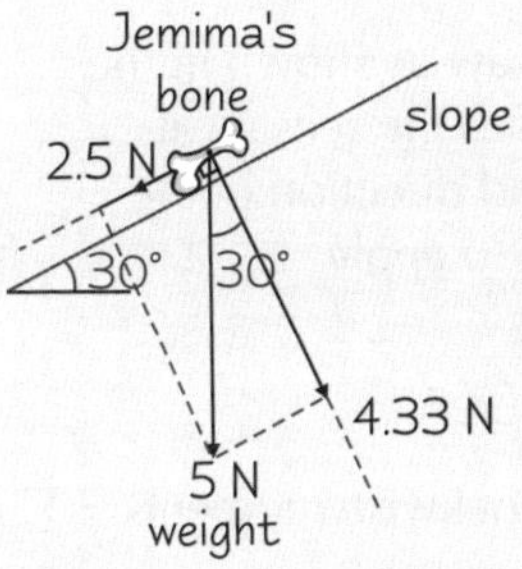

The component of the bone's weight down the slope is 2.5 N so you'd need 2.5 N of friction to stop it sliding away.

Practice Questions

Q1 Sketch a free-body force diagram for an ice hockey puck moving across the ice (assuming no friction).

Q2 What are the horizontal and vertical components of the force F?

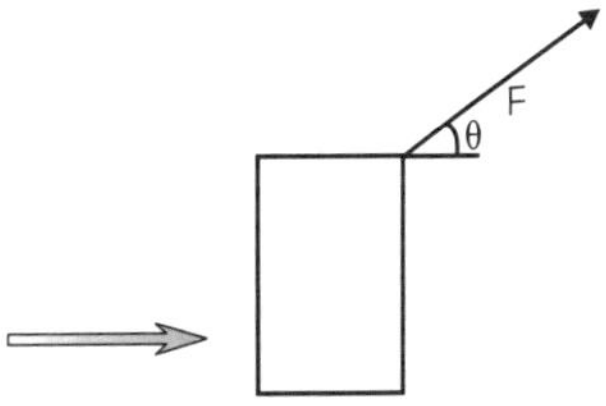

Exam Questions

Q1 A picture is suspended from a hook as shown in the diagram. Calculate the tension force, *T*, in the string.

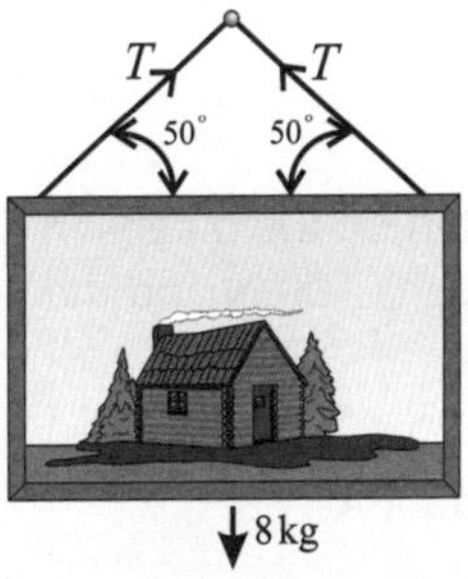

[2 marks]

Q2 Two elephants pull a tree trunk as shown in the diagram. Calculate the resultant force on the tree trunk.

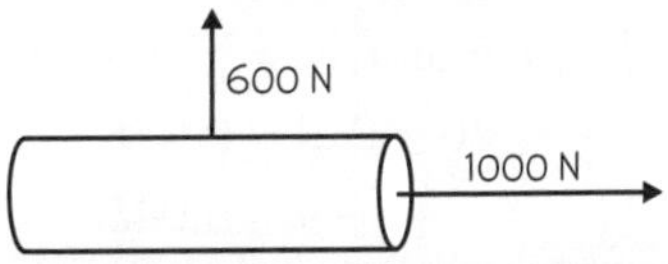

[2 marks]

Free-body force diagram — sounds like something you'd get with a dance mat...

Remember those F cos θ and F sin θ bits. Write them on bits of paper and stick them to your wall. Scrawl them on your pillow. Tattoo them on your brain. Whatever it takes — you just ***have to learn them****.*

Moments and Torques

*This is not a time for jokes. There is not a moment to lose. The time for torquing is over. Oh ho ho ho ho *bang*. (Ow.)*

A Moment is the Turning Effect of a Force

The **moment**, or **torque**, of a **force** depends on the **size** of the force and **how far** the force is applied from the **turning point**:

moment of a force (in Nm) = **force** (in N) × **perpendicular distance from pivot** (in m)

In symbols, that's: $M = F \times d$

Moments must be Balanced or the Object will Turn

The **principle of moments** states that for a body to be in **equilibrium**, the **sum of the clockwise moments** about any point **equals** the **sum of the anticlockwise moments** about the same point.

Example

Two children sit on a seesaw as shown in the diagram. An adult balances the seesaw at one end. Find the size and direction of the force that the adult needs to apply.

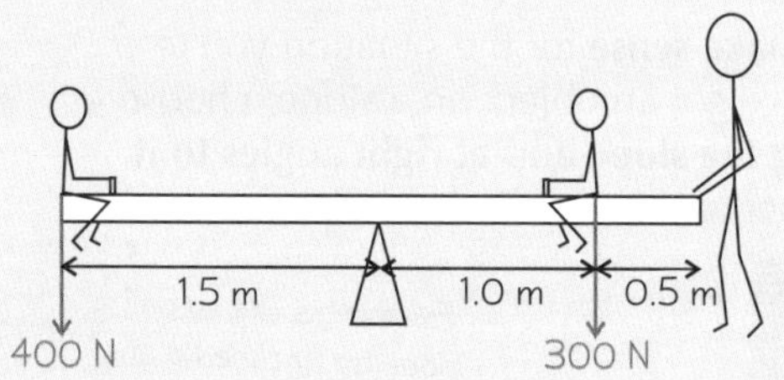

In equilibrium, $\sum$ anticlockwise moments = $\sum$ clockwise moments

$400 \times 1.5 = 300 \times 1 + 1.5F$

$600 = 300 + 1.5F$

Final answer: F = 200 N downwards

$\sum$ means "the sum of"

Muscles, Bones and Joints Act as Levers

1) In a lever, an **effort force** (in this case from a muscle) acts against a **load force** (e.g. the weight of your arm) by means of a **rigid object** (the bone) rotating around a **pivot** (the joint).
2) You can use the **principle of moments** to answer lever questions:

Example

Find the force exerted by the biceps in holding a bag of gold still. The bag of gold weighs 100 N and the forearm weighs 20 N.

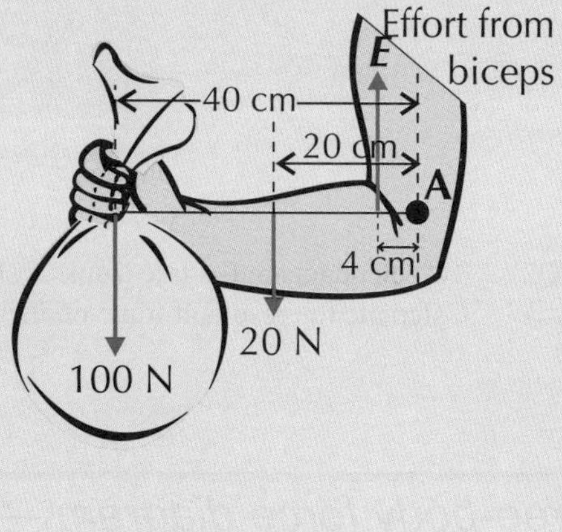

Take moments about **A**.

In equilibrium:

$\sum$ anticlockwise moments = $\sum$ clockwise moments

$(100 \times 0.4) + (20 \times 0.2) = 0.04E$

$40 + 4 = 0.04E$

Final answer: E = 1100 N

Moments and Torques

A Couple is a Pair of Forces

1) A couple is a **pair** of **forces** of **equal size** which act **parallel** to each other, but in **opposite directions**.
2) A couple doesn't cause any resultant linear force, but **does** produce a **turning force** (usually called a **torque** rather than a moment).

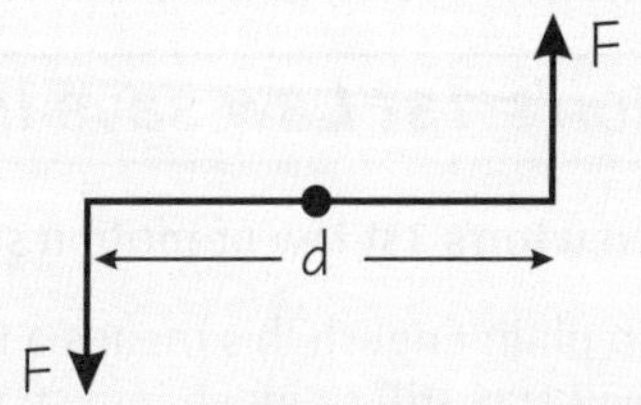

The **size** of this **torque** depends on the **size** of the **forces** and the **distance** between them.

Torque of a couple (in Nm) = **size of one of the forces** (in N) × **perpendicular distance between the forces** (in m)

In symbols, that's: $T = F \times d$

Example

A cyclist turns a sharp right corner by applying equal but opposite forces of 20 N to the ends of the handlebars.

The length of the handlebars is 0.6 m.
Calculate the torque applied to the handlebars.

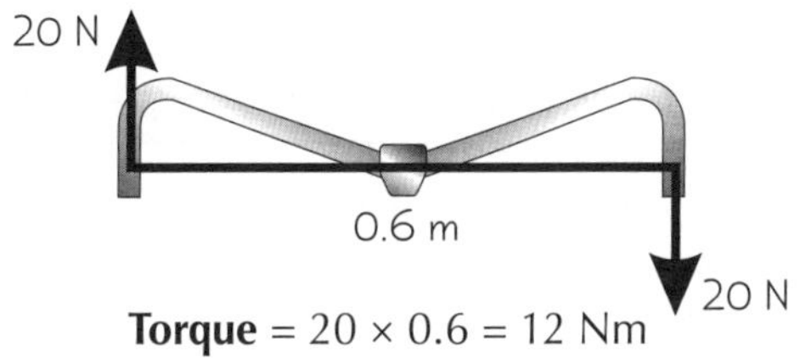

Torque = 20 × 0.6 = 12 Nm

Practice Questions

Q1 A girl of mass 40 kg sits 1.5 m from the middle of a seesaw.
Show that her brother, mass 50 kg, must sit 1.2 m from the middle if the seesaw is to balance.

Q2 What is meant by the word 'couple'?

Exam Questions

Q1 A driver is changing his flat tyre. The torque required to undo the nut is 60 Nm.
He uses a 0.4 m long double-ended wheel wrench.
Calculate the force that he must apply at each end of the spanner. [2 marks]

Q2 A diver of mass 60 kg stands on the end of a diving board 2 m from the pivot point.
Calculate the upward force exerted on the retaining spring 30 cm from the pivot.

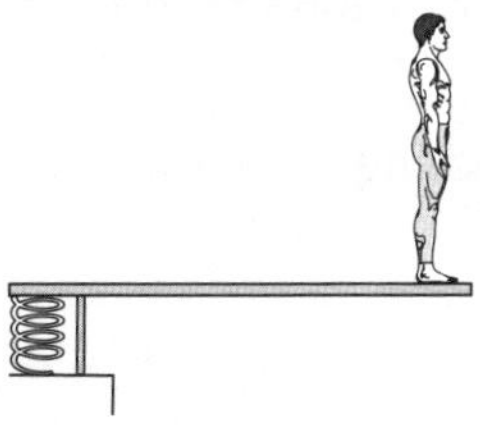

[2 marks]

It's all about balancing — just ask a tightrope walker...

*They're always seesaw questions aren't they. It'd be nice if once, **just once**, they'd have a question on... I don't know, rotating knives or something. Just something unexpected... anything. It'd make physics a lot more fun, I'm sure. *sigh**

Newton's Laws of Motion

You did most of this at GCSE, but that doesn't mean you can just skip over it now. You'll be kicking yourself if you forget this stuff in the exam — easy marks...

Newton's **1st Law** says that a **Force** is Needed to Change Velocity

1) **Newton's 1st law of motion** states that the **velocity** of an object will **not change** unless a **resultant force** acts on it.
2) In plain English this means a body will stay still or move in a **straight line** at a **constant speed**, unless there's a **resultant force** acting on it.

An apple sitting on a table won't go anywhere because the **forces** on it are **balanced**.

reaction (***R***) (force of table pushing apple up) = **weight** (mg) (force of gravity pulling apple down)

3) If the forces **aren't balanced**, the **overall resultant force** will make the body **accelerate**. This could be a change in **direction**, or **speed**, or both. (See Newton's 2nd law, below.)

Newton's **2nd Law** says that **Acceleration** is **Proportional** to the Force

...which can be written as the well-known equation:

resultant force (N) = mass (kg) × acceleration (ms^{-2})

$$F = m \times a$$

Learn this — it crops up all over the place in AS Physics. And learn what it means too:

1) It says that the **more force** you have acting on a certain mass, the **more acceleration** you get.
2) It says that for a given force the **more mass** you have, the **less acceleration** you get.

REMEMBER:
1) The **resultant force** is the **vector sum** of all the forces.
2) The force is **always** measured in **newtons**.
3) The **mass** is always measured in **kilograms**.
4) The **acceleration** is always in the **same direction** as the **resultant force** and is measured in **ms^{-2}**.

Galileo said: **All Objects Fall** at the **Same Rate** (if you **Ignore Air Resistance**)

You need to understand **why** this is true. Newton's 2nd law explains it neatly — consider two balls dropped at the same time — ball **1** being heavy, and ball **2** being light. Then use Newton's 2nd law to find their acceleration.

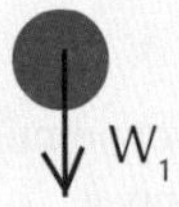

mass = m_1
resultant force = F_1
acceleration = a_1
By Newton's Second Law:

$$F_1 = m_1 a_1$$

Ignoring air resistance, the only force acting on the ball is weight, given by $W_1 = m_1 g$ (where g = gravitational field strength = 9.81 Nkg^{-1}).

So: $F_1 = m_1 a_1 = W_1 = m_1 g$

So: $m_1 a_1 = m_1 g$, then m_1 cancels out to give: $a_1 = g$

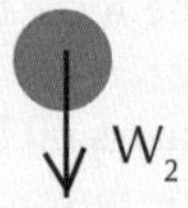

mass = m_2
resultant force = F_2
acceleration = a_2
By Newton's Second Law:

$$F_2 = m_2 a_2$$

Ignoring air resistance, the only force acting on the ball is weight, given by $W_2 = m_2 g$ (where g = gravitational field strength = 9.81 Nkg^{-1}).

So: $F_2 = m_2 a_2 = W_2 = m_2 g$

So: $m_2 a_2 = m_2 g$, then m_2 cancels out to give: $a_2 = g$

... in other words, the **acceleration** is **independent of the mass**. It makes **no difference** whether the ball is **heavy or light**. And I've kindly **hammered home the point** by showing you two almost identical examples.

Newton's Laws of Motion

Newton's **3rd Law** says each Force has an **Equal, Opposite Reaction Force**

There are a few different ways of stating Newton's 3rd law, but the clearest way is:

If an object A EXERTS a FORCE on object B, then object B exerts AN EQUAL BUT OPPOSITE FORCE on object A.

You'll also hear the law as "every action has an equal and opposite reaction". But this confuses people who wrongly think the forces are both applied to the same object. (If that were the case, you'd get a resultant force of zero and nothing would ever move anywhere...)

The two forces actually represent the **same interaction**, just seen from two **different perspectives**:

1) If you **push against a wall**, the wall will **push back** against you, **just as hard**. As soon as you stop pushing, so does the wall. Amazing...
2) If you **pull a cart**, whatever force **you exert** on the rope, the rope exerts the **exact opposite** pull on you (unless the rope's stretching).
3) When you go **swimming**, you push **back** against the water with your arms and legs, and the water pushes you **forwards** with an equal-sized force.

Newton's 3rd law applies in **all situations** and to all **types of force**. But the pairs of forces are always the **same type**, e.g. both gravitational or both electrical.

This looks like Newton's 3rd law...

But it's <u>NOT</u>.

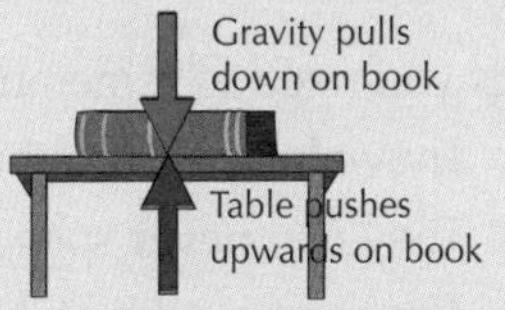

...because both forces are acting on the book, and they're not of the same type. This is two separate interactions. The forces are equal and opposite, resulting in zero acceleration, so this is showing Newton's 1st law.

Practice Questions

Q1 State Newton's 1st, 2nd and 3rd laws of motion, and explain what they mean.

Q2 What are the two equal and opposite forces acting between an orbiting satellite and the Earth?

Exam Questions

Q1 Draw diagrams to show the forces acting on a parachutist:
(a) accelerating downwards. [1 mark]
(b) having reached terminal velocity. [1 mark]

Q2 A boat is moving across a river. The engines provide a force of 500 N at right angles to the flow of the river and the boat experiences a drag of 100 N in the opposite direction. The force on the boat due to the flow of the river is 300 N. The mass of the boat is 250 kg.
(a) Calculate the magnitude of the resultant force acting on the boat. [2 marks]
(b) Calculate the magnitude of the acceleration of the boat. [2 marks]

Q3 This question asks you to use Newton's second law to explain three situations.
(a) Two cars have different maximum accelerations. What are the only two overall factors that determine the acceleration a car can have? [2 marks]
(b) Michael can always beat his younger brother Tom in a sprint, however short the distance. Give two possible reasons for this. [2 marks]
(c) Michael and Tom are both keen on diving. They notice that they seem to take the same time to drop from the diving board to the water. Explain why this is the case. (Assume no air resistance.) [3 marks]

<u>*Newton's three incredibly important laws of motion...*</u>

These laws may not really fill you with a huge amount of excitement (and I could hardly blame you if they don't)... but it was pretty fantastic at the time — suddenly people actually understood how forces work, and how they affect motion. I mean arguably it was one of the most important scientific discoveries ever...

Drag and Terminal Velocity

If you jump out of a plane at 2000 ft, you want to know that you're not going to be accelerating all the way.

Friction is a ***Force*** that ***Opposes Motion***

There are two main types of friction:

1) **Contact friction** between **solid surfaces** (which is what we usually mean when we just use the word 'friction'). You don't need to worry about that too much for now.
2) **Fluid friction** (known as **drag** or fluid resistance or air resistance).

Fluid Friction or Drag:

1) 'Fluid' is a word that means either a **liquid or a gas** — something that can **flow**.
2) The force depends on the thickness (or **viscosity**) of the fluid.
3) It **increases** as the **speed increases** (for simple situations it's directly proportional, but you don't need to worry about the mathematical relationship).
4) It also depends on the **shape** of the object moving through it — the larger the **area** pushing against the fluid, the greater the resistance force.

Things you need to remember about frictional forces:

1) They **always** act in the **opposite direction** to the **motion** of the object.
2) They can **never** speed things up or start something moving.
3) They convert **kinetic energy** into **heat**.

Terminal Velocity — when the ***Friction*** Force ***Equals*** the ***Driving Force***

You will reach a **terminal velocity** at some point, if you have:

1) a **driving force** that stays the **same** all the time
2) a **frictional** or **drag force** (or collection of forces) that increases with speed

There are **three main stages** to reaching terminal velocity:

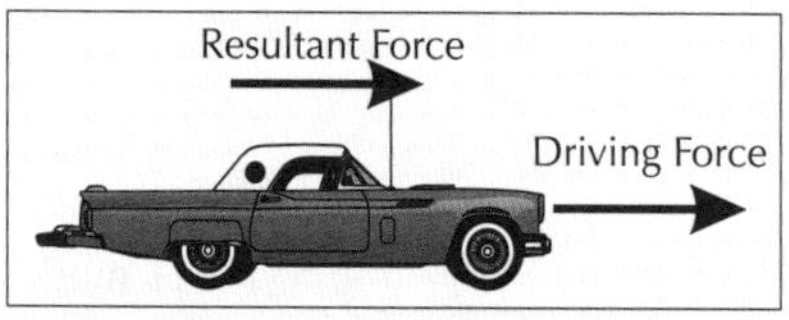

The car **accelerates** from **rest** using a constant driving force.

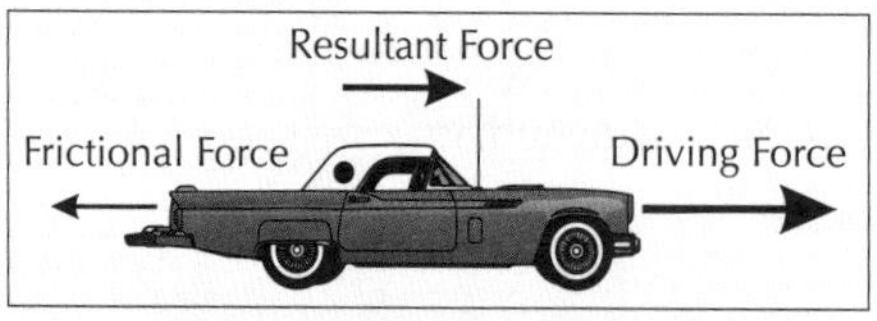

As the **velocity increases**, the **resistance forces increase** (because of things like turbulence — you don't need the details). This **reduces the resultant force** on the car and hence **reduces its acceleration**.

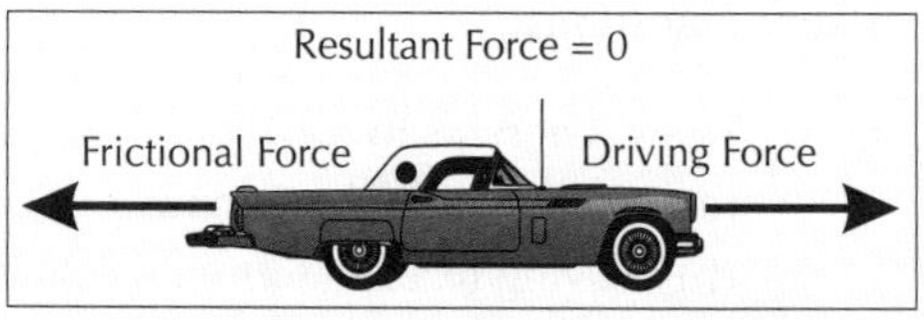

Eventually the car reaches a velocity at which the **resistance forces are equal to the driving force**. There is now **no resultant force** and **no acceleration**, so the car carries on at **constant velocity**.

Sketching a ***Graph*** for ***Terminal Velocity***

You need to be able to **recognise** and **sketch** the graphs for **velocity against time** and **acceleration against time** for the **terminal velocity** situation.

Nothing for it but practice — shut the book and sketch them from memory. Keep doing it till you get them right every time.

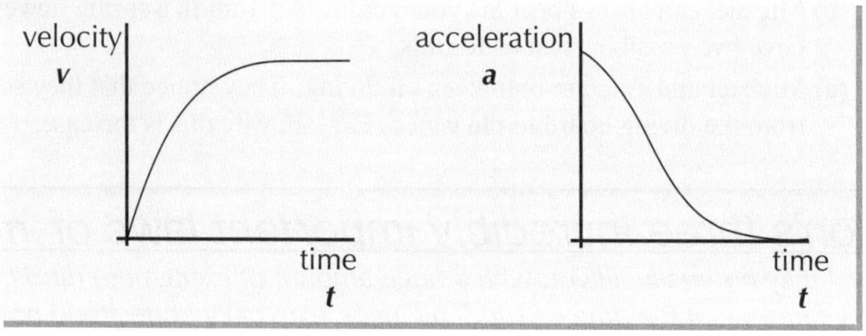

Drag and Terminal Velocity

*Things **Falling** through **Air** or **Water** Reach a **Terminal Velocity** too*

When something's falling through **air**, the **weight** of the object is a **constant force** accelerating the object downwards. **Air resistance** is a **frictional force** opposing this motion, which **increases** with **speed**.
So before a parachutist opens the parachute, exactly the same thing happens as with the car example:

1) A skydiver leaves a plane and will **accelerate** until the **air resistance** equals his **weight**.

2) He will then be travelling at a **terminal velocity**.

But... the terminal velocity of a person in free fall is too great to land without dying a horrible death.
The **parachute increases** the **air resistance massively**, which slows him down to a lower terminal velocity:

3) Before reaching the ground he will **open his parachute**, which immediately **increases the air resistance** so it is now **bigger** than his **weight**.

4) This **slows him down** until his speed has dropped enough for the **air resistance** to be **equal to his weight** again. This new terminal velocity is small enough to survive landing.

The v-t graph is a bit different, because you have a new terminal velocity being reached after the parachute is opened:

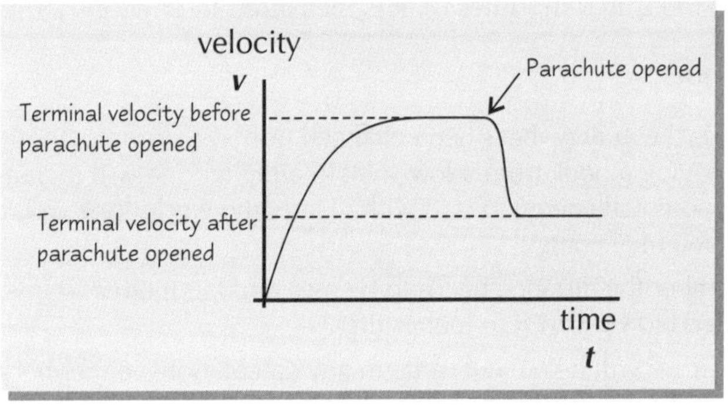

Practice Questions

Q1 What forces limit the speed of a skier going down a slope?

Q2 What conditions cause a terminal velocity to be reached?

Q3 Sketch a graph to show how the velocity changes with time for an object falling through air.

Exam Question

Q1 A space probe free-falls towards the surface of a planet.
The graph on the right shows the velocity data recorded by the probe as it falls.

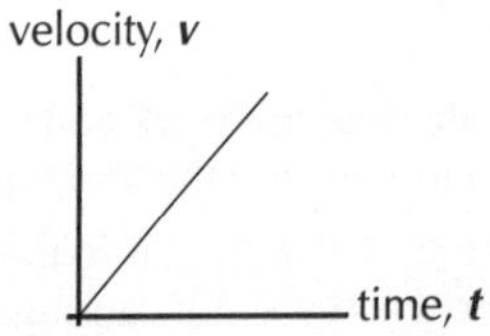

(a) The planet does not have an atmosphere. Explain how you can tell this from the graph. [2 marks]

(b) Sketch the velocity-time graph you would expect to see if the planet did have an atmosphere. [2 marks]

(c) Explain the shape of the graph you have drawn. [3 marks]

You'll never understand this without going parachuting...*

When you're doing questions about terminal velocity, remember the frictional forces reduce acceleration, not speed. They usually don't slow an object down, apart from in the parachute example, where the skydiver is travelling faster when the parachute opens than the terminal velocity for the parachute-skydiver combination.

* No. 37 in a series of the 100 least convincing excuses for an interesting holiday.

Work and Power

As everyone knows, work in Physics isn't like normal work. It's harder. Work also has a specific meaning that's to do with movement and forces. You'll have seen this at GCSE — it just comes up in more detail for AS level.

Work *is Done Whenever* ***Energy*** *is* ***Transferred***

This table gives you some examples of **work being done** and the **energy changes** that happen.

1) Usually you need a force to move something because you're having to **overcome another force**.
2) The thing being moved has **kinetic energy** while it's **moving**.
3) The kinetic energy is transferred to **another form of energy** when the movement stops.

ACTIVITY	WORK DONE AGAINST	FINAL ENERGY FORM
Lifting up a box.	gravity	gravitational potential energy
Pushing a chair across a level floor.	friction	heat
Pushing two magnetic north poles together.	magnetic force	magnetic energy
Stretching a spring.	stiffness of spring	elastic potential energy

The word **'work'** in Physics means the **amount of energy transferred** from one form to another when a force causes a movement of some sort.

Work = Force × Distance

When a car tows a caravan, it applies a force to the caravan to moves it to where it's wanted.
To **find out** how much **work** has been **done**, you need to use the **equation**:

work done (W) = **force causing motion** (F) × **distance moved** (s)
...where W is measured in joules (J), F is measured in newtons (N) and s is measured in metres (m).

Points to remember:

1) **Work** is the **energy** that's been **changed** from one form to another — it's not necessarily the **total** energy. E.g. moving a book from a low shelf to a higher one will increase its gravitational potential energy, but it had some potential energy to start with. Here, the **work done** would be the **increase** in potential energy, **not the total** potential energy.
2) Remember the distance needs to be measured in metres — if you have **distance in centimetres or kilometres**, you need to **convert** it to metres first.
3) The force F will be a **fixed** value in any calculations, either because it's **constant** or because it's the **average** force.
4) The equation assumes that the **direction of the force** is the **same** as the **direction of movement**.
5) The equation gives you the **definition** of the joule (symbol J): 'One joule is the work done when a force of 1 newton moves an object through a distance of 1 metre'.

The ***Force*** *isn't always in the* ***Same Direction*** *as the* ***Movement***

Sometimes the **direction of movement** is **different** from the **direction of the force**.

Example

1) To **calculate the work done** in a situation like the one on the right, you need to consider the **horizontal** and **vertical components** of the **force**.
2) The only **movement** is in the **horizontal** direction. This means the **vertical force** is not causing any motion (and hence not doing any work) — it's just **balancing** out some of the **weight**, meaning there's a **smaller reaction force**.

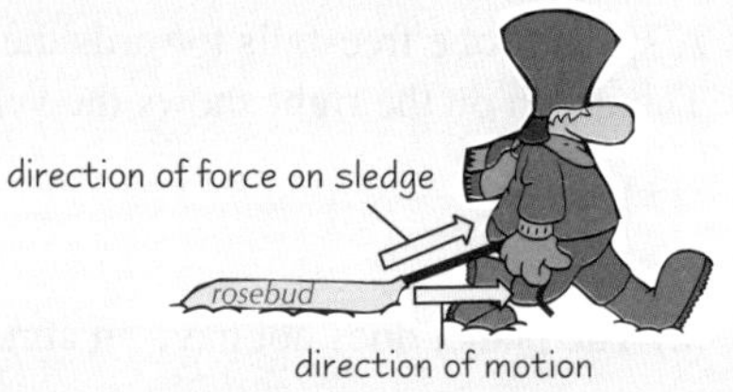

3) The horizontal force is causing the motion — so to **calculate** the **work done**, this is the **only force** you need to consider. Which means we get:

$$W = Fs\cos\theta$$

Where θ is the **angle** between the **direction of the force** and the **direction of motion**. See page 52 for more on resolving forces.

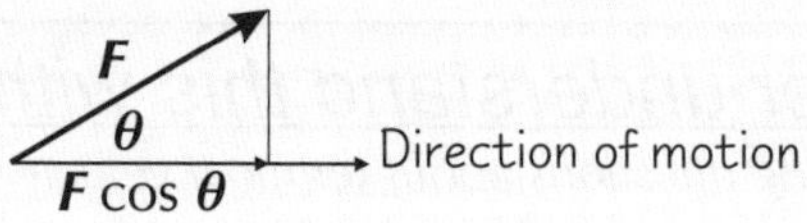

Work and Power

Power = Work Done per Second

Power means many things in everyday speech, but in physics (of course!) it has a special meaning. Power is the **rate of doing work** — in other words it is the **amount of energy transformed** from one form to another **per second.**

You **calculate power** from this equation:

> **Power (P) = work done (W) / time (t)**
> ...where P is measured in watts (W), W is measured in joules (J) and t is measured in seconds (s)

The **watt** (symbol W) is defined as a **rate of energy transfer** equal to **1 joule per second** (Js^{-1}).

Yep, that's another **equation and definition** for you to **learn**.

Power is also Force × Velocity ($P = Fv$)

Sometimes, it's **easier** to use **this version** of the power equation. This is how you get it:

1) You **know** $P = W/t$.
2) You also **know** $W = Fs$, which gives $P = Fs/t$.
3) But $v = s/t$, which you can substitute into the above equation to give $P = Fv$.
4) It's easier to use this if you're given the **speed** in the question.
   Learn this equation as a **shortcut** to link **power** and **speed**.

Example

A car is travelling at a speed of $10\,ms^{-1}$ and is kept going against the frictional force by a driving force of 500 N in the direction of motion. Find the power supplied by the engine to keep the car moving.

Use the shortcut $P = Fv$, which gives:

$P = 500 \times 10 = 5000$ W

If the force and motion are in different directions, you can replace F with $F\cos\theta$ to get: $P = Fv\cos\theta$

You **aren't** expected to **remember** this equation, but it's made up of bits that you **are supposed to know**, so be ready for the possibility of calculating **power** in a situation where the **direction of the force and direction of motion are different**.

Practice Questions

Q1 Write down the equation used to calculate work if the force and motion are in the same direction.

Q2 Write down the equation for work if the force is at an angle to the direction of motion.

Q3 Write down the equations relating (i) power and work and (ii) power and speed.

Exam Questions

Q1 A traditional narrowboat is drawn by a horse walking along the towpath. The horse pulls the boat at a constant speed between two locks which are 1500 m apart. The tension in the rope is 100 N at 40° to the direction of motion.

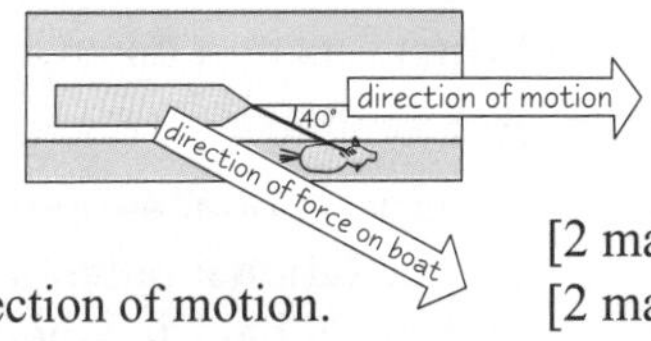

(a) How much work is done on the boat? [2 marks]

(b) The boat moves at $0.8\,ms^{-1}$. Calculate the power supplied to the boat in the direction of motion. [2 marks]

Q2 A motor is used to lift a 20 kg load a height of 3 m. (Take $g = 9.81\,Nkg^{-1}$.)

(a) Calculate the work done in lifting the load. [2 marks]

(b) The speed of the load during the lift is $0.25\,ms^{-1}$. Calculate the power delivered by the motor. [2 marks]

Work — there's just no getting away from it...

Loads of equations to learn. Well, that's what you came here for, after all. Can't beat a good bit of equation-learning, as I've heard you say quietly to yourself when you think no one's listening. Aha, can't fool me. Ahahahahahahahahahahahaha.

Conservation of Energy

Energy *can never be* ***lost****. I repeat —* ***energy*** *can* ***never*** *be lost. Which is basically what I'm about to take up two whole pages saying. But that's, of course, because you need to do exam questions on this as well as understand the principle.*

Learn the Principle of Conservation of Energy

The **principle of conservation of energy** says that:

Energy **cannot be created** or **destroyed**. Energy **can be transferred** from one form to another but the total amount of energy in a closed system will not change.

Example

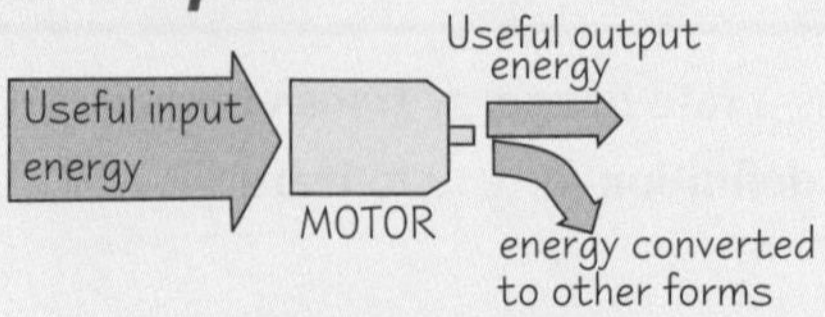

Total energy in = Total energy out

You can talk about how well energy is transferred in terms of **efficiency**.

$$\text{Efficiency} = \frac{\text{useful power output}}{\text{power input}}$$

You need it for Questions about Kinetic and Potential Energy

The principle of conservation of energy nearly always comes up when you're doing questions about changes between kinetic and potential energy.

A quick reminder:

1) **Kinetic energy** is energy of anything **moving**, which you work out from $E_k = \frac{1}{2}mv^2$, where v is the velocity it's travelling at and m is its mass.
2) There are **different types of potential energy** — e.g. gravitational and elastic.
3) **Gravitational potential energy** is the energy something gains if you lift it up. You work it out using: $\Delta E_p = mg\Delta h$, where m is the mass of the object, Δh is the height it is lifted and g is the gravitational field strength (9.81 Nkg^{-1} on Earth).
4) **Elastic potential energy** (elastic stored energy) is the energy you get in, say, a stretched rubber band or spring. You work this out using $E = \frac{1}{2}ke^2$, where **e** is the extension of the spring and k is the stiffness constant.

Examples These pictures show you three **examples** of changes between kinetic and potential energy.

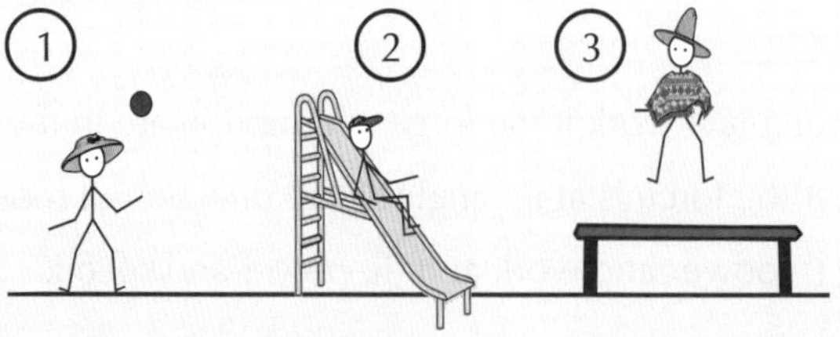

1) As Becky throws the **ball upwards, kinetic energy** is converted into **gravitational potential energy**. When it **comes down** again, that **gravitational potential** energy is **converted back** into **kinetic** energy.
2) As Dominic goes **down the slide, gravitational potential energy** is converted to **kinetic energy**.
3) As Simon bounces upwards from the trampoline, **elastic potential energy** is converted to **kinetic energy**, to **gravitational potential energy**. As he comes back down again, that **gravitational potential** energy is **converted back** to **kinetic** energy, to **elastic potential** energy, and so on.

In **real life** there are also **frictional forces** — Simon would have to use some **force** from his **muscles** to keep **jumping** to the **same height** above the trampoline each time. Each time the trampoline **stretches**, some **heat** is generated in the trampoline material. You're usually told to **ignore friction** in exam questions — this means you can **assume** that the **only forces** are those that provide the **potential or kinetic energy** (in this example that's **Simon's weight** and the **tension** in the springs and trampoline material).
If you're ignoring friction, you can say that the **sum of the kinetic and potential energies is constant**.

Conservation of Energy

Use Conservation of Energy to **Solve Problems**

You need to be able to **use** conservation of mechanical energy (change in potential energy = change in kinetic energy) to solve problems. The classic example is the **simple pendulum**.

In a simple pendulum, you assume that all the mass is in the **bob** at the end.

Example

A simple pendulum has a mass of 700 g and a length of 50 cm. It is pulled out to an angle of 30° from the vertical.

(a) Find the gravitational potential energy stored in the pendulum bob.

Start by drawing a diagram.

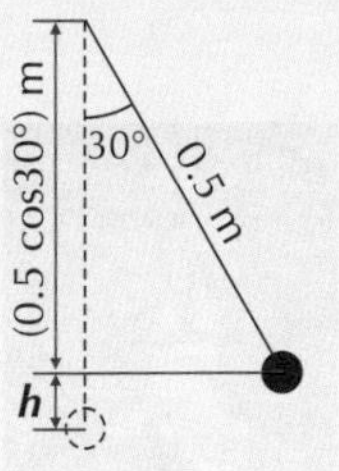

You can work out the increase in height, ***h***, of the end of the pendulum using trig.

Gravitational potential energy = mgh
$= 0.7 \times 9.81 \times (0.5 - 0.5 \cos 30°)$
$= 0.46$ J

(b) The pendulum is released. Find the maximum speed of the pendulum bob as it passes the vertical position.

To find the *maximum* speed, assume no air resistance, then $mgh = \frac{1}{2}mv^2$.

So $\frac{1}{2}mv^2 = 0.46$

rearrange to find $v = \sqrt{\frac{2 \times 0.46}{0.7}} = 1.15 \text{ ms}^{-1}$

OR

Cancel the ***m***s and rearrange to give:
$v^2 = 2gh$
$= 2 \times 9.81 \times (0.5 - 0.5 \cos 30°)$
$= 1.31429...$
$v = 1.15 \text{ ms}^{-1}$

You could be asked to apply this stuff to just about any situation in the exam. **Rollercoasters** are a bit of a favourite.

Practice Questions

Q1 State the principle of conservation of energy.

Q2 What are the equations for calculating kinetic energy and gravitational potential energy?

Q3 Show that, if there's no air resistance and the mass of the string is negligible, the speed of a pendulum is independent of the mass of the bob.

Exam Questions

Q1 A skateboarder is on a half-pipe. He lets the board run down one side of the ramp and up the other. The height of the ramp is 2 m. Take **g** as 9.81 Nkg^{-1}.

(a) If you assume that there is no friction, what would be his speed at the lowest point of the ramp? [3 marks]

(b) How high will he rise up the other side? [1 mark]

(c) Real ramps are not frictionless, so what must the skater do to reach the top on the other side? [1 mark]

Q2 A 20 g rubber ball is released from a height of 8 m. (Assume that the effect of air resistance is negligible.)

(a) Find the kinetic energy of the ball just before it hits the ground. [2 marks]

(b) The ball strikes the ground and rebounds to a height of 6.5 m. How much energy is converted to heat and sound in the impact with the ground? [2 marks]

Energy is never lost — it just sometimes prefers the scenic route...

Remember to check your answers — I can't count the number of times I've forgotten to square the velocities or to multiply by the ½... I reckon it's definitely worth the extra minute to check.

Hooke's Law

Hooke's law doesn't apply to all materials, and only works for the rest up to a point, but it's still pretty handy.

Hooke's Law Says that Extension is Proportional to Force

If a **metal wire** is supported at the top and then a weight attached to the bottom, it **stretches**. The weight pulls down with force ***F***, producing an equal and opposite force at the support.

The material will only deform (stretch, bend, twist etc.) if there's a pair of opposite forces acting on it.

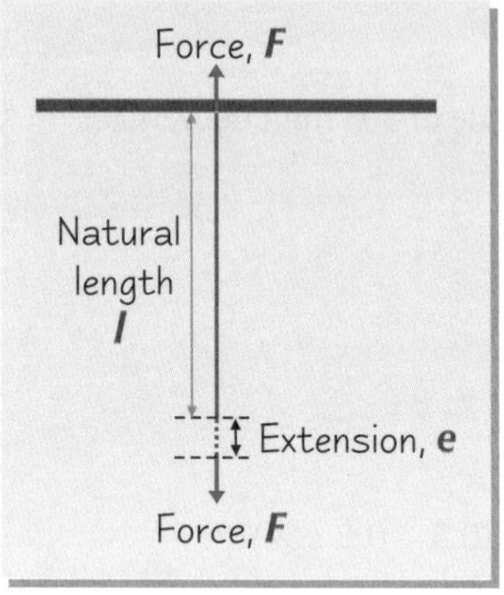

1) **Robert Hooke** discovered in 1676 that the extension of a stretched wire, **e**, is proportional to the load or force, ***F***.
   This relationship is now called **Hooke's law**.
2) Hooke's law can be written:

$$F = ke$$

Where **k** is a constant that depends on the material being stretched.
k is called the **stiffness constant**.

Hooke's law Also Applies to Springs

A metal spring also changes length when you apply a **pair of opposite forces**.

1) The **extension** or **compression** of a spring is **proportional** to the **force** applied — so Hooke's law applies.
2) For springs, **k** in the formula ***F* = ke** is usually called the **spring stiffness** or **spring constant**.

> Hooke's law works just as well for **compressive** forces as **tensile** forces. For a spring, **k** has the **same value** whether the forces are tensile or compressive (that's not true for all materials).

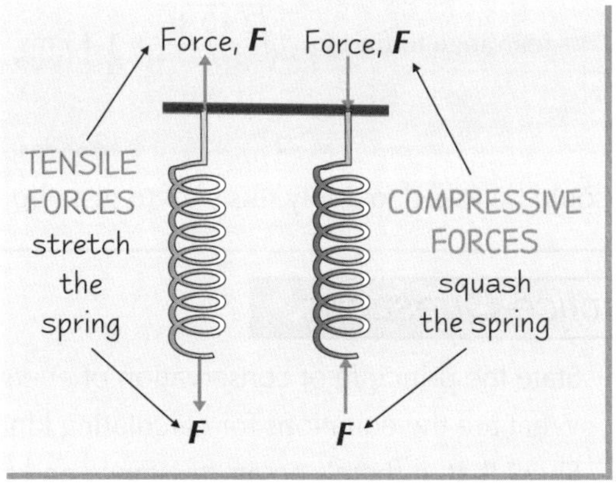

Hooke's law Stops Working when the Load is Great Enough

There's a **limit** to the force you can apply for Hooke's law to stay true.

1) The graph shows load against extension for a **typical metal wire**.
2) The first part of the graph shows Hooke's law being obeyed — there's a **straight-line relationship** between **load** and **extension**.
3) When the load becomes great enough, the graph starts to **curve**. The point marked E on the graph is called the **elastic limit**.
4) If you increase the load past the elastic limit, the material will be **permanently stretched**. When all the force is removed, the material will be **longer** than at the start.
5) **Metals** generally obey Hooke's law up to the limit of proportionality, which is very near the elastic limit.
6) Be careful — there are some materials, like **rubber**, that only obey Hooke's law for **really small** extensions.

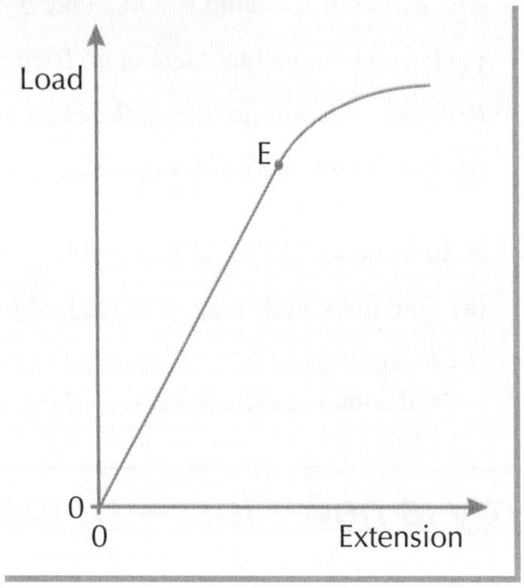

Hooke's Law

So basically...

A Stretch can be **Elastic** or **Plastic**

Elastic

If a **deformation** is **elastic**, the material returns to its **original shape** once the forces are removed.

1) When the material is put under **tension**, the **atoms** of the material are **pulled apart** from one another.
2) Atoms can **move** small distances relative to their **equilibrium positions**, without actually changing position in the material.
3) Once the **load** is **removed**, the atoms **return** to their **equilibrium** distance apart.

For a metal, elastic deformation happens as long as **Hooke's law** is obeyed.

Plastic

If a deformation is **plastic**, the material is **permanently stretched.**

1) Some atoms in the material move position relative to one another.
2) When the load is removed, the **atoms don't return** to their original positions.

A metal stretched **past its elastic limit** shows plastic deformation.

Practice Questions

Q1 State Hooke's law.

Q2 Define tensile forces and compressive forces.

Q3 Explain what is meant by the elastic limit of a material.

Q4 From studying the force-extension graph for a material as it is loaded and unloaded, how can you tell:
(a) if Hooke's law is being obeyed,
(b) if the elastic limit has been reached?

Q5 What is plastic behaviour of a material under load?

Exam Questions

Q1 A metal guitar string stretches 4.0 mm when a 10 N force is applied.

(a) If the string obeys Hooke's law, how far will the string stretch with a 15 N force? [1 mark]

(b) Calculate the stiffness constant for this string in Nm^{-1}. [2 marks]

(c) The string is tightened beyond its elastic limit. What would be noticed about the string? [1 mark]

Q2 A rubber band is 6.0 cm long. When it is loaded with 2.5 N, its length becomes 10.4 cm.
Further loading increases the length to 16.2 cm when the force is 5.0 N.

Does the rubber band obey Hooke's law when the force on it is 5.0 N?
Justify your answer with a suitable calculation. [2 marks]

Sod's Law — if you don't learn it, it'll be in the exam...

Three things you didn't know about Robert Hooke — he was the first person to use the word 'cell' (in terms of biology, not prisons), he helped Christopher Wren with his designs for St. Paul's Cathedral and no-one knows quite what he looked like. I'd like to think that if I did all that stuff, then someone would at least remember what I looked like — poor old Hooke.

Stress and Strain

How much a material stretches for a particular applied force depends on its dimensions.
If you want to compare the properties of two different materials, you need to use stress and strain instead.
A stress-strain graph is the same for any sample of a particular material — the size of the sample doesn't matter.

A Stress Causes a Strain

A material subjected to a pair of **opposite forces** might **deform**, i.e. **change shape**. If the forces **stretch** the material, they're **tensile**. If the forces **squash** the material, they're **compressive**.

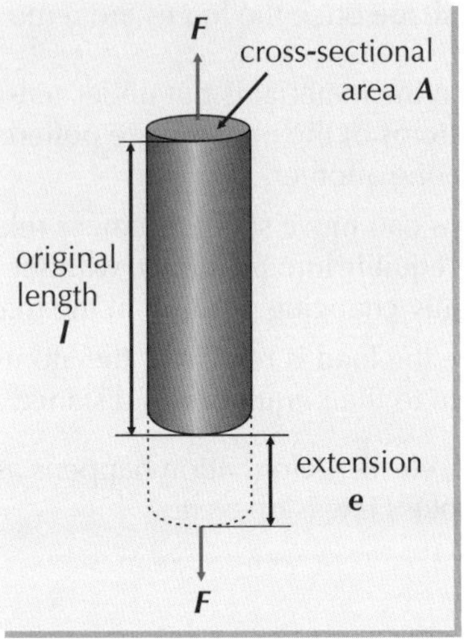

1) **Tensile stress** is defined as the **force applied**, *F*, divided by the **cross-sectional area**, *A*:

$$\text{stress} = \frac{F}{A}$$

The **units** of stress are **Nm^{-2}** or pascals, **Pa**.

2) **Tensile strain** is defined as the **change in length**, i.e. the **extension**, divided by the **original length** of the material:

$$\text{strain} = \frac{e}{l}$$

Strain has **no units** — it's just a **number**.

3) It doesn't matter whether the forces producing the **stress** and **strain** are **tensile** or **compressive** — the **same equations** apply. The only difference is that you tend to think of **tensile** forces as **positive**, and **compressive** forces as **negative**.

A Stress Big Enough to Break the Material is Called the Breaking Stress

As a greater and greater tensile **force** is applied to a material, the **stress** on it **increases**.

1) The effect of the **stress** is to start to **pull** the **atoms apart** from one another.
2) Eventually the stress becomes **so great** that atoms **separate completely**, and the **material breaks**. This is shown by point **B** on the graph. The stress at which this occurs is called the **breaking stress**.
3) The point marked **UTS** on the graph is called the **ultimate tensile stress**. This is the **maximum stress** that the material can withstand.
4) **Engineers** have to consider the **UTS** and **breaking stress** of materials when designing a **structure**.

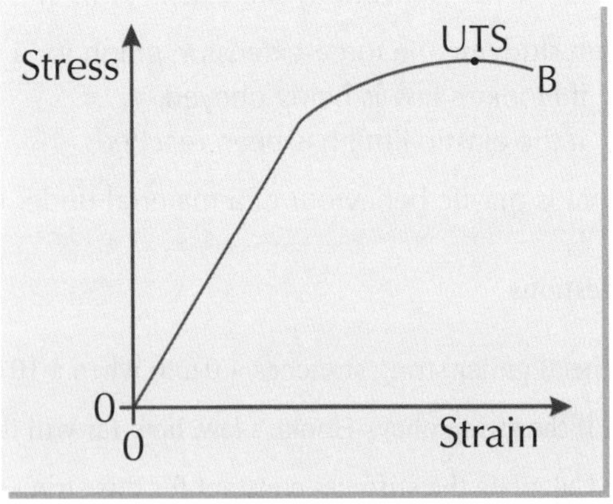

Elastic Strain Energy is the Energy Stored in a Stretched Material

When a material is **stretched**, **work** has to be done in stretching the material.

1) **Before** the **elastic limit**, **all** the **work done** in stretching is **stored** as **potential energy** in the material.
2) This stored energy is called **elastic strain energy**.
3) On a **graph** of **force against extension**, the elastic strain energy is given by the **area under the graph**.

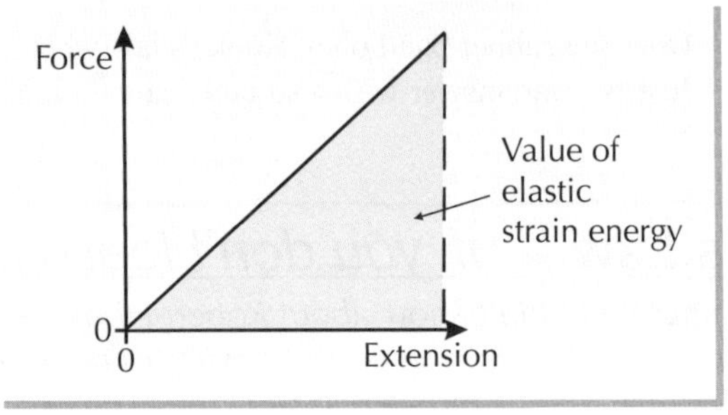

Stress and Strain

You can Calculate the *Energy Stored* in a *Stretched Wire*

Provided a material obeys Hooke's law, the **potential energy** stored inside it can be **calculated** quite easily.

1) The work done on the wire in stretching it is equal to the energy stored.
2) **Work done** equals **force × displacement**.
3) However, the **force** on the material **isn't constant**. It rises from zero up to force ***F***.
   To calculate the **work done**, use the average force between zero and ***F***, i.e. ½***F***.

   work done = ½***F*** × **e**

4) Then the **elastic strain energy**, ***E***, is:

   E = ½***Fe***

   This is the triangular area under the force-extension graph — see previous page.

5) Because Hooke's law is being obeyed, ***F*** = ***ke***,
   which means ***F*** can be replaced in the equation to give:

   E = ½**ke²**

6) If the material is stretched beyond the **elastic limit**, some work is done separating atoms.
   This will **not** be **stored** as strain energy and so isn't available when the force is released.

Practice Questions

Q1 Write a definition for tensile stress.

Q2 Explain what is meant by the tensile strain on a material.

Q3 What is meant by the breaking stress of a material?

Q4 How can the elastic strain energy be found from the force against extension graph of a material under load?

Q5 The work done is usually calculated as force multiplied by displacement.
Explain why the work done in stretching a wire is ½*Fe*.

Exam Questions

Q1 A steel wire is 2.00 m long. When a 300 N force is applied to the wire, it stretches 4.0 mm.
The wire has a circular cross-section with a diameter of 1.0 mm.

(a) What is the cross-sectional area of the wire? [1 mark]

(b) Calculate the tensile stress in the wire. [1 mark]

(c) Calculate the tensile strain of the wire. [1 mark]

Q2 A copper wire (which obeys Hooke's law) is stretched by 3.0 mm when a force of 50 N is applied.

(a) Calculate the stiffness constant for this wire in Nm^{-1}. [2 marks]

(b) What is the value of the elastic strain energy in the stretched wire? [1 mark]

Q3 A pinball machine contains a spring which is used to fire a small, 12 g metal ball to start the game.
The spring has a stiffness constant of 40.8 Nm^{-1}. It is compressed by 5 cm and then released to fire the ball.

Calculate the maximum possible speed of the ball. [4 marks]

UTS a laugh a minute, this stuff...

-lere endeth the proper physics for this section — the rest of it's materials science (and I don't care what your exam ɔoards say). It's all a bit "useful" for my liking. Calls itself a physics course... grumble... grumble... wasn't like this in my lay... But to be fair — some of it's quite interesting, and there are some lovely graphs coming up on page 70.

The Young Modulus

Busy chap, Thomas Young. He did this work on tensile stress as something of a sideline. Light was his main thing. He proved that light behaved like a wave, explained how we see in colour and worked out what causes astigmatism.

The **Young Modulus** is Stress ÷ Strain

When you apply a **load** to stretch a material, it experiences a **tensile stress** and a **tensile strain**.

1) Up to a point called the **limit of proportionality**, the stress and strain of a material are proportional to each other.
2) So below this limit, for a particular material, stress divided by strain is a constant. This constant is called the **Young modulus**, ***E***.

$$E = \frac{\text{tensile stress}}{\text{tensile strain}} = \frac{F/A}{e/l} = \frac{Fl}{eA}$$

Where, F = force in N, A = cross-sectional area in m^2, l = initial length in m and e = extension in m.

3) The **units** of the Young modulus are the same as stress (**Nm^{-2}** or pascals), since strain has no units.
4) The Young modulus is used by **engineers** to make sure their materials can withstand sufficient forces.

To **Find** the Young Modulus, You need a **Very Long Wire**

This is the experiment you're most likely to do in class:

Mum moment: if you're doing this experiment, wear safety goggles — if the wire snaps, it could get very messy...

The Young Modulus

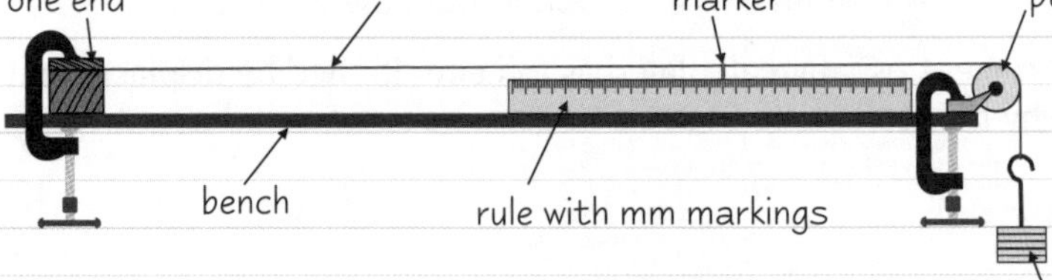

The test wire should be thin, and as long as possible. The **longer and thinner** the wire, the more it **extends** for the same force.

Start with the **smallest weight** necessary to straighten the wire.

Measure the **distance** between the **fixed end of the wire** and the **marker** — this is your unstretched length.

If you then increase the weight, the **wire stretches** and the **marker moves**.

Increase the **weight** by steps, recording the marker reading each time — the **extension** is the **difference** between this reading and the **unstretched length**.

Once you've taken all your readings, use a **micrometer** to measure the **diameter** of the wire in several places. Take an average of your measurements, and use that to work out the average **cross-sectional area** of the wire.

The other standard way of measuring the Young modulus in the lab is using **Searle's apparatus**. This is a bit more accurate, but it's harder to do and the equipment's more complicated.

The Young Modulus

Use a Stress-Strain Graph to Find E

You can plot a **graph** of **stress against strain** from your results.

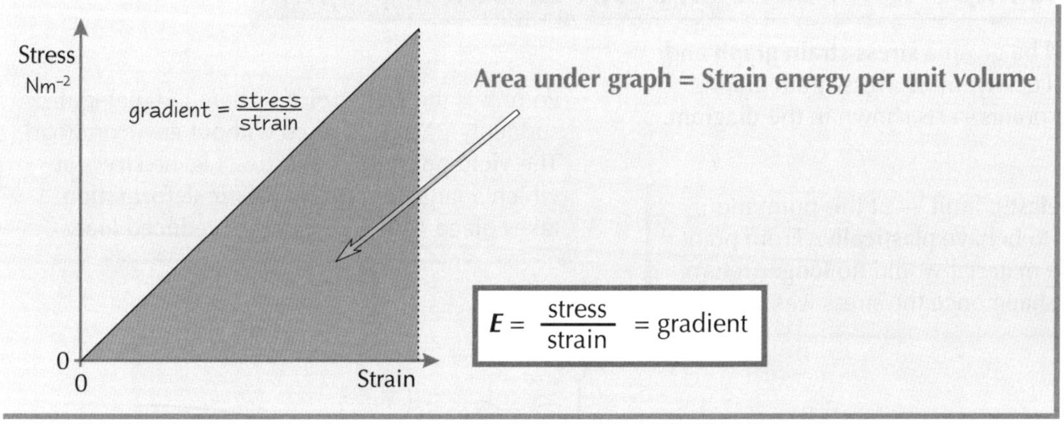

1) The **gradient** of the graph gives the Young modulus, **E**.
2) The **area under the graph** gives the **strain energy** (or energy stored) per unit volume i.e. the energy stored per 1 m^3 of wire.
3) The stress-strain graph is a **straight line** provided that Hooke's law is obeyed, so you can also calculate the energy per unit volume as:

energy = ½ × stress × strain

Practice Questions

Q1 Define the Young modulus for a material.

Q2 What are the units for the Young modulus?

Q3 Explain why a thin test wire is used to find the Young modulus.

Q4 What is given by the area contained under a stress-strain graph?

Exam Questions

Q1 A steel wire is stretched elastically. For a load of 80 N, the wire extends by 3.6 mm. The original length of the wire was 2.50 m and its average diameter is 0.6 mm.

(a) Calculate the cross-sectional area of the wire in m^2. [1 mark]

(b) Find the tensile stress applied to the wire. [1 mark]

(c) Calculate the tensile strain of the wire. [1 mark]

(d) What is the value of the Young modulus for steel? [1 mark]

Q2 The Young modulus for copper is 1.3×10^{11} Nm^{-2}.

(a) If the stress on a copper wire is 2.6×10^{8} Nm^{-2}, what is the strain? [2 marks]

(b) If the load applied to the copper wire is 100 N, what is the cross-sectional area of the wire? [1 mark]

(c) Calculate the strain energy per unit volume for this loaded wire. [1 mark]

Learn that experiment — it's important...

Getting back to the good Dr Young... As if ground-breaking work in light, the physics of vision and materials science wasn't enough, he was also a well-respected physician, a linguist and an Egyptologist. He was one of the first to try to decipher the Rosetta stone (he didn't get it right, but nobody's perfect). Makes you feel kind of inferior, doesn't it. Best get learning.

Interpreting Stress-Strain Curves

I hope the stresses and strains of this section aren't getting to you too much. Don't worry, though — there's just these two pages to go before you're on to the wonderful world of waves.

*There are **Three Important Points** on a **Stress-Strain Graph***

In the exam you could be given a **stress-strain graph** and asked to **interpret** it. Luckily, most stress-strain graphs share **three** important points — as shown in the **diagram**.

Point **Y** is the **yield point** — here the material suddenly starts to **stretch** without any extra load. The **yield point** (or yield stress) is the **stress** at which a large amount of **plastic deformation** takes place with a **constant** or **reduced load.**

Point **E** is the **elastic limit** — at this point the material starts to behave **plastically**. From point E onwards, the material would **no longer** return to its **original shape** once the stress was removed.

Point **P** is the **limit of proportionality** — after this, the graph is no longer a straight line but starts to **bend**. At this point, the material **stops** obeying **Hooke's law**, but would still **return** to its **original shape** if the stress was removed.

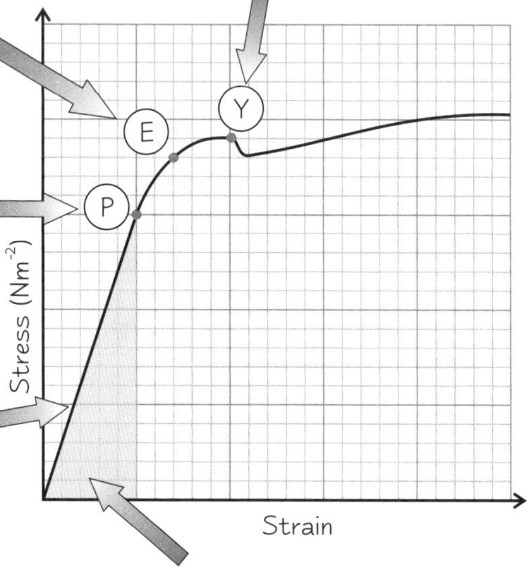

Before point **P**, the graph is a **straight line** through the **origin**. This shows that the material is obeying **Hooke's law** (page 64). The **gradient** of the line is constant — its the **Young modulus** (see page 68).

The **area** under the first part of the graph gives the **energy stored** in the **material per unit volume** (see page 69).

Stress-Strain** Graphs for **Brittle** Materials **Don't Curve

The graph shown below is typical of a **brittle** material.

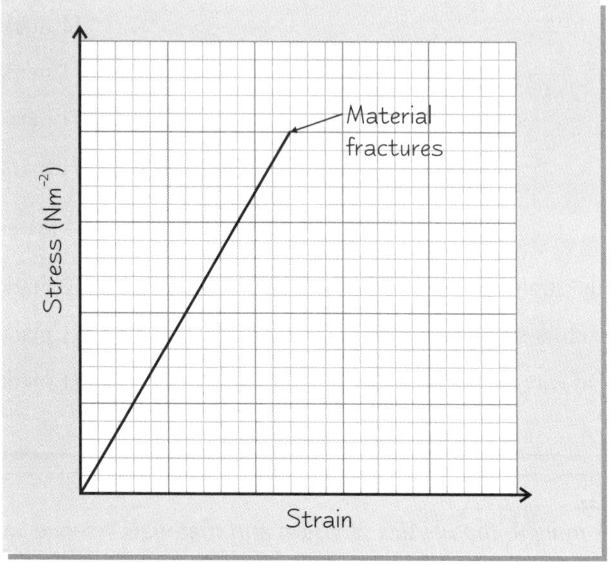

The graph starts the same as the one above — with a **straight line through the origin**. So brittle materials also obey **Hooke's law**.

However, when the **stress** reaches a certain point, the material **snaps** — it doesn't deform plastically.

Hooke's law — it's the pirates' code... yarr

Interpreting Stress-Strain Curves

Brittle Materials Break Suddenly Without Deforming Plastically

As shown on the previous page, if you apply a **force** to a **brittle material**, it won't **deform plastically** (see p 65), but will suddenly **snap** when the force gets to a certain size. Brittle materials can also be quite **weak** if they have **cracks** in them.

A **chocolate bar** is an example of a brittle material — you can break chunks of chocolate off the bar without the whole thing changing shape.
Ceramics (e.g. **glass** and **pottery**) are brittle too — they tend to shatter.

Brittle Materials Have a Rigid Structure

1) Ceramics are made by melting certain materials, and then letting them cool.
2) The arrangement of atoms in a ceramic can be **crystalline** or **polycrystalline** — where there are many regions (or **grains**) of crystalline structure. The atoms in each grain line up in a different direction.
3) However they're arranged, the atoms in a ceramic are bonded in a **giant rigid structure**. The **strong bonds** between the atoms make them **stiff**, while the **rigid** structure means that ceramics are very **brittle** materials.

A Rigid Structure Makes Brittle Materials Fracture

When **stress** is applied to a brittle material any **tiny cracks** at the material's surface get **bigger** and **bigger** until the material **breaks** completely. This is called **brittle fracture**.

The **cracks** in **brittle** materials are able to **grow** because these materials have a **rigid structure**. Other materials, like most metals, are not brittle because the **atoms** within them can **move** to **prevent** any **cracks** getting bigger.

Practice Questions

Q1 What is the difference between the limit of proportionality and the elastic limit?

Q2 Describe what happens at the yield point.

Q3 What happens to a brittle material when a large load is applied?

Exam Questions

Q1 Hardened steel is a hard, brittle form of steel made by heating it up slowly and then quenching it in cold water.

(a) What is meant by the term *brittle*? [1 mark]

(b) Sketch a stress-strain graph for hardened steel. [2 marks]

(c) Explain what happens to hardened steel when stress is applied to it. [2 marks]

Q2 The diagram shows a stress-strain graph for a nylon thread.

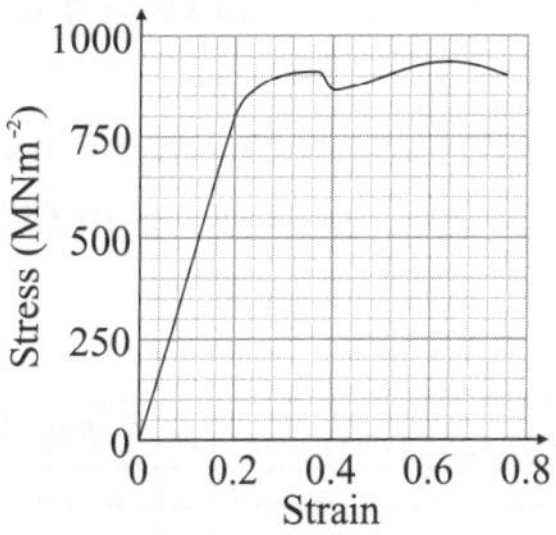

(a) Use the graph to find the Young modulus for nylon. [2 marks]

(b) How much energy was stored in the thread at the limit of proportionality? [2 marks]

My sister must be brittle — she's always snapping...

Yes, it's the end of the section on materials, but don't shed a tear — there's some good stuff coming up in the next one. You can always come back and re-read these pages too — it's a good idea, even if you could do all of the practice questions.

The Nature of Waves

Aaaah... playing with slinky springs and waggling ropes about. It's all good clean fun as my mate Richard used to say...

A **Wave Transfers Energy** Away from Its Source

A **progressive** (moving) wave carries **energy** from one place to another **without transferring any material**. Here are some ways you can tell waves carry energy:

1) Electromagnetic waves cause things to **heat up**.
2) **X-rays** and **gamma rays** knock electrons out of their orbits, causing **ionisation**.
3) Loud **sounds** make things **vibrate**.
4) **Wave power** can be used to **generate electricity**.
5) Since waves carry energy away, the **source** of the wave **loses energy**.

Here are all the **bits** of a **Wave** you Need to Know

1) **Displacement**, ***X***, metres — how far a **point** on the wave has **moved** from its **undisturbed position**.
2) **Amplitude**, **a**, metres — **maximum displacement**.
3) **Wavelength**, **λ**, metres — the **length** of **one whole wave**, from **crest** to **crest** or **trough** to **trough**.

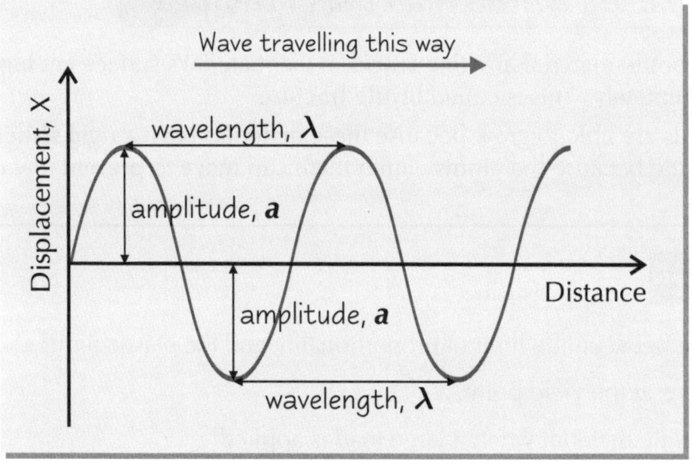

4) **Period**, ***T***, seconds — the **time taken** for a **whole vibration**.
5) **Frequency**, ***f***, hertz — the **number** of **vibrations per second** passing a given **point**.
6) **Phase difference** — the amount by which **one** wave **lags behind another** wave.
Measured in **degrees** or **radians**. See page 80.

Waves Can Be **Reflected** and **Refracted**

Reflection — the wave is **bounced back** when it **hits a boundary**. E.g. you can see the reflection of light in mirrors. The reflection of water waves can be demonstrated in a ripple tank.

Refraction — the wave **changes direction** as it enters a **different medium**.
The change in direction is a result of the wave slowing down or speeding up.

The **Frequency** is the **Inverse** of the **Period**

$$\textit{Frequency} = \frac{1}{\textit{period}}$$

It's that simple.
Get the **units** straight: **1 Hz = 1 s^{-1}**.

The Nature of Waves

Wave Speed, Frequency and Wavelength are Linked by the Wave Equation

Wave speed can be measured just like the speed of anything else:

$$\text{Speed } (v) = \frac{\text{distance moved } (d)}{\text{time taken } (t)}$$

Remember, you're not measuring how fast a physical point (like one molecule of rope) moves. You're measuring how fast a point on the **wave pattern** moves.

Learn the **Wave Equation...**

Speed of wave (v) = wavelength (λ) × frequency (f)

$$v = \lambda f$$

You need to be able to rearrange this equation for v, λ or f.

... and How to **Derive** it

You can work out the **wave equation** by imagining **how long** it takes for the **crest** of a wave to **move** across a **distance** of **one wavelength**. The **distance travelled** is λ. **By definition**, the **time taken** to travel **one whole wavelength** is the **period** of the wave, which is equal to **1/*f***.

$$\text{Speed } (v) = \frac{\text{distance moved } (d)}{\text{time taken } (t)} \Rightarrow \text{Speed } (v) = \frac{\text{distance moved } (\lambda)}{\text{time taken } (1/f)}$$

Learn to recognise when to use $v = \lambda f$ and when to use $v = d/t$. Look at which variables are mentioned in the question.

Practice Questions

Q1 Does a wave carry matter **or** energy from one place to another?

Q2 Diffraction and interference are two wave properties. Write down two more.

Q3 Write down the relationship between the frequency of a wave and its time period.

Q4 Give the units of frequency, displacement and amplitude.

Q5 Write down the equation connecting v, λ and f.

Exam Question

Q1 A buoy floating on the sea takes 6 seconds to rise and fall once (complete a full period of oscillation). The difference in height between the buoy at its lowest and highest points is 1.2 m, and waves pass it at a speed of 3 ms^{-1}.

(a) How long are the waves? [2 marks]

(b) What is the amplitude of the waves? [1 mark]

Learn the wave equation and its derivation — pure poetry...

This isn't too difficult to start you off — most of it you'll have done at GCSE anyway. But once again, it's a whole bunch of equations to learn, and you won't get far without learning them. Yada yada.

Longitudinal and Transverse Waves

There are different types of wave — and the difference is easiest to see using a slinky. Try it — you'll have hours of fun.

*In **Transverse Waves** the **Vibration** is at **Right Angles** to the **Direction** of Travel*

All **electromagnetic waves** are **transverse**. Other examples of transverse waves are **ripples** on water and waves on **ropes**.

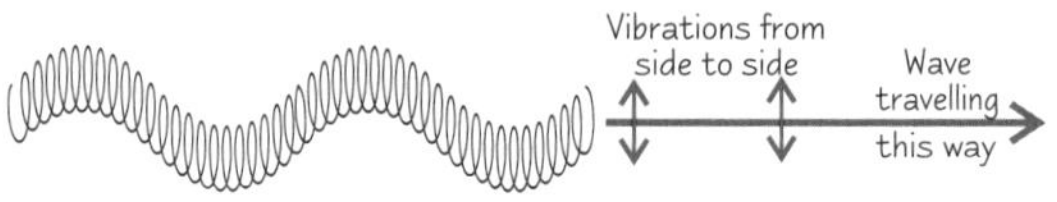

There are **two** main ways of **drawing** transverse waves:

1) They can be shown as **graphs of displacement** against **distance along the path of the wave**.

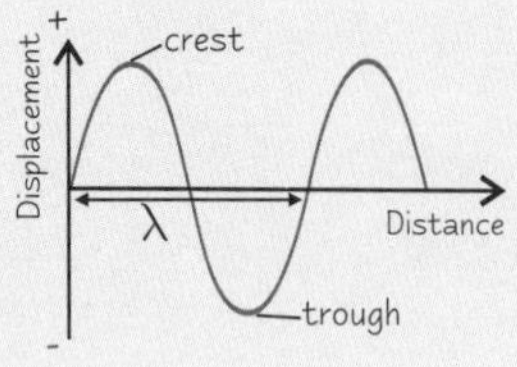

2) Or, they can be shown as graphs of **displacement against time** for a point as the wave passes.

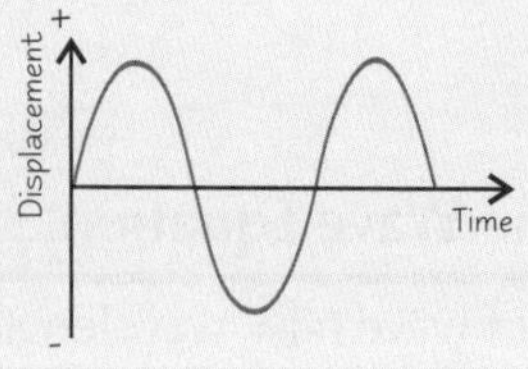

Both sorts of graph often give the **same shape**, so make sure you check out the label on the **x-axis**. Displacements **upwards** from the centre line are given a **+ sign**. Displacements downwards are given a **– sign**.

*In **Longitudinal Waves** the **Vibrations** are **Along** the Direction of Travel*

The most **common** example of a **longitudinal wave** is **sound**. A sound wave consists of alternate **compressions** and **rarefactions** of the **medium** it's travelling through. (That's why sound can't go through a vacuum.) Some types of **earthquake shock waves** are also longitudinal.

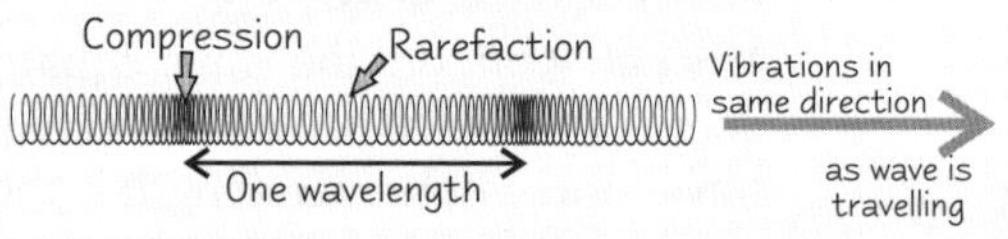

It's hard to **represent** longitudinal waves **graphically**. You'll usually see them plotted as **displacement** against **time**. These can be **confusing** though, because they look like a **transverse wave**.

*A **Polarised Wave** only **Oscillates** in One Direction*

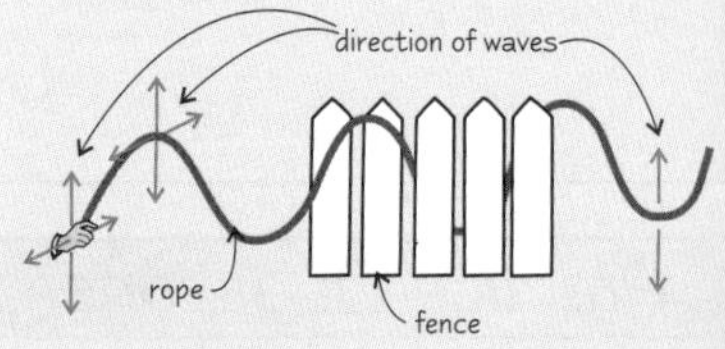

1) If you **shake a rope** to make a **wave** you can move your hand **up and down** or **side to side** or in a **mixture** of directions — it still makes a **transverse wave**.
2) But if you try to pass **waves in a rope** through a **vertical fence**, the wave will only get through if the **vibrations** are **vertical**. The fence filters out vibration in other directions. This is called **polarising** the wave.
3) Ordinary **light waves** are a mixture of **different directions** of **vibration**. (The things vibrating are electric and magnetic fields.) A **polarising filter** only transmits vibrations in one direction.
4) If you have two polarising filters at **right angles** to each other, then **no** light will get through.
5) Polarisation **can only happen** for **transverse** waves. The fact that you can polarise light is one **proof** that it's a transverse wave.

*When **Light Reflects** it is **Partially Polarised***

1) Rotating a **polarising filter** in a beam of light shows the fraction of the light that is vibrating in each **direction**.
2) If you direct a beam of unpolarised light at a reflective surface then view the **reflected ray** through a polarising filter, the intensity of light leaving the filter **changes** with the **orientation** of the filter.
3) The intensity changes because light is **partially polarised** when it is **reflected**.
4) This effect is used to remove **unwanted reflections** in photography and in **Polaroid sunglasses** to remove **glare**.

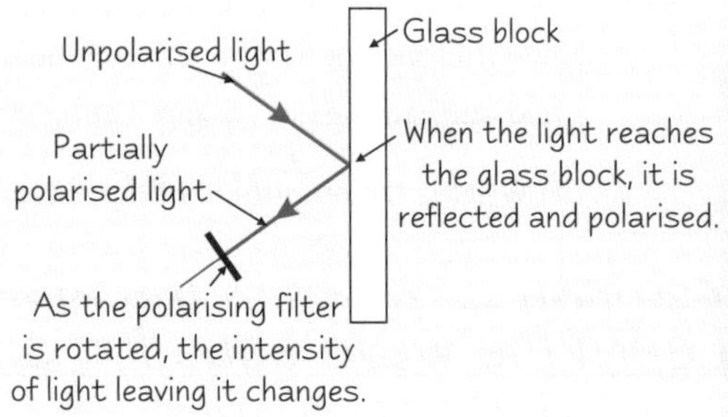

Longitudinal and Transverse Waves

Television and Radio Signals are Polarised

If you walk down the street and look up at the **TV aerials** on people's houses, you'll see that the **rods** (the sticky-out bits) on them are all **horizontal**. The reason for this is that **TV signals** are **polarised** by the orientation of the **rods** on the **broadcasting aerial**.
To receive a strong signal, you have to **line up** the **rods** on the **receiving aerial** with the **rods** on the **transmitting aerial** — if they aren't aligned, the signal strength will be lower.

It's the **same** with **radio** — if you try **tuning a radio** and then **moving** the **aerial** around, your signal will **come and go** as the transmitting and receiving aerials go in and out of **alignment**.

The rods on this broadcasting aerial are horizontal.

Materials Can Rotate the Plane of Polarisation

The **plane** in which a wave moves and **vibrates** is called the **plane of polarisation** — e.g. the rope on the last page was polarised in the **vertical** plane by the fence. Some **materials** (e.g. crystals) **rotate** the plane of polarised light. You can **measure** how much a material rotates the plane of polarised light using two **polarising filters**:

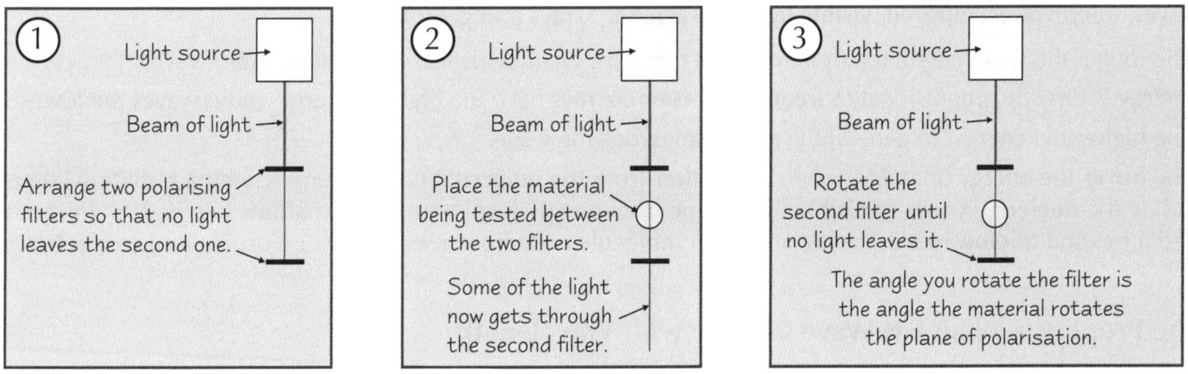

Practice Questions

Q1 Give an example of a transverse wave and a longitudinal wave.

Q2 What is a polarised wave?

Q3 How can you polarise a wave?

Q4 Why do you have to line up transmitting and receiving television aerials?

Exam Questions

Q1 In an experiment, light is shone through a disc of a crystal called "Iceland spar". The beam of light is less bright when it emerges from the crystal than when it enters. Next, a second identical disc of Iceland spar is placed in front of the first. The first disc is held steady while the second is rotated (in the plane of the disc). The intensity of light emerging changes as the second disc rotates. At two points in each rotation, no light gets through at all.

Explain the results of these experiments. You may use a diagram to help your answer. [5 marks]

Q2 Give one example of an application of polarisation and explain how it works. [2 marks]

Caution — rotating the plane may cause nausea...

Right, there's lots to learn on these two pages so I won't hold you up with chat. Don't panic though — a lot of this stuff will be familiar from GCSE, so it's not like you're starting from scratch. One last thing — I know television is on this page, but it doesn't mean you can tune in and call it revision — it won't help. Get the revision done, then take a break.

The Electromagnetic Spectrum

There's nothing really deep and meaningful to understand on this page — just a load of facts to learn I'm afraid.

All **Electromagnetic Waves** Have Some **Properties** In Common

1) They travel in a **vacuum** at a **speed** of **2.998×10^8 ms^{-1}**, and at slower speeds in other media.
2) They are **transverse** waves consisting of **vibrating electric** and **magnetic fields**. The **electric** and **magnetic** fields are at **right angles** to each other and to the **direction of travel**.
3) Like all waves, EM waves can be **reflected**, **refracted** and **diffracted** and can undergo **interference**.
4) Like all waves, EM waves obey $v = f\lambda$ (v = velocity, f = frequency, λ = wavelength).
5) Like all progressive waves, progressive EM waves **carry energy**.
6) Like all transverse waves, EM waves can be **polarised**.

Some **Properties Vary** Across the **EM Spectrum**

EM waves with different wavelengths behave differently in some respects. The spectrum is split into seven categories: **radio waves**, **microwaves**, **infrared**, **visible light**, **ultraviolet**, **X-rays** and **gamma rays**.

1) The longer the wavelength, the more **obvious** the wave characteristics — e.g., long radio waves diffract round hills.
2) **Energy** is directly proportional to **frequency**. **Gamma rays** have the **highest energy**; **radio waves** the **lowest**.
3) The **higher** the **energy**, in general the more **dangerous** the wave.
4) The **lower the energy** of an EM wave, the **further from the nucleus** it comes from. **Gamma radiation** comes from inside the **nucleus**. **X-rays to visible light** come from energy-level transitions in **atoms** (see p. 20). **Infrared** radiation and **microwaves** are associated with **molecules**. **Radio waves** come from oscillations in **electric fields**.

The **Properties** of an **EM Wave** Change with **Wavelength**

Type	Approximate wavelength / m	Penetration	Uses
Radio waves	10^{-1} — 10^{6}	Pass through matter.	Radio transmissions.
Microwaves	10^{-3} — 10^{-1}	Mostly pass through matter, but cause some heating.	Radar. Microwave cookery. TV transmissions.
Infrared (IR)	7×10^{-7} — 10^{-3}	Mostly absorbed by matter, causing it to heat up.	Heat detectors. Night-vision cameras. Remote controls. Optical fibres.
Visible light	4×10^{-7} — 7×10^{-7}	Absorbed by matter, causing some heating effect.	Human sight. Optical fibres.
Ultraviolet (UV)	10^{-8} — 4×10^{-7}	Absorbed by matter. Slight ionisation.	Sunbeds. Security markings that show up in UV light.
X-rays	10^{-13} — 10^{-8}	Mostly pass through matter, but cause ionisation as they pass.	To see damage to bones and teeth. Airport security scanners. To kill cancer cells.
Gamma rays	10^{-16} — 10^{-10}	Mostly pass through matter, but cause ionisation as they pass.	Irradiation of food. Sterilisation of medical instruments. To kill cancer cells.

The Electromagnetic Spectrum

Different Types of EM Wave Have Different **Effects** on the **Body**

Type	Production	Effect on human body
Radio waves	Oscillating electrons in an aerial	No effect.
Microwaves	Electron tube oscillators. Masers.	Absorbed by water — danger of cooking human body*.
Infrared (IR)	Natural and artificial heat sources.	Heating. Excess heat can harm the body's systems.
Visible light	Natural and artificial light sources.	Used for sight. Too bright a light can damage eyes.
Ultraviolet (UV)	e.g. the Sun.	Tans the skin. Can cause skin cancer and eye damage.
X-rays	Bombarding metal with electrons.	Cancer due to cell damage. Eye damage.
Gamma rays	Radioactive decay of the nucleus.	Cancer due to cell damage. Eye damage.

1) UV radiation is split into categories based on frequency — **UV-A**, **UV-B** and **UV-C**.
2) **UV-A** is the **least damaging**, although it's thought to be a major cause of **skin aging**.
3) **UV-B** is more dangerous — it's responsible for **sunburn** and can lead to **cancer** too.
4) **UV-C** is **ionising** — it can cause **cell mutation** or **destruction**, and **cancer**. It's almost **entirely blocked** by the ozone layer, though.
5) **Dark** skin gives some protection from UV rays, stopping them reaching more vulnerable tissues below — so **tanning** is a protection mechanism.
6) **Sunscreens** provide some protection from UV in sunlight. They're designed to **limit** the amount of **UV-A** and **UV-B** radiation reaching your skin.

* Or small animals.

Practice Questions

Q1 What are the main practical uses of infrared radiation?

Q2 Which types of electromagnetic radiation have the highest and lowest energies?

Q3 What is the significance of the speed 2.998×10^8 ms^{-1}?

Q4 Why are microwaves dangerous?

Q5 How does the energy of an EM wave vary with frequency?

Exam Questions

Q1 In a vacuum, do X-rays travel faster, slower or at the same speed as visible light? Explain your answer. [2 marks]

Q2 (a) Describe briefly the physics behind a practical use of X rays. [2 marks]

(b) What is the difference between gamma rays and X-rays? [2 marks]

Q3 Give an example of a type of electromagnetic wave causing a hazard to health. [2 marks]

I've got UV hair...

No really I have. It's great. It's purple. And it's got shiny glittery white bits in it. Aaaanyway... moving swiftly on. Loads of facts to learn on these pages. You probably know most of this from GCSE anyway, but make sure you know it well enough to answer a question on it in the exam. Not much fun, but... there you go.

Refractive Index

The stuff on the next two pages explains why your legs look short in a swimming pool.

The **Refractive Index** of a Material Measures **How Much** It Slows Down Light

Light goes fastest in a **vacuum**. It **slows down** in other materials, because it **interacts** with the particles in them. The more **optically dense** a material is, the more light slows down when it enters it.

The **absolute refractive index** of a material, n, is the **ratio** between the **speed of light in a vacuum**, c, and the speed of light in that **material**, v.

$$n = \frac{c}{v}$$

$c = 3 \times 10^8\ \text{ms}^{-1}$

The **relative** refractive index **between two materials**, ${}_1n_2$, is the ratio of the speed of light **in material 1** to the speed of light **in material 2**.

$${}_1n_2 = \frac{v_1}{v_2}$$

Combining the two equations gives:

$${}_1n_2 = \frac{n_2}{n_1}$$

The speed of light in air is only a tiny bit smaller than c. So you can assume the refractive index of air is 1.

1) The **absolute refractive index** of a material is a **property** of that material only. But a **relative refractive index** is a property of the **interface** between two materials. It's different for **every possible pair**.
2) Because you can assume $n_{air} = 1$, you can assume the refractive index for an **air to glass boundary** equals the **absolute refractive index** of the glass.

Snell's Law uses **Angles** to Calculate the Refractive Index

1) The **angle** the **incoming light** makes to the **normal** is called the **angle of incidence**, i. The **angle** the **refracted ray** makes with the **normal** is the **angle of refraction**, r.
2) When light enters an optically denser medium it is refracted **towards** the normal.
3) n, i and r are related by **Snell's law**.

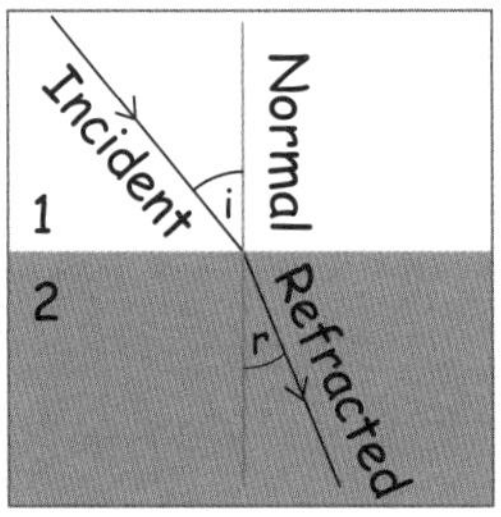

$$n_1 \sin i = n_2 \sin r$$

You don't need to know the geometry of why Snell's law works, but you need to be able to use it.

4) You can use a device called a **refractometer** to accurately measure the refractive index of a material. The machine shines a beam of light at the sample. You then view the refracted beam through a **microscope** and measure its angle of refraction.

Sometimes You Need to Know The Refractive Index **Very Accurately**

1) **Making lenses** — There are several different types of glass that can be used to make lenses, and they all have different refractive indices. The refractive index of the glass affects how thick and how curved the lens has to be to get the same magnifying power.
It's important to know the refractive index accurately so you can calculate the exact shape of lens to fit, say, someone's prescription for a pair of glasses.
2) **In forensic investigations** — Identifying broken bits of glass and plastic can be vital when tracking down crime suspects. Once you know the exact refractive index of the glass from the crime scene, you can compare it with shards found on the clothes or shoes of a suspect. Most shards are too small for a refractometer, so you need to measure the refractive index differently. You use the fact that if two materials with the same refractive index are side by side, you can't see the boundary between them.

Refractive Index

Light Leaving an Optically Denser Material is Refracted Away from the Normal

When light **goes from** an optically denser material into an optically **less dense** material (e.g. glass to air), interesting things can start to happen.

Shine a ray of light at a **glass to air** boundary, then gradually **increase** the angle of incidence. As you increase the angle of incidence, the angle of **refraction** gets closer and closer to **90°**. Eventually ***i*** reaches a **critical angle *C*** for which ***r* = 90°**. The light is **refracted** along the **boundary**.

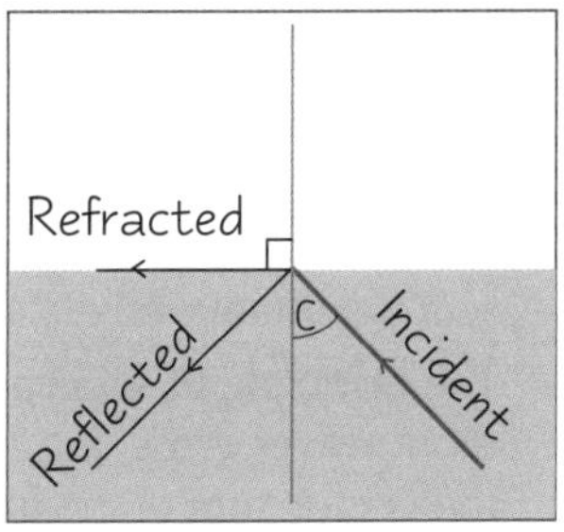

At this **critical angle**, Snell's law $n_{glass} \sin i = n_{air} \sin r$ becomes:

$n \sin C = 1$ so: $$n = \frac{1}{\sin C}$$

That's because the refractive index of air is 1 and sin 90° = 1

At angles of incidence **greater than *C*** refraction is **impossible**.
That means **all** the light is reflected back into the material. This effect is called **total internal reflection**.

Optical Fibres Use Total Internal Reflection

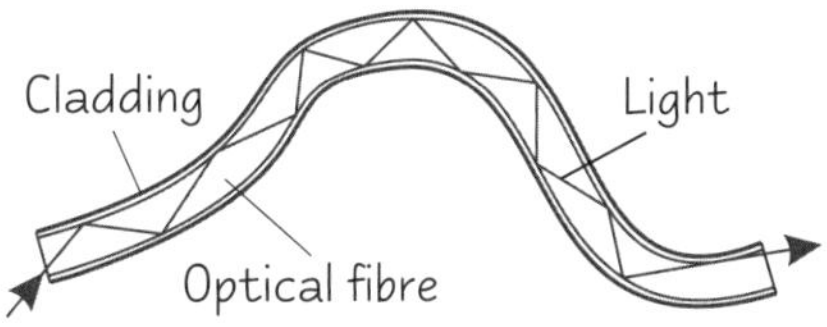

1) An optical fibre is a very **thin flexible tube** of **glass** or **plastic** fibre that can carry **light signals** over long distances and round corners.
2) The optical fibres themselves have a **high refractive index** but are surrounded by **cladding** with a lower refractive index.
3) Light is shone in at **one end** of the fibre. The fibre is so **narrow** that the light always **hits the boundary** between fibre and cladding at an **angle bigger** than the **critical angle**.
4) So all the light is **totally internally reflected** from boundary to boundary until it reaches the other end.

Light running through optical fibres is used to transmit **phone and cable TV signals**.
It beats the **old system** of using **electricity** flowing through **copper cables**:

1) The signal can carry **more information** because light has a high frequency.
2) The light **doesn't heat up the fibre** — so almost no energy is lost as heat.
3) There is no electrical **interference**.

Practice Questions

Q1 Why does light go fastest in a vacuum and slow down in other media?

Q2 What is the formula for the critical angle for a ray of light at a water/air boundary?

Exam Questions

Q1 (a) Light travels in diamond at 1.24×10^8 ms^{-1}. What is the refractive index of diamond? [1 mark]

(b) What is the angle of refraction if light strikes a facet of a diamond ring at an angle of 50° to the normal of the air/diamond boundary? [2 marks]

Q2 An adjustable underwater spotlight is placed on the floor of an aquarium tank. When the light points upwards at a steep angle a beam comes through the surface of the water into the air, and the tank is dimly lit. When the spotlight is placed at a shallower angle, no light comes up through the water surface, and the tank is brightly lit.

(a) Explain what is happening. [2 marks]

(b) It is found that the beam into the air disappears when the spotlight is pointed at any angle of less than 41.25° to the floor. Calculate the refractive index of water. [2 marks]

I don't care about expensive things — all I care about is wave speed...

AS Physics examiners are always saying how candidates do worst in the waves bit of the exam. You'd think they'd have something more important to worry about — third world poverty, war, Posh & Becks... But no.

Superposition and Coherence

When two waves get together, it can be either really impressive or really disappointing.

Superposition Happens When **Two** or **More** Waves **Pass Through** Each Other

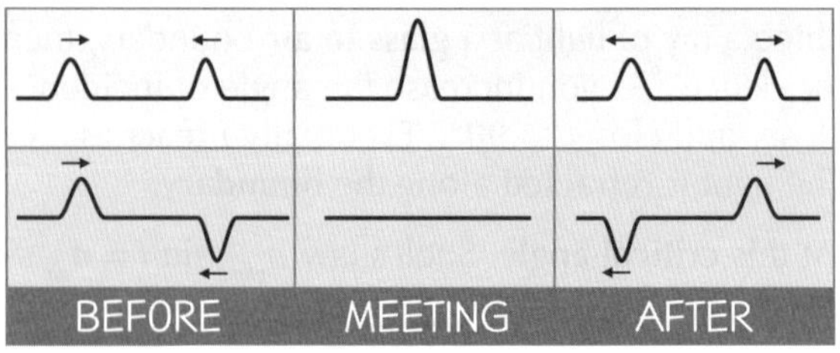

1) At the **instant** the waves **cross**, the **displacements** due to each wave **combine**. Then **each wave** goes on its merry way. You can **see** this if **two pulses** are sent **simultaneously** from each end of a rope.
2) The **principle of superposition** says that when two or more **waves cross**, the **resultant** displacement equals the **vector sum** of the **individual** displacements.

"Superposition" means "one thing on top of another thing". You can use the same idea in reverse — a complex wave can be separated out mathematically into several simple sine waves of various sizes.

Interference can be **Constructive** or **Destructive**

1) A **crest** plus a **crest** gives a **big crest**. A **trough** plus a **trough** gives a **big trough**. These are both examples of **constructive interference**.
2) A **crest** plus a **trough** of **equal size** gives... **nothing**. The two displacements **cancel each other out** completely. This is called **destructive interference**.
3) If the **crest** and the **trough** aren't the **same size**, then the destructive interference **isn't total**. For the interference to be **noticeable**, the two **amplitudes** should be **nearly equal**.

Graphically, you can superimpose waves by adding the individual displacements at each point along the x-axis, and then plotting them.

In Phase Means In **Step** — Two Points **In Phase** Interfere **Constructively**

1) Two points on a wave are **in phase** if they are both at the **same point** in the **wave cycle**. Points in phase have the **same displacement** and **velocity**.
   On the graph, points **A** and **B** are **in phase**; points **A** and **C** are **out of phase**.

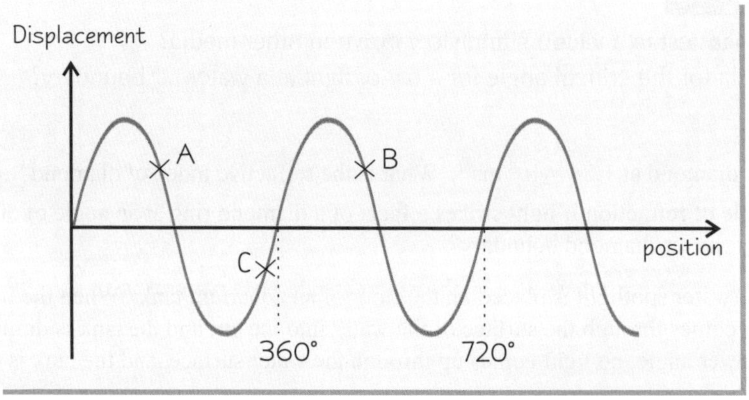

2) It's mathematically **handy** to show one **complete cycle** of a wave as an **angle of 360° (2π radians)**. **Two points** with a **phase difference** of **zero** or a **multiple of 360°** are **in phase**. **Points** with a **phase difference** of **odd-number multiples** of **180° (π radians)** are **exactly out of phase**.
3) You can also talk about two **different waves** being **in phase**. **In practice** this happens because **both** waves came from the **same oscillator**. In **other** situations there will nearly always be a **phase difference** between two waves.

Superposition and Coherence

To Get **Interference Patterns** the **Two Sources** Must Be **Coherent**

Interference **still happens** when you're observing waves of **different wavelength** and **frequency** — but it happens in a **jumble**. In order to get clear **interference patterns**, the two or more sources must be **coherent**.

Two sources are **coherent** if they have the **same wavelength** and **frequency** and a **fixed phase difference** between them.

In exam questions at AS, the 'fixed phase difference' is almost certainly going to be zero. The two sources will be in phase.

Constructive or **Destructive** Interference Depends on the **Path Difference**

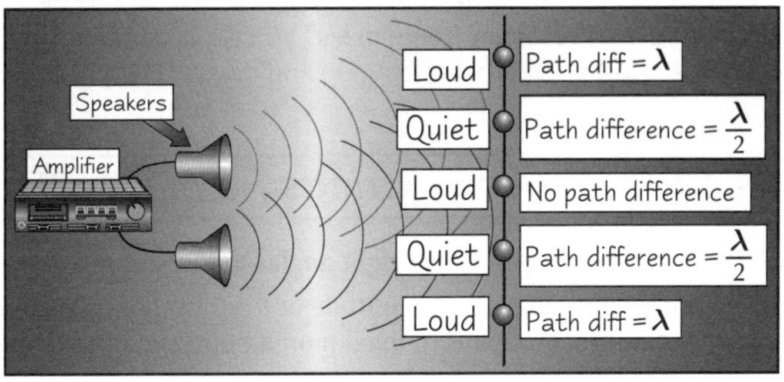

1) Whether you get **constructive** or **destructive** interference at a **point** depends on how **much further one wave** has travelled than the **other wave** to get to that point.
2) The **amount** by which the path travelled by one wave is **longer** than the path travelled by the other wave is called the **path difference.**
3) At **any point an equal distance** from both sources you will get **constructive interference.** You also get constructive interference at any point where the **path difference** is a **whole number of wavelengths**. At these points the two waves are **in phase** and **reinforce** each other. But at points where the path difference is **half a wavelength**, **one and a half** wavelengths, **two and a half** wavelengths etc., the waves arrive **out of phase** and you get **destructive interference**.

Constructive interference occurs when: path difference $= n\lambda$ (where n is an integer)

Destructive interference occurs when: path difference $= \dfrac{(2n + 1)\lambda}{2} = (n + ½)\lambda$

Practice Questions

Q1 Why does the principle of superposition deal with the **vector** sum of two displacements?

Q2 What happens when a crest meets a slightly smaller trough?

Q3 If two points on a wave have a phase difference of 1440°, are they in phase?

Exam Questions

Q1 (a) Two sources are coherent.
What can you say about their frequencies, wavelengths and phase difference? [2 marks]

(b) Suggest why you might have difficulty in observing interference patterns in an area affected by two waves from two sources even though the two sources are coherent. [1 mark]

Q2 Two points on an undamped wave are exactly out of phase.

(a) What is the phase difference between them, expressed in degrees? [1 mark]

(b) Compare the displacements and velocities of the two points. [2 marks]

Learn this and you'll be in a super position to pass your exam...

...I'll get my coat.

There are a few really crucial concepts here: a) interference can be constructive or destructive, b) constructive interference happens when the path difference is a whole number of wavelengths, c) the sources must be coherent.

Stationary Waves

Standing waves are waves that... er... stand still... well, not still exactly... I mean, well...they don't go anywhere... um...

You get Standing Waves When a **Progressive Wave** is **Reflected** at a **Boundary**

A standing wave is the **superposition** of **two progressive waves** with the **same wavelength**, moving in **opposite directions.**

1) Unlike progressive waves, **no energy** is transmitted by a standing wave.
2) You can demonstrate standing waves by setting up a **driving oscillator** at one end of a **stretched string** with the other end fixed. The wave generated by the oscillator is **reflected** back and forth.
3) For most frequencies the resultant **pattern** is a **jumble**. However, if the oscillator happens to produce an **exact number of waves** in the time it takes for a wave to get to the **end** and **back again**, then the **original** and **reflected** waves **reinforce** each other.
4) At these **"resonant frequencies"** you get a **standing wave** where the **pattern doesn't move** — it just sits there, bobbing up and down. Happy, at peace with the world...

A sitting wave.

Standing Waves in **Strings** Form **Oscillating "Loops"** Separated by **Nodes**

1) Each particle vibrates at **right angles** to the string.
2) **Nodes** are where the **amplitude** of the vibration is **zero**. **Antinodes** are points of **maximum amplitude**.
3) At resonant frequencies, an **exact number** of **half wavelengths** fits onto the string.

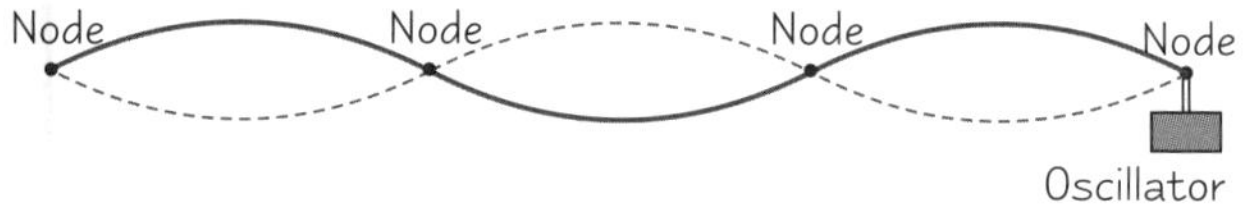

Fundamental Frequency

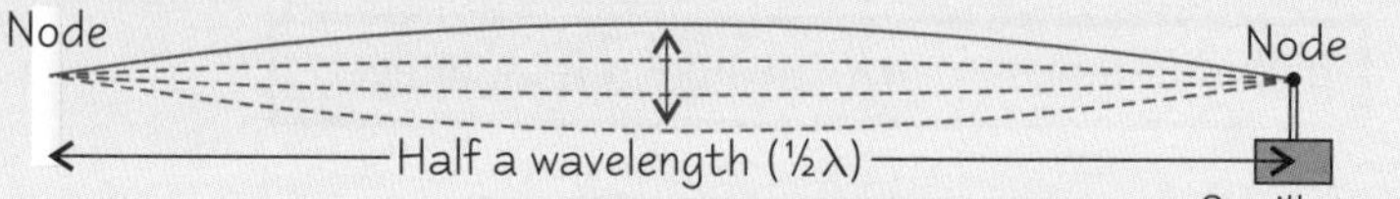

This standing wave is vibrating at the **lowest possible** resonant frequency (the **fundamental frequency**). It has **one "loop"** with a **node at each end**.

Second Harmonic

This is the **second harmonic** (or **first overtone**). It is **twice** the fundamental frequency. There are two **"loops"** with a **node** in the **middle** and **one at each end**.

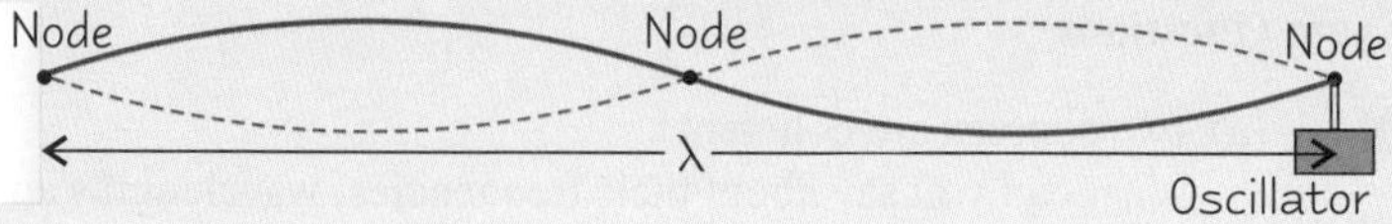

Third Harmonic

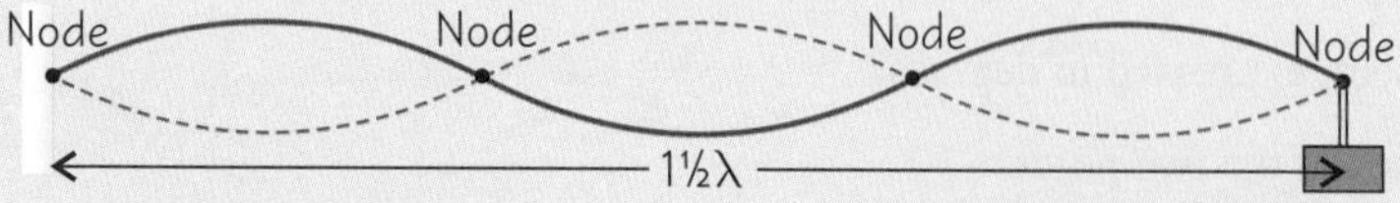

The **third harmonic** (or **second overtone**) is **three times** the fundamental frequency. **1½ wavelengths** fit on the string.

Stationary Waves

Fundamental Frequency of a String Depends on *Length, Weight* and *Tension*

1) The **longer** the string, the **lower** the note — because the **half-wavelength** at the natural frequency is longer.
2) The **heavier** (i.e. the more mass per length) the string, the **lower** the note — because waves travel more **slowly** down the string. For a given **length** a **lower** velocity, ***v*** makes a **lower** frequency, ***f***.
3) The **looser** the string the **lower** the note — again because waves travel more **slowly** down a **loose** string.

You can *Demonstrate Standing Waves* with Microwaves and Sounds

Microwaves Reflected Off a Metal Plate Set Up a Standing Wave

Microwave standing wave apparatus ⟶

You can find the **nodes** and **antinodes** by moving the **probe** between **transmitter** and **reflecting** plate.

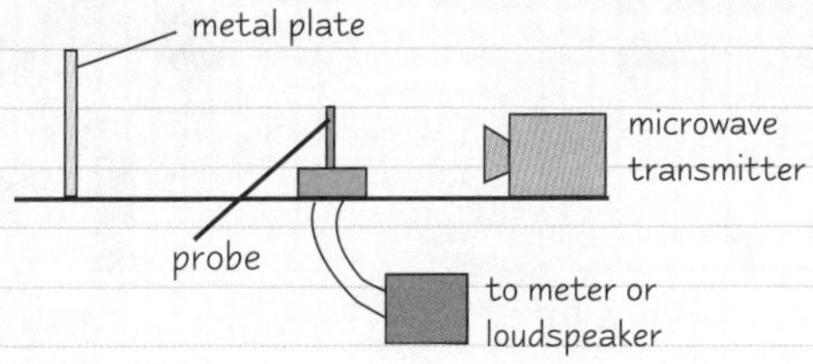

Powder can Show Standing Waves in a Tube of Air

Stationary sound waves are produced in the **glass tube**.

The **lycopodium powder** (don't worry, you don't need to know what that is) laid along the bottom of the tube is **shaken away** from the **antinodes** but left **undisturbed** at the **nodes**.

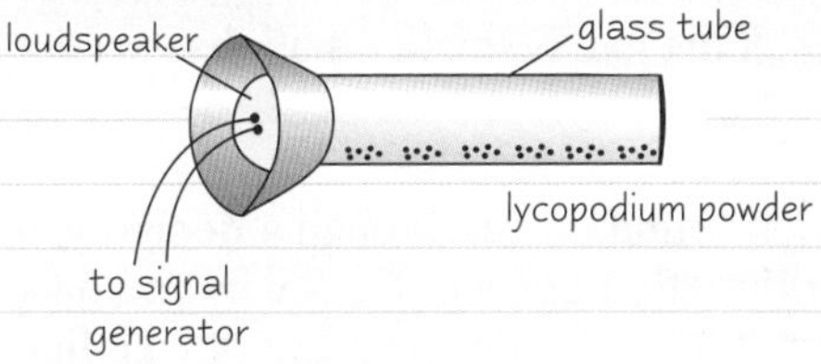

Practice Questions

Q1 How do standing waves form?

Q2 At four times the fundamental frequency, how many half wavelengths fit on a violin string?

Q3 Describe an experiment to investigate standing waves in a column of air.

Exam Question

Q1 (a) A standing wave of three times the fundamental frequency is formed on a stretched string of length 1.2m. Sketch a diagram showing the form of the wave. [2 marks]

(b) What is the wavelength of the standing wave? [1 mark]

(c) Explain how the amplitude varies along the string. How is that different from the amplitude of a progressive wave? [2 marks]

(d) At a given moment, how does the displacement of a particle at one antinode compare to the displacement of a particle at the next antinode? [2 marks]

Just get hold of a guitar and try it out...

Exam questions often get you to compare standing waves to progressive waves.
Remember the fundamental (ho ho) difference is that standing waves don't transmit energy.

Diffraction

Ripple tanks, ripple tanks — yeah.

*Waves Go **Round Corners** and **Spread out** of **Gaps***

The way that **waves spread out** as they come through a **narrow gap** or go round obstacles is called **diffraction**. **All** waves diffract, but it's not always easy to observe.

Use a **Ripple Tank** to Show Diffraction of **Water Waves**

You can make diffraction patterns in ripple tanks.

The **amount** of diffraction depends on the **wavelength** of the wave compared with the **size of the gap**.

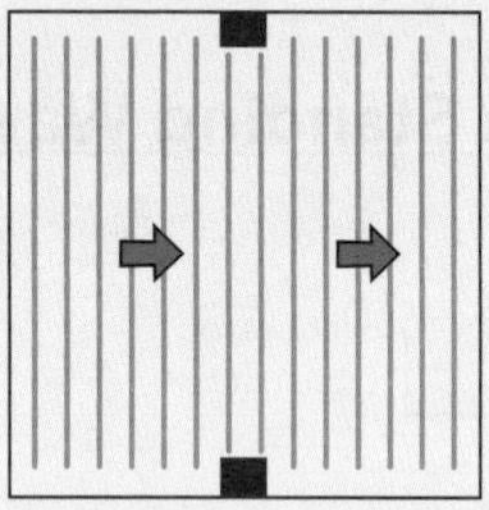

When the gap is **a lot bigger** than the **wavelength**, diffraction is **unnoticeable**.

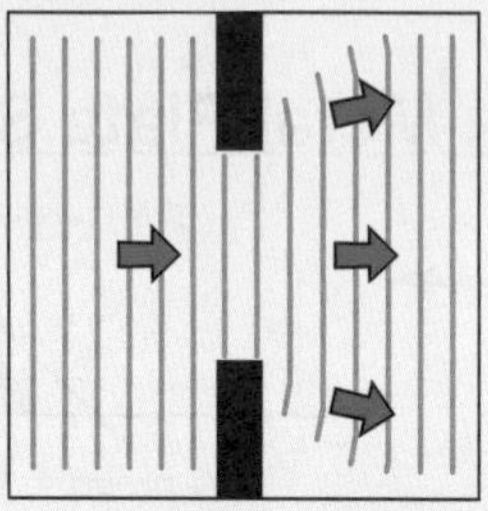

You get **noticeable diffraction** through a gap **several** wavelengths wide.

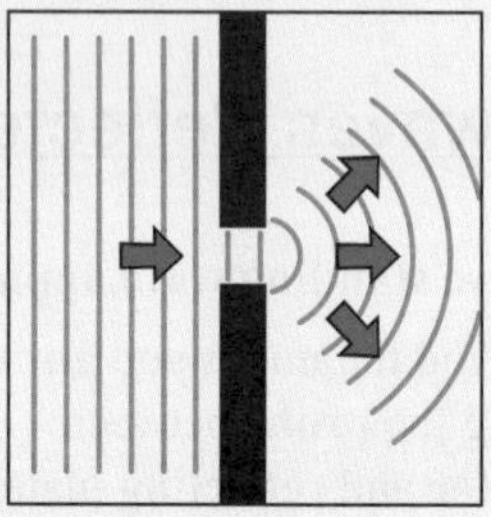

You get the **most** diffraction when the gap is **the same** size as the **wavelength**.

If the gap is **smaller** than the wavelength, the waves are mostly just **reflected back**.

When **sound** passes through a **doorway**, the **size of gap** and the **wavelength** are usually roughly **equal**, so **a lot** of **diffraction** occurs. That's why you have no trouble **hearing** someone through an **open door** to the next room, even if the other person is out of your **line of sight**. The reason that you can't **see** him or her is that when **light** passes through the doorway, it is passing through a **gap** around a **hundred million times bigger** than its wavelength — the amount of diffraction is **tiny**.

Demonstrate **Diffraction** in **Light** Using **Laser Light**

1) Diffraction in **light** can be demonstrated by shining a **laser light** through a very **narrow slit** onto a screen (see next page). You can alter the amount of diffraction by changing the width of the slit.
2) You can do a similar experiment using a **white light** source instead of the laser (which is monochromatic) and a set of **colour filters**. The size of the slit can be kept constant while the **wavelength** is varied by putting different **colour filters** over the slit.

Warning. Use of coloured filters may result in excessive fun.

*You Get a **Similar** Effect Around an **Obstacle***

When a wave meets an **obstacle**, you get diffraction around the edges.

Behind the obstacle is a '**shadow**', where the wave is blocked. The **wider** the obstacle compared with the wavelength of the wave, the less diffraction you get, and so the **longer** the shadow.

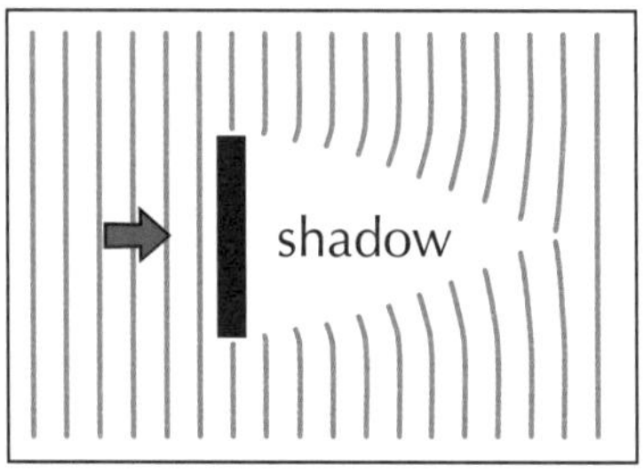

Diffraction

With **Light Waves** you get a **Pattern** of **Light** and **Dark Fringes**

1) If the **wavelength** of a **light wave** is about the same size as the **aperture**, you get a **diffraction pattern** of light and dark fringes.
2) The pattern has a **bright central fringe** with alternating **dark and bright fringes** on either side of it.

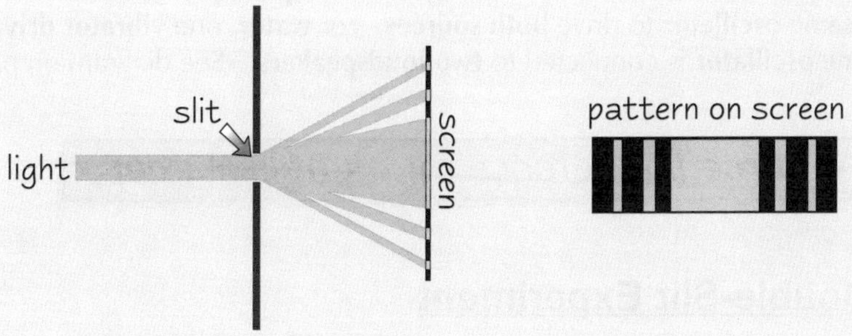

You need to use a coherent light source for this experiment.

3) The **narrower** the slit, the **wider** the diffraction pattern.

You Get a **Similar Pattern** with **Electrons**

1) It's not just with **light** that you get diffraction patterns.
2) In **1927**, two American physicists, **Clinton Davisson** and **Lester Germer**, succeeded in diffracting **electrons**.
3) This was a **huge** discovery. A few years earlier, **Louis de Broglie** had **hypothesised** that electrons would show **wave-like** properties (in the same way that light can show particle-like properties — see page 22), but this was the first **direct evidence** for it.

Electron diffraction patterns look like this

Practice Questions

Q1 What is diffraction?

Q2 Sketch what happens when plane waves meet an obstacle about as wide as one wavelength.

Q3 For a long time some scientists argued that light couldn't be a wave because it did not seem to diffract. Suggest why they might have got this impression.

Q4 Do all waves diffract?

Exam Question

Q1 A mountain lies directly between you and a radio transmitter.

Explain using diagrams why you can pick up long-wave radio broadcasts from the transmitter but not short-wave radio broadcasts. [4 marks]

Even hiding behind a mountain, you can't get away from long-wave radio...

*Unfortunately "Bay FM" don't transmit using long wave radio. So as I'm giving the singing-in-the-car performance of my life, I go over a hill and the signal cuts out. Where's diffraction when I need it then hmm? Diffraction crops up again in particle physics, quantum physics and astronomy, so make sure you **really** understand it.*

Two-Source Interference

Yeah, I know, fringe spacing doesn't really sound like a Physics topic — just trust me on this one, OK.

*Demonstrating Two-Source Interference in **Water** and **Sound** is Easy*

1) It's **easy** to demonstrate **two-source interference** for either **sound** or **water** because they've got **wavelengths** of a handy **size** that you can **measure**.
2) You need **coherent** sources, which means the **wavelength** and **frequency** have to be the **same**. The trick is to use the **same oscillator** to drive **both sources**. For **water**, one **vibrator drives two dippers**. For sound, **one oscillator** is connected to **two loudspeakers**. (See diagram on page 81.)

*Demonstrating **Two-Source** Interference for **Light** is Harder*

Young's Double-Slit Experiment

1) You **can't** arrange **two separate coherent light sources** because **light** from **each source** is emitted in **random bursts**. Instead a **single** laser is shone through **two slits**.
2) Laser light is **coherent** and **monochromatic** (there's only **one wavelength** present).
3) The slits have to be narrower than the wavelength of the laser light so that it is **diffracted** — then the light from the slits is equivalent to **two coherent point sources**.

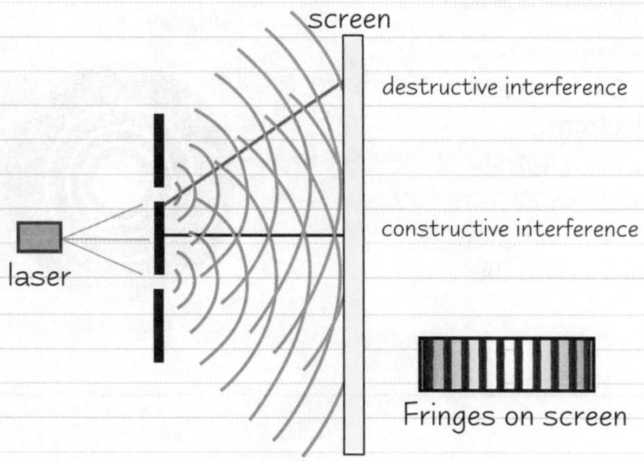

You need to be aware of the safety issues of using lasers. It's not rocket science: laser light damages your retina, so never look directly at the beam. And be careful not to point one at a reflective surface (for obvious reasons).

4) You get a pattern of light and dark **fringes**, depending on whether constructive or destructive **interference** is taking place. Thomas Young — the first person to do this experiment (with a lamp rather than a laser) — came up with an **equation** to **work out** the **wavelength** of the **light** from this experiment (see next page).

*You Can Do a **Similar** Experiment with **Microwaves***

1) To see interference patterns with **microwaves**, you can **replace** the laser and slits with two microwave **transmitter cones** attached to the **same** signal generator.
2) You also need to replace the screen with a microwave **receiver probe** (like the one used in the standing waves experiment on page 83).
3) If you move the probe along the path of the green arrow, you'll get an **alternating pattern** of **strong** and **weak** signals — just like the light and dark fringes on the screen.

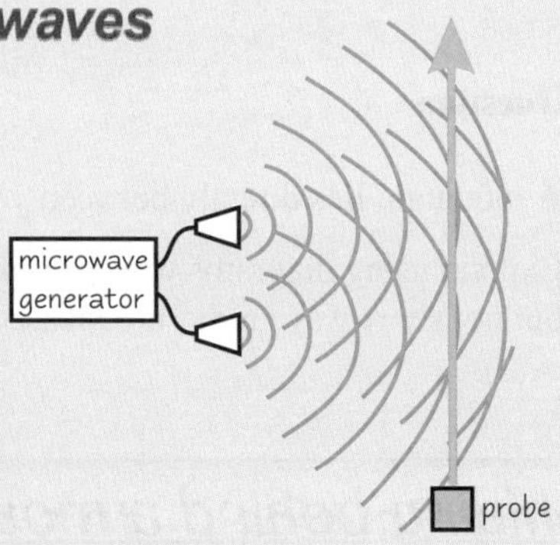

Two-Source Interference

Work Out the Wavelength with **Young's Double-Slit Formula**

1) The fringe spacing (***X***), wavelength (**λ**), spacing between slits (***d***) and the distance from slits to screen (***D***) are all related by **Young's double-slit formula**, which works for all waves (you need to know it, but not derive it).

$$\text{Fringe spacing, } X = \frac{D\lambda}{d}$$

"Fringe spacing" means the distance from the centre of one minimum to the centre of the next minimum or from the centre of one maximum to the centre of the next maximum.

Always check your fringe spacing.

2) Since the wavelength of light is so small you can see from the formula that a high ratio of ***D / d*** is needed to make the fringe spacing **big enough to see**.
3) Rearranging, you can use **λ = *Xd / D*** to **calculate the wavelength** of light.
4) The fringes are **so tiny** that it's very hard to get an **accurate value of X**. It's easier to measure across **several** fringes then **divide** by the number of **fringe widths** between them.

Young's Experiment was ***Evidence*** for the ***Wave Nature*** of Light

1) Towards the end of the **17th century**, two important **theories of light** were published — one by Isaac Newton and the other by a chap called Huygens. **Newton's** theory suggested that light was made up of tiny particles, which he called "**corpuscles**". And **Huygens** put forward a theory using **waves**.
2) The **corpuscular theory** could explain **reflection** and **refraction**, but **diffraction** and **interference** are both **uniquely** wave properties. If it could be **shown** that light showed interference patterns, that would help settle the argument once and for all.
3) **Young's** double-slit experiment (over 100 years later) provided the necessary evidence. It showed that light could both **diffract** (through the narrow slits) and **interfere** (to form the interference pattern on the screen).

Of course, this being Physics, nothing's ever simple — give it another 100 years or so and the debate would be raging again.

Practice Questions

Q1 In Young's experiment, why do you get a bright fringe at a point equidistant from both slits?

Q2 What does Young's experiment show about the nature of light?

Q3 Write down Young's double-slit formula.

Exam Questions

Q1 (a) The diagram on the right shows waves from two coherent light sources, S_1 and S_2. Sketch the interference pattern, marking on constructive and destructive interference. [2 marks]

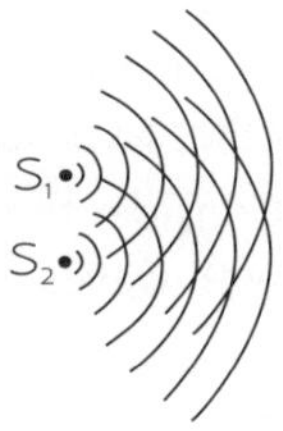

(b) In practice if interference is to be observed, S_1 and S_2 must be slits in a screen behind which there is a source of laser light. Why? [2 marks]

Q2 In an experiment to study sound interference, two loudspeakers are connected to an oscillator emitting sound at 1320 Hz and set up as shown in the diagram below. They are 1.5 m apart and 7 m away from the line AC. A listener moving from A to C hears minimum sound at A and C and maximum sound at B.

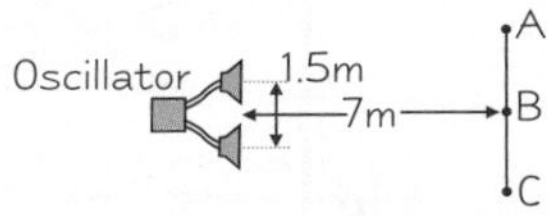

(a) Calculate the wavelength of the sound waves if the speed of sound in air is taken to be 330 ms^{-1}. [1 mark]

(b) Calculate the separation of points A and C. [2 marks]

Carry on Physics — this page is far too saucy...

*

Be careful when you're calculating the fringe width by averaging over several fringes. Don't just divide by the number of bright lines. Ten bright lines will only have nine fringe-widths between them, not ten. It's an easy mistake to make, but you have been warned... mwa ha ha ha (felt necessary, sorry).

* Sadly, Barbara Windsor was unavailable.

Diffraction Gratings

Ay... starting to get into some pretty funky stuff now. I like light experiments.

*Interference Patterns Get **Sharper** When You Diffract Through **More Slits***

1) You can repeat **Young's double-slit** experiment (see p. 86) with **more than two equally spaced** slits. You get basically the **same shaped** pattern as for two slits — but the **bright bands** are **brighter** and **narrower** and the **dark areas** between are **darker**.
2) When **monochromatic light** (one wavelength) is passed through a **grating** with **hundreds** of slits per millimetre, the interference pattern is **really sharp** because there are so **many beams reinforcing** the **pattern**.
3) Sharper fringes make for more **accurate** measurements.

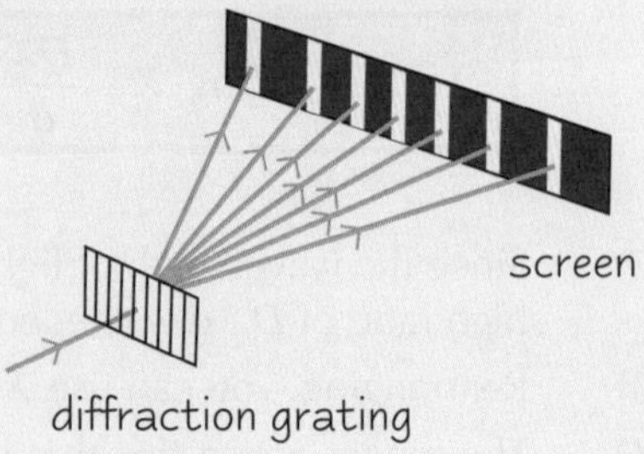

Monochromatic Light** on a **Diffraction Grating** gives **Sharp Lines

1) For **monochromatic** light, all the **maxima** are sharp lines. (It's different for white light — see next page.)
2) There's a line of **maximum brightness** at the centre called the **zero order** line.
3) The lines just **either side** of the central one are called **first order lines**. The **next pair out** are called **second order** lines and so on.
4) For a grating with slits a distance d apart, the angle between the **incident beam** and **the nth order maximum** is given by:

$$d \sin \theta = n\lambda$$

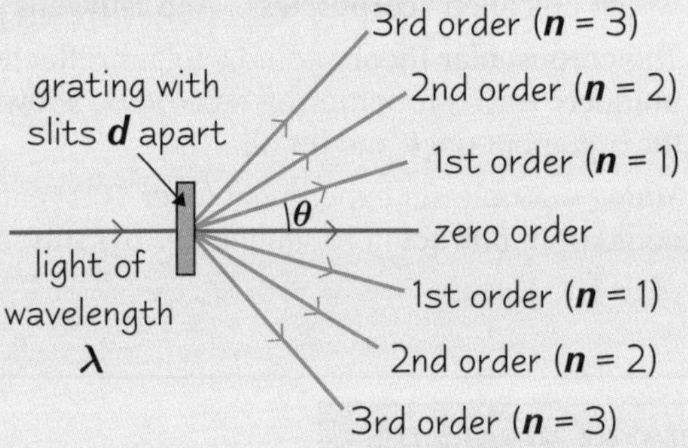

5) So by observing d, θ and n you can **calculate the wavelength** of the light.

If the grating has N slits per metre, then the slit spacing, d, is just 1/N metres.

DERIVING THE EQUATION:

1) At **each slit**, the incoming waves are **diffracted**. These diffracted waves then **interfere** with each other to produce an **interference pattern**.
2) Consider the **first order maximum**. This happens at the **angle** when the waves from one slit line up with waves from the **next slit** that are **exactly one wavelength** behind.

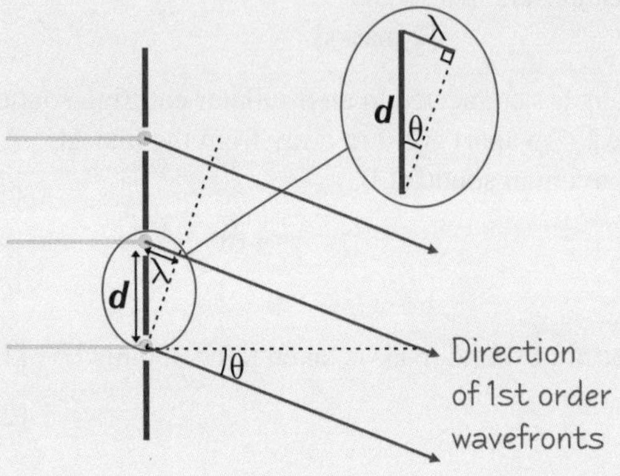

3) Call the **angle** between the **first order maximum** and the **incoming light** θ.
4) Now, look at the **triangle** highlighted in the diagram. The angle is θ (using basic geometry), d is the slit spacing and the **path difference** is λ.
5) So, for the first maximum, using trig:
$$d \sin \theta = \lambda$$
6) The other maxima occur when the path difference is 2λ, 3λ, 4λ, etc. So to make the equation **general**, just replace λ with $n\lambda$, where n is an integer — the **order** of the maximum.

Diffraction Gratings

You can Draw General Conclusions from d sin θ = nλ

1) If λ is **bigger**, **sin θ** is **bigger**, and so θ is **bigger**. This means that the larger the **wavelength**, the more the pattern will **spread out**.
2) If d is **bigger**, **sin θ** is **smaller**. This means that the **coarser** the **grating**, the **less** the pattern will **spread out**.
3) Values of **sin θ** greater than **1** are **impossible**. So if for a certain n you get a result of **more than 1** for **sin θ** you know that that order **doesn't exist**.

Shining White Light Through a Diffraction Grating Produces Spectra

1) **White light** is really a **mixture** of **colours**. If you **diffract** white light through a **grating** then the patterns due to **different wavelengths** within the white light are **spread out** by **different** amounts.
2) Each **order** in the pattern becomes a **spectrum**, with **red** on the **outside** and **violet** on the **inside**. The **zero order maximum** stays **white** because all the wavelengths just pass straight through.

Astronomers and **chemists** often need to study spectra to help identify elements. They use diffraction gratings rather than prisms because they're **more accurate**.

Practice Questions

Q1 How is the diffraction grating pattern for white light different from the pattern for laser light?

Q2 What difference does it make to the pattern if you use a finer grating?

Q3 What equation is used to find the angle between the nth order maximum and the incident beam for a diffraction grating?

Q4 Derive the equation you quoted in Q3.

Exam Questions

Q1 Yellow laser light of wavelength 600 nm (6×10^{-7} m) is transmitted through a diffraction grating of 4×10^5 lines per metre.

(a) At what angle to the normal are the first and second order bright lines seen? [4 marks]

(b) Is there a fifth order line? [1 mark]

Q2 Visible, monochromatic light is transmitted through a diffraction grating of 3.7×10^5 lines per metre. The first order maximum is at an angle of 14.2° to the incident beam.

Find the wavelength of the incident light. [2 marks]

Ooooooooooooo — pretty patterns...

Derivation — ouch. At least it's not a bad one though. As long as you learn the diagram, its just geometry and a bit of trig from there. Make sure you ***learn*** *the equation — that way, you know what you're aiming for. As for the rest of the page, remember that the more slits you have, the sharper the image — and white light makes a pretty spectrum.*

Error Analysis

Science is all about getting good evidence to test your theories... and part of that is knowing how good the results from an experiment are. Physicists always have to include the uncertainty in a result, so you can see the range the actual value probably lies within. Dealing with error and uncertainty is an important skill, so those pesky examiners like to sneak in a couple of questions about it... but if you know your stuff you can get some easy marks.

Nothing is Certain

1) **Every** measurement you take has an **experimental uncertainty**. Say you've done something outrageous like measure the length of a piece of wire with a centimetre ruler. You might think you've measured its length as 30 cm, but at **best** you've probably measured it to be 30 **± 0.5** cm. And that's without taking into account any other errors that might be in your measurement...
2) The ± bit gives you the **range** in which the **true** length (the one you'd really like to know) probably lies — 30 ± 0.5 cm tells you the true length is very likely to lie in the range of 29.5 to 30.5 cm.
3) The smaller the uncertainty, the nearer your value must be to the true value, so the more **accurate** your result.
4) There are **two types** of **error** that cause experimental uncertainty:

Random errors

1) No matter how hard you try, you **can't get rid** of random errors.
2) They can just be down to **noise**, or be that you're measuring a **random process** such as nuclear radiation emission.
3) You get random error in **any** measurement. If you measured the length of a wire 20 times, the chances are you'd get a **slightly different** value each time, e.g. due to your head being in a slightly different position when reading the scale.
4) It could be that you just can't keep controlled variables **exactly** the same throughout the experiment.
5) Or it could just be the wind was blowing in the wrong direction at the time...

Systematic errors

1) You get systematic errors not because you've made a mistake in a measurement — but because of the **apparatus** you're using, or your experimental method, e.g. using an inaccurate clock.
2) The problem is often that you **don't know they're there**. You've got to spot them first to have any chance of correcting for them.
3) Systematic errors usually **shift** all of your results to be too high or too low by the **same amount**. They're annoying, but there are things you can do to reduce them if you manage to spot them...

You Need to Know How to Improve Measurements

There are a few different ways you can **reduce** the uncertainty in your results:

Repeating measurements — by repeating a measurement **several times** and **averaging**, you reduce the **random uncertainty** in your result. The **more** measurements you average over, the **less error** you're likely to have.

Use higher precision apparatus — the **more precisely** you can measure something, the **less random error** there is in the measurement. So if you use more precise equipment — e.g. swapping a millimetre ruler for a micrometer to measure the diameter of a wire — you can instantly cut down the **random error** in your experiment.

Calibration — you can calibrate your apparatus by measuring a **known value**. If there's a **difference** between the **measured** and **known** value, you can use this to **correct** the inaccuracy of the apparatus, and so reduce your **systematic error**.

Lorraine thought getting an uncertainty of ± 0.1 A deserved a victory dance.

Error Analysis

You can Calculate the Percentage Uncertainty in a Measurement

1) You might get asked to work out the percentage uncertainty in a measurement.
2) It's just working out a percentage, so nothing too tricky. It's just that sometimes you can get **the fear** as soon as you see the word uncertainty... but just keep your cool and you can pick up some easy marks.

Example

Tom finds the resistance of a filament lamp to be **5.0 ± 0.4 Ω.**
Find the percentage uncertainty in Tom's measurement.

Just divide the **absolute uncertainty** by the value for the resistance:

$$\text{Percentage uncertainty} = \frac{0.4}{5.0} \times 100 = \mathbf{8\%}$$

You can Estimate Values by Averaging

You might be given a graph of information showing the results for many **repetitions** of the **same** experiment, and asked to estimate the true value and give an uncertainty in that value. Yuk.

Here's how to go about it:

1) Estimate the true value by **averaging** the results you've been given.
(Make sure you state whatever average it is you take, otherwise you might not get the mark.)
2) To get the uncertainty, you just need to look **how far away** from your average value the maximum and minimum values in the graph you've been given are.

Example — Estimating the resistance of a component

A class measure the resistance of a component and record their results on the bar chart shown on the right.

Estimate the resistance of the component, giving a suitable range of uncertainty in your answer.

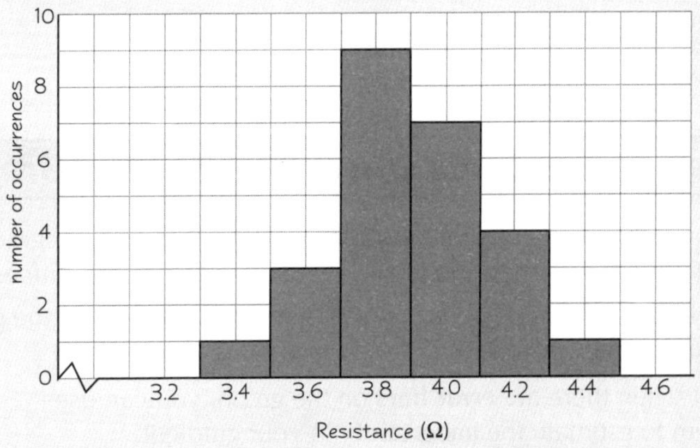

There were 25 measurements, so taking the **mean**:

$$\frac{(3.4+(3.6\times3)+(3.8\times9)+(4.0\times7)+(4.2\times4)+4.4)}{25} = \frac{97.6}{25} = 3.90 \text{ (3 s.f.)}$$

The maximum value found was 4.4 Ω, the minimum value was 3.4.

Both values are about 0.5 Ω from the average value, so the answer is **3.9 ± 0.5 Ω.**

Random error in your favour — collect £200...

These pages should give you a fair idea of how to deal with errors... which are rather annoyingly in everything. Even if you're lucky enough to not get tested on this sort of thing in the exam, it's really useful to know for your lab coursework. Have a look at the last page for how to show errors on graphs...

Errors on Graphs

No, you can't use errors and uncertainty as an excuse for getting your graph wrong.

Error Bars *to Show* ***Uncertainty*** *on a* ***Graph***

1) Most of the time in science, you work out the uncertainty in your **final result** using the uncertainty in **each measurement** you make.
2) When you're plotting a graph, you show the uncertainty in a value by using **error bars** to show the range the point is likely to lie in.
3) You probably won't get asked to **plot** any error bars (phew...) — but you might need to **read off** a graph that has them.

Example

Use the graph below to find the error in measuring the extension of material X.

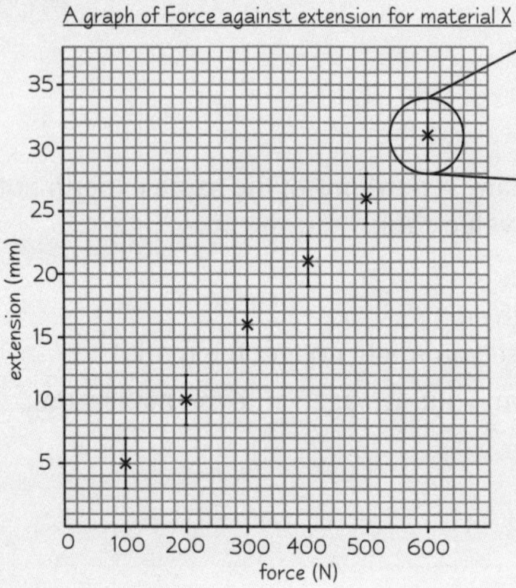

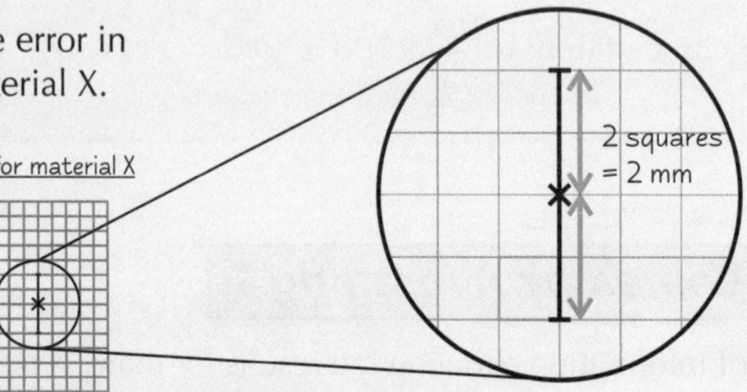

The error bars extend 2 squares above and below each measurement, which is equivalent to 2 mm.

So, the uncertainty in each measurement is **± 2 mm.**

Be careful — sometimes error bars are calculated using a set percentage of uncertainty for each measurement — so their size will change depending on the size of the measurement.

You can ***Estimate*** *the* ***Uncertainty*** *of the* ***Graph's Gradient***

1) Normally when you draw a graph you'll want to find the **gradient** or the **intercept**. E.g. for a force-extension graph, the gradient's 1/***k***, the stiffness constant of the material.
2) To find the value of ***k***, you draw a nice line of best fit on the graph and calculate your answer from that. No problem there.
3) But when there are **error bars** on the graph, you can use them to estimate the **uncertainty** in your gradient.

- Draw the **maximum** and **minimum** slopes possible for the data through **all** of the error bars.
- Calculate the value of the gradient (or intercept) for these slopes to find maximum and minimum values.
- The true answer is likely to lie between these values, so they give you the **uncertainty** in your answer.

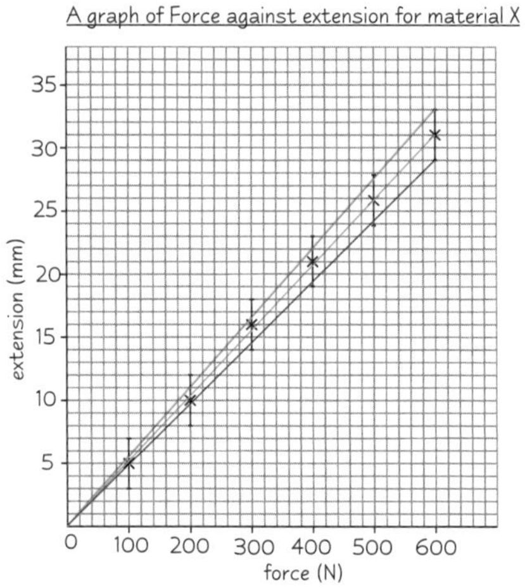

Hedgehog nougat — error bar...

Errors and uncertainty aren't a Physicist's cop-out — they're a really useful way of saying how confident you are that you've got the right answer. Getting slightly different results from your mate doesn't mean you've done something wrong, it probably just means there's some factor you can't control that's causing random errors in your results.

A2-Level Physics

Exam Board: AQA A

Momentum and Impulse

These pages are about linear momentum — that's momentum in a straight line (not a circle).

Understanding **Momentum** helps you do **Calculations** on **Collisions**

The **momentum** of an object depends on two things — its **mass** and **velocity**.
The **product** of these two values is the momentum of the object.

momentum = mass × velocity or in symbols: $\boldsymbol{p}$ (in kg ms^{-1}) = $\boldsymbol{m}$ (in kg) × $\boldsymbol{v}$ (in ms^{-1})

Remember, velocity is a vector quantity, so it has size and direction.

Momentum is always **Conserved**

1) Assuming **no external forces** act, momentum is always **conserved**.
2) This means the **total momentum** of two objects **before** they collide **equals** the total momentum **after** the collision.
3) This is really handy for working out the **velocity** of objects after a collision (as you do...):

Example A skater of mass 75 kg and velocity 4 ms^{-1} collides with a stationary skater of mass 50 kg. The two skaters join together and move off in the same direction. Calculate their velocity after impact.

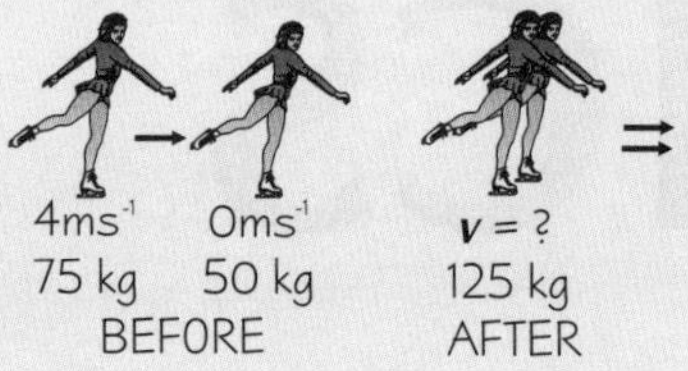

Before you start a momentum calculation, always draw a quick sketch.

Momentum of skaters before = Momentum of skaters after
$(75 \times 4) + (50 \times 0) = 125\boldsymbol{v}$
$300 = 125\boldsymbol{v}$
So $\boldsymbol{v} = 2.4\ ms^{-1}$

4) The same principle can be applied in **explosions**. E.g. if you fire an **air rifle**, the **forward momentum** gained by the pellet **equals** the **backward momentum** of the rifle, and you feel the rifle recoiling into your shoulder.

Example A bullet of mass 0.005 kg is shot from a rifle at a speed of 200 ms^{-1}. The rifle has a mass of 4 kg. Calculate the velocity at which the rifle recoils.

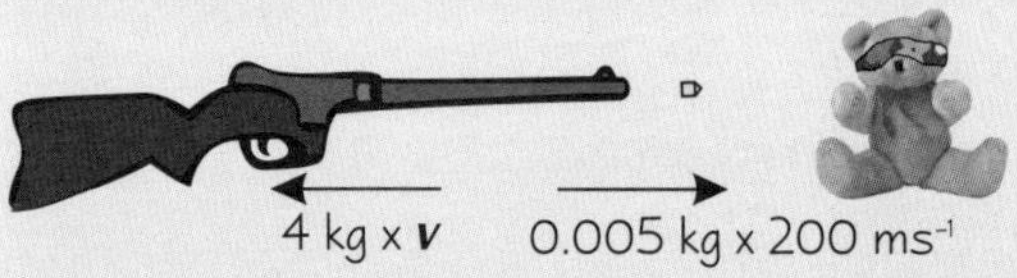

Momentum before explosion = Momentum after explosion
$0 = (0.005 \times 200) + (4 \times \boldsymbol{v})$
$0 = 1 + 4\boldsymbol{v}$
$\boldsymbol{v} = -0.25\ ms^{-1}$

Collisions can be **Elastic** or **Inelastic**

An **elastic collision** is one where **momentum** is **conserved** and **kinetic energy** is **conserved** — i.e. no energy is dissipatec as heat, sound, etc. If a collision is **inelastic** it means that some of the kinetic energy is converted into other forms durin the collision. But **momentum is always conserved.**

Example A toy lorry (mass 2 kg) travelling at 3 ms^{-1} crashes into a smaller toy car (mass 800 g), travelling in the same direction at 2 ms^{-1}. The velocity of the lorry after the collision is 2.6 ms^{-1} in the same direction. Calculate the new velocity of the car and the total kinetic energy before and after the collision.

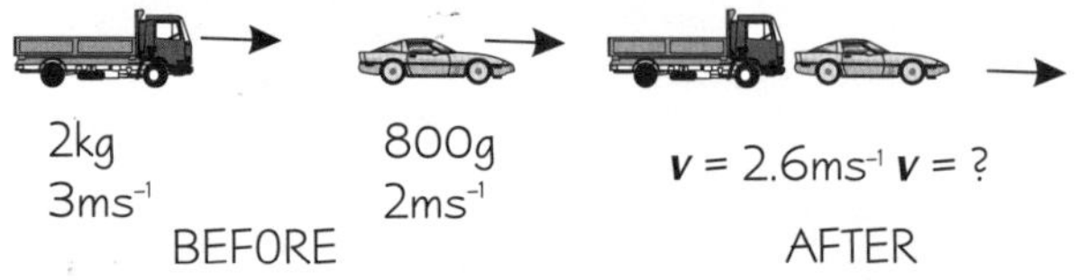

Momentum before collision = Momentum after collision
$(2 \times 3) + (0.8 \times 2) = (2 \times 2.6) + (0.8\boldsymbol{v})$
$7.6 = 5.2 + 0.8\boldsymbol{v}$
$2.4 = 0.8\boldsymbol{v}$
$\boldsymbol{v} = 3\ ms^{-1}$

Kinetic Energy before = KE of lorry + KE of car
$= \frac{1}{2}mv^2$ (lorry) + $\frac{1}{2}mv^2$ (car)
$= \frac{1}{2}(2 \times 3^2) + \frac{1}{2}(0.8 \times 2^2)$
$= 9 + 1.6$
$= 10.6$ J

Kinetic Energy after $= \frac{1}{2}(2 \times 2.6^2) + \frac{1}{2}(0.8 \times 3^2)$
$= 6.76 + 3.6$
$= 10.36$ J

The difference in the two values is the amount of kinetic energy dissipated as heat or sound, or in damaging the vehicles — so this is an inelastic collision.

Momentum and Impulse

Newton's 2nd Law says that Force is the **Rate of Change in Momentum...**

> *"The **rate of change of momentum** of an object is **directly proportional** to the **resultant force** which acts on the object."*

so $F = \dfrac{\Delta mv}{\Delta t}$

If mass is constant, this can be written as the well-known equation:

resultant force (F) = mass (m) × acceleration (a)

Learn this — it crops up all over the place in A2 Physics. And learn what it means too:

1) It says that the **more force** you have acting on a certain mass, the **more acceleration** you get.
2) It says that for a given force the **more mass** you have, the **less acceleration** you get.

REMEMBER:
1) The **resultant force** is the **vector sum** of all the forces.
2) The force is **always** measured in **newtons**. **Always.**
3) The **mass** is always measured in **kilograms**.
4) **a** is the **acceleration** of the object as a result of F. It's **always** measured in **metres per second per second** (ms^{-2}).
5) The **acceleration** is always in the **same direction** as the **resultant force**.

Impulse = Change in ***Momentum***

1) Newton's second law says **force = rate of change of momentum**, or $F = (mv - mu) \div t$
2) **Rearranging** Newton's 2nd law gives: ⟹ Where **impulse** is defined as **force × time**, ***Ft***. The units of impulse are **newton seconds**, Ns.

   $Ft = mv - mu$
   (where v is the final velocity and u is the initial velocity)
   so ***impulse*** = ***change of momentum***

3) **Impulse** is the **area under** a **force-time graph** — this is really handy for solving problems where the force changes.

   Example The graph shows the resultant force acting on a toy car. If the car is initially at rest, what is its momentum after 3 seconds?

   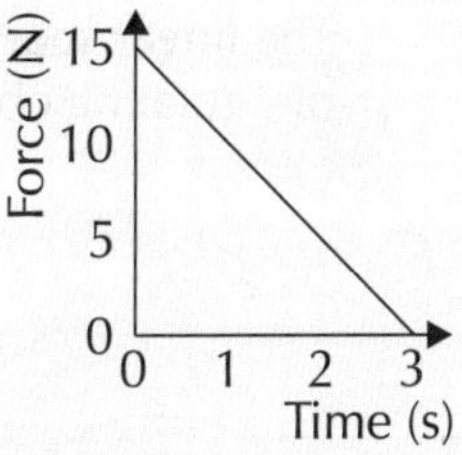

   Impulse = **change of momentum** = ***mv – mu***. The **initial momentum** (***mu***) is **zero** because the toy car is stationary to begin with. So, **impulse** = ***mv***.
   Impulse is the **area under the graph**, so to find the **momentum** of the car after 3 seconds, you need to find the **area under the graph** between 0 and 3 seconds.
   Area under graph = (15 × 3) ÷ 2 = **22.5 Ns**

4) The **force** of an impact can be **reduced** by **increasing the time** of the impact.

Practice Questions

Q1 Give two examples of conservation of momentum in practice.

Q2 Describe what happens when a tiny object makes an elastic collision with a massive object, and why.

Exam Questions

Q1 A ball of mass 0.6 kg moving at 5 ms^{-1} collides with a larger stationary ball of mass 2 kg. The smaller ball rebounds in the opposite direction at 2.4 ms^{-1}.
(a) Calculate the velocity of the larger ball immediately after the collision. [3 marks]
(b) Is this an elastic or inelastic collision? Explain your answer. [3 marks]

Q2 A toy train of mass 0.7 kg, travelling at 0.3 ms^{-1}, collides with a stationary toy carriage of mass 0.4 kg. The two toys couple together. Calculate their new velocity. [3 marks]

Momentum will never be an endangered species — it's always conserved...

It seems a bit of a contradiction to say that momentum's always conserved then tell you that impulse is the change in momentum. The difference is that impulse is only talking about the change of momentum of one of the objects, whereas conservation of momentum is talking about the ***whole*** *system.*

Circular Motion

*It's probably worth putting a bookmark in here — this stuff is needed **all over** the place.*

Angles *can be Expressed in* ***Radians***

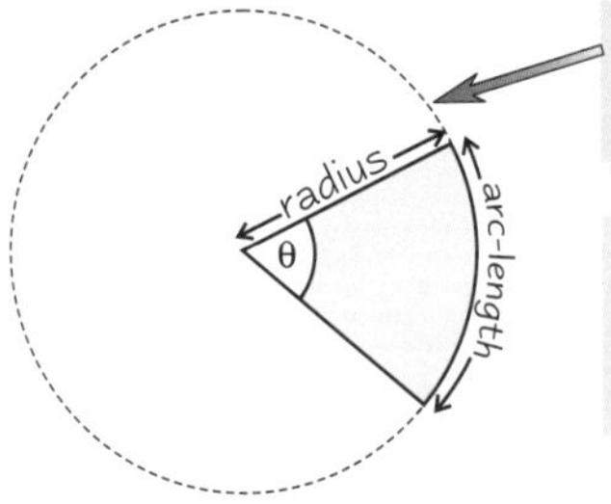

The angle in **radians**, θ, is defined as the **arc-length** divided by the radius of the circle.

For a **complete circle** (360°), the arc-length is just the circumference of the circle ($2\pi \boldsymbol{r}$). Dividing this by the radius ($\boldsymbol{r}$) gives 2π. So there are 2π radians in a complete circle.

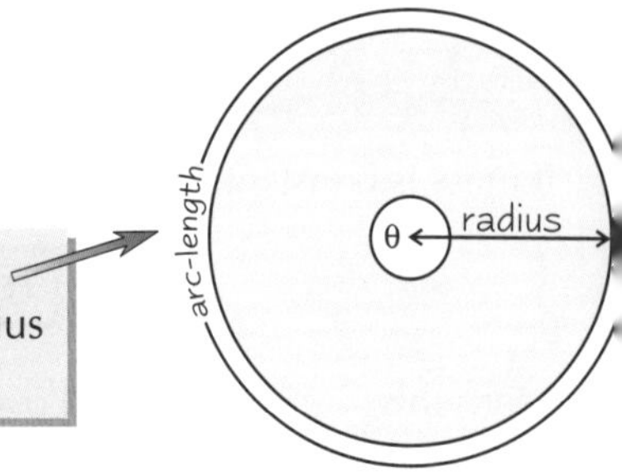

Some common angles:

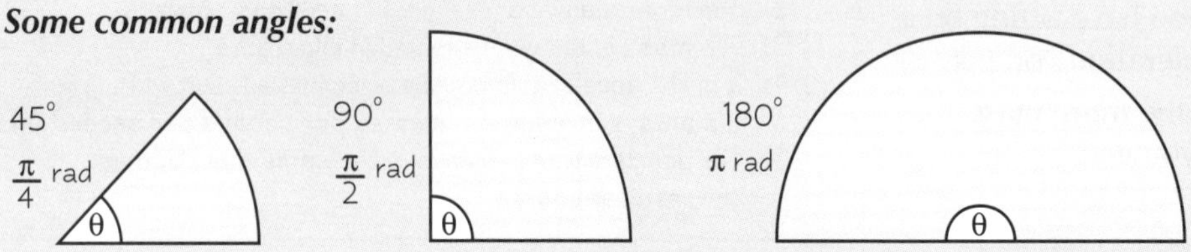

angle in radians $= \frac{2\pi}{360} \times$ angle in degrees

1 radian is about 57°

The ***Angular Speed*** *is the* ***Angle*** *an Object Rotates Through* ***per Second***

1) Just as **linear speed**, $\boldsymbol{v}$, is defined as distance ÷ time, the **angular speed**, ω, is defined as **angle ÷ time**. The unit is rad s^{-1} — radians per second.

$$\omega = \frac{\theta}{t}$$

ω = angular speed (rad s^{-1}) — the symbol for angular speed is the little Greek 'omega', not a w.
$\boldsymbol{\theta}$ = angle (radians) turned through in a time, $\boldsymbol{t}$ (seconds)

2) The **linear speed**, $\boldsymbol{v}$, and **angular speed**, ω, of a rotating object are linked by the equation:

$$v = r\omega$$

$\boldsymbol{v}$ = linear speed (ms^{-1}), $\boldsymbol{r}$ = radius of the circle (m), ω = angular speed (rad s^{-1})

Example ***Beam of Particles in a Cyclotron*** *(a type of particle accelerator)*

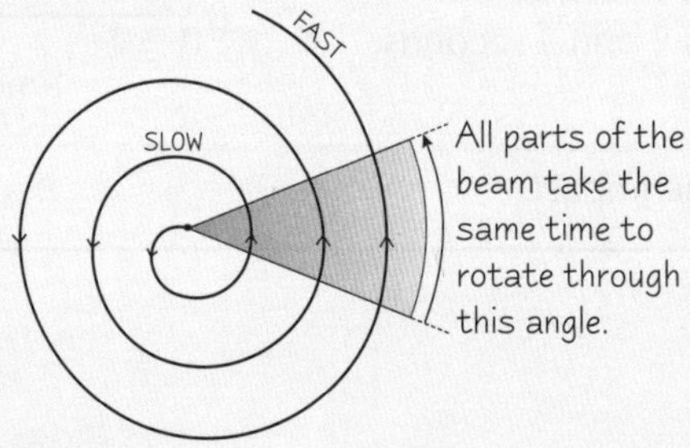

1) Different parts of the particle beam are rotating at **different linear speeds**, $\boldsymbol{v}$. (The linear speed is sometimes called **tangential velocity**.)
2) But all the parts **rotate** through the **same angle** in the **same time** — so they have the **same angular speed**.

Circular Motion has a ***Frequency*** *and* ***Period***

1) The frequency, $\boldsymbol{f}$, is the number of complete **revolutions per second** (rev s^{-1} or hertz, Hz).
2) The period, $\boldsymbol{T}$, is the **time taken** for a complete revolution (in seconds).
3) Frequency and period are **linked** by the equation:

$$f = \frac{1}{T}$$

$\boldsymbol{f}$ = frequency in rev s^{-1}, $\boldsymbol{T}$ = period in s

4) For a complete circle, an object turns through 2π radians in a time $\boldsymbol{T}$, so frequency and period are related to ω by

$$\omega = 2\pi f \quad \text{and} \quad \omega = \frac{2\pi}{T}$$

$\boldsymbol{f}$ = frequency in rev s^{-1}, $\boldsymbol{T}$ = period in s, ω = angular speed in rad s^{-1}

Circular Motion

Objects Travelling in Circles are **Accelerating** since their **Velocity is Changing**

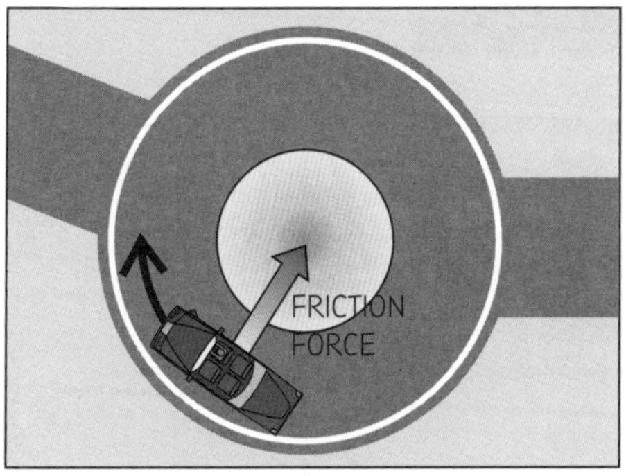

1) Even if the car shown is going at a **constant speed**, its **velocity** is changing since its **direction** is changing.
2) Since acceleration is defined as the **rate of change of velocity**, the car is accelerating even though it isn't going any faster.
3) This acceleration is called the **centripetal acceleration** and is always directed towards the **centre of the circle**.

There are two formulas for centripetal acceleration:

$$a = \frac{v^2}{r} \quad \text{and} \quad a = \omega^2 r$$

a = centripetal acceleration in ms^{-2}
v = linear speed in ms^{-1}
ω = angular speed in rad s^{-1}
r = radius in m

The **Centripetal Acceleration** is produced by a **Centripetal Force**

From Newton's laws, if there's a **centripetal acceleration**, there must be a **centripetal force** acting towards the **centre of the circle**.

Since $F = ma$, the centripetal force must be:

$$F = \frac{mv^2}{r} \quad \text{and} \quad F = m\omega^2 r$$

The centripetal force is what keeps the object moving in a circle — remove the force and the object would fly off at a tangent.

Men cowered from the force of the centripede.

Practice Questions

Q1 How many radians are there in a complete circle?
Q2 How is angular speed defined and what is the relationship between angular speed and linear speed?
Q3 Define the period and frequency of circular motion. What is the relationship between period and angular speed?
Q4 In which direction does the centripetal force act, and what happens when this force is removed?

Exam Questions

Q1 (a) At what angular speed does the Earth orbit the Sun? (1 year = 3.2×10^7 s) [2 marks]

(b) Calculate the Earth's linear speed. (Assume radius of orbit = 1.5×10^{11} m) [2 marks]

(c) Calculate the centripetal force needed to keep the Earth in its orbit. (Mass of Earth = 6.0×10^{24} kg) [2 marks]

(d) What is providing this force? [1 mark]

Q2 A bucket full of water, tied to a rope, is being swung around in a vertical circle (so it is upside down at the top of the swing). The radius of the circle is 1 m.

(a) By considering the acceleration due to gravity at the top of the swing, what is the minimum frequency with which the bucket can be swung without any water falling out? [3 marks]

(b) The bucket is now swung with a constant angular speed of 5 rad s^{-1}. What will be the tension in the rope when the bucket is at the top of the swing if the total mass of the bucket and water is 10 kg? [2 marks]

I'm spinnin' around, move out of my way...

*"Centripetal" just means "centre-seeking". The centripetal force is what actually causes circular motion. What you **feel** when you're spinning, though, is the reaction (centrifugal) force. Don't get the two mixed up.*

Simple Harmonic Motion

Something simple at last — I like the sound of this. And colourful graphs too — you're in for a treat here.

SHM is Defined in terms of Acceleration and Displacement

1) An object moving with **simple harmonic motion** (SHM) **oscillates** to and fro, either side of a **midpoint**.
2) The distance of the object from the midpoint is called its **displacement**.
3) There is always a **restoring force** pulling or pushing the object back **towards** the **midpoint**.
4) The **size** of the **restoring force** depends on the **displacement**, and the force makes the object **accelerate** towards the midpoint:

> **SHM:** an oscillation in which the **acceleration** of an object is **directly proportional** to its **displacement** from the **midpoint**, and is directed **towards the midpoint**.

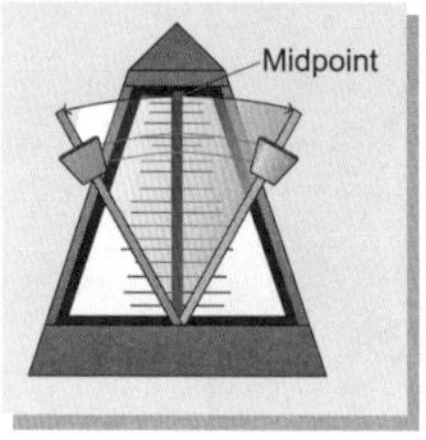

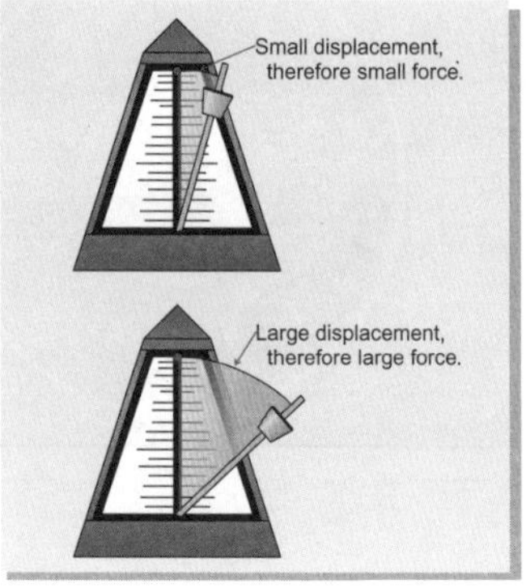

The Restoring Force makes the Object Exchange PE and KE

1) The **type** of **potential energy** (PE) depends on **what it is** that's providing the **restoring force**. This will be **gravitational PE** for pendulums and **elastic PE** (elastic stored energy) for masses on springs.
2) As the object moves **towards the midpoint**, the restoring force **does work** on the object and so **transfers** some **PE** to **KE**. When the object is moving **away from the midpoint**, all that KE is transferred **back to PE** again.
3) At the **midpoint**, the object's **PE** is **zero** and its **KE** is **maximum**.
4) At the **maximum displacement** (the **amplitude**) on both sides of the midpoint, the object's **KE** is **zero** and its **PE** is **maximum**.

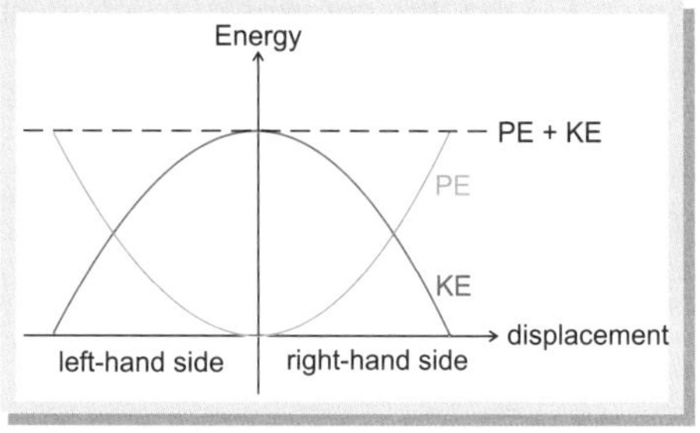

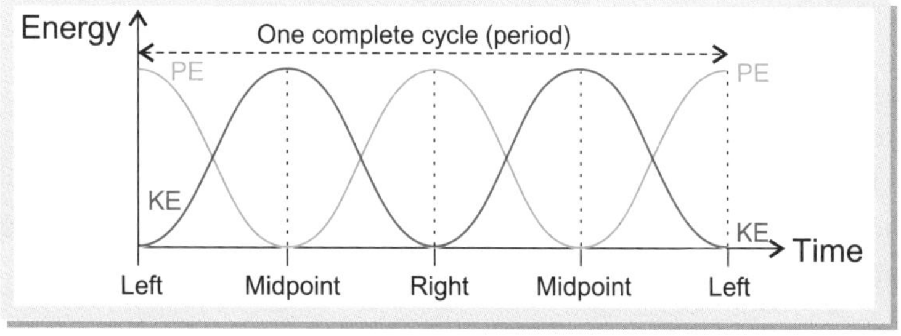

5) The **sum** of the **potential** and **kinetic** energy is called the **mechanical energy** and **stays constant** (as long as the motion isn't damped — see p. 102-103).
6) The **energy transfer** for one complete cycle of oscillation (see graph) is: PE to KE to PE to KE to PE ... and then the process repeats...

You can Draw Graphs to Show Displacement, Velocity and Acceleration

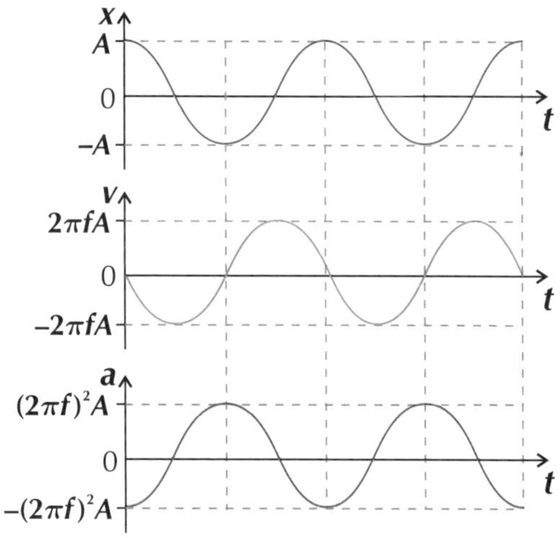

Displacement, x, varies as a cosine or sine wave with a maximum value, A (the amplitude).

Velocity, v, is the gradient of the displacement-time graph. It has a maximum value of $(2\pi f)A$ (where f is the frequency of the oscillation) and is a quarter of a cycle in front of the displacement.

Acceleration, a, is the gradient of the velocity-time graph. It has a maximum value of $(2\pi f)^2 A$, and is in antiphase with the displacement.

Simple Harmonic Motion

The **Frequency** and **Period** don't depend on the **Amplitude**

1) From **maximum positive displacement** (e.g. maximum displacement to the right) to **maximum negative displacement** (e.g. maximum displacement to the left) and **back again** is called a **cycle** of oscillation.
2) The **frequency**, f, of the SHM is the number of cycles per second (measured in Hz).
3) The **period**, T, is the **time** taken for a complete cycle (in seconds).

In SHM, the **frequency** and **period** are independent of the **amplitude** (i.e. constant for a given oscillation). So a pendulum clock will keep ticking in regular time intervals even if its swing becomes very small.

Learn the SHM Equations

I know it looks like there are loads of complicated equations to learn here, but don't panic — it's not that bad really. You'll be given these formulas in the exam, so just make sure you know what they mean and how to use them.

1) According to the definition of SHM, the **acceleration**, a, is directly proportional to the **displacement**, x. The **constant of proportionality** depends on the **frequency**, and the acceleration is always in the **opposite direction** from the displacement (so there's a minus sign in the equation).

$$a = -(2\pi f)^2 x$$

$$\text{Max acceleration: } a_{max} = (2\pi f)^2 A$$

Don't forget, A is the maximum displacement — it's not acceleration.

Jeremy was investigating swinging as a form of simple harmonic motion.

2) The **velocity** is **positive** if the object's moving **away** from the **midpoint**, and **negative** if it's moving **towards** the midpoint — that's why there's a ± sign.

$$v = \pm 2\pi f\sqrt{A^2 - x^2}$$

$$\text{Max velocity: } v_{max} = 2\pi f A$$

3) The **displacement** varies with time according to the equation on the right. To use this equation you need to start timing when the pendulum is at its **maximum displacement** — i.e. when $t = 0$, $x = A$.

$$x = A\cos(2\pi f t)$$

Practice Questions

Q1 Sketch a graph of how the velocity of an object oscillating with SHM varies with time.

Q2 What is the special relationship between the acceleration and the displacement in SHM?

Q3 Given the amplitude and the frequency, how would you work out the maximum acceleration?

Exam Questions

Q1 (a) Define *simple harmonic motion.* [2 marks]

(b) Explain why the motion of a ball bouncing off the ground is not SHM. [1 mark]

Q2 A pendulum is pulled a distance 0.05 m from its midpoint and released.
It oscillates with simple harmonic motion with a frequency of 1.5 Hz. Calculate:

(a) its maximum velocity [1 mark]

(b) its displacement 0.1 s after it is released [2 marks]

(c) the time it takes to fall to 0.01 m from the midpoint after it is released [2 marks]

"Simple" harmonic motion — hmmm, I'm not convinced...

The basic concept of SHM is simple enough (no pun intended). Make sure you can remember the shapes of all the graphs on page 98 and the equations from this page, then just get as much practice at using the equations as you can.

Simple Harmonic Oscillators

There are a couple more equations to learn on this page I'm afraid. The experiment described at the bottom of the page shows where they come from, though, so that should help you remember them.

A **Mass** on a **Spring** is a **Simple Harmonic Oscillator (SHO)**

1) When the mass is **pushed to the left** or **pulled to the right** of the **equilibrium position**, there's a **force** exerted on it.

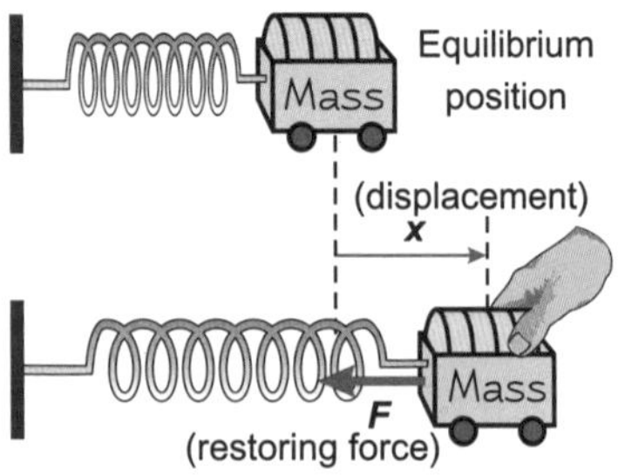

2) The size of this force is:

$$F = -kx$$

where k is the **spring constant** (stiffness) of the spring in Nm^{-1} and x is the displacement in m.

3) After a bit of jiggery-pokery involving Newton's second law and some of the ideas on the previous page, you get the **formula for the period of a mass oscillating on a spring**:

$$T = 2\pi\sqrt{\frac{m}{k}}$$

where T = period of oscillation in seconds
m = mass in kg
k = spring constant in Nm^{-1}

A simple theory of how atoms in a lattice (i.e. a solid) behave can be worked out by considering them as masses oscillating on springs. So there you go.

You Can **Check the Formula Experimentally**

As promised, this experiment shows you where the equations come from.
It's not too tricky — you just have to change **one variable at a time** and see what happens.

Investigating the Mass-Spring System

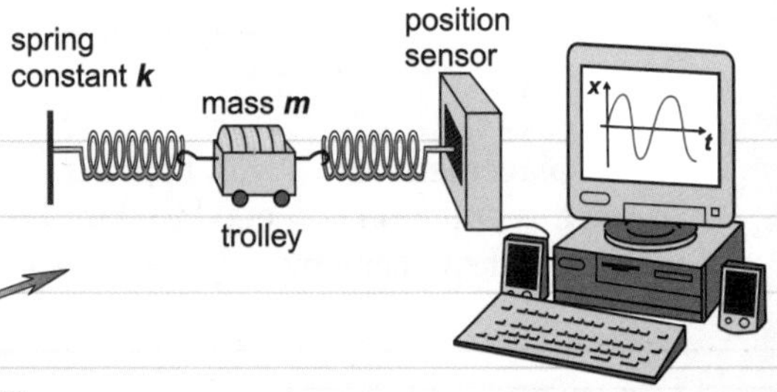

1) You could measure the **period**, T, by getting a computer to plot a **displacement-time graph** from a **data logger**.

2) Attach a **trolley** between two **springs**, pull it to one side by a certain amount and then let go. The trolley will **oscillate** back and forth as the springs pull it in each direction.

3) Change the **mass**, m, by loading the trolley with **masses** — don't forget to include the mass of the trolley in your calculations.

4) Change the **spring stiffness**, k, by using different combinations of springs.

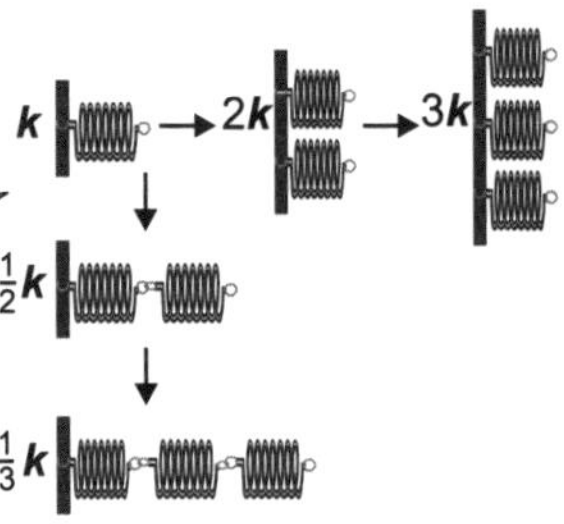

5) Change the **amplitude**, **A**, by pulling the trolley across by different amounts.

6) You'll get the following **results**: *(∝ means "is proportional to")*

a) $T \propto \sqrt{m}$ so $T^2 \propto m$

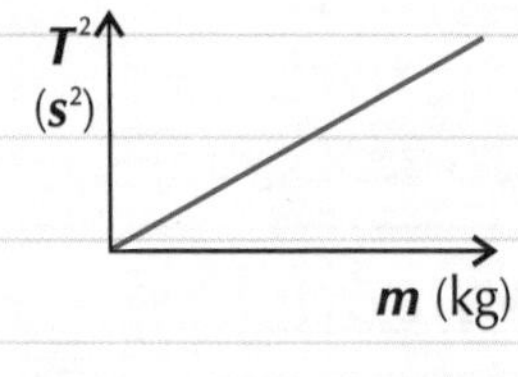

b) $T \propto \sqrt{\frac{1}{k}}$ so $T^2 \propto \frac{1}{k}$

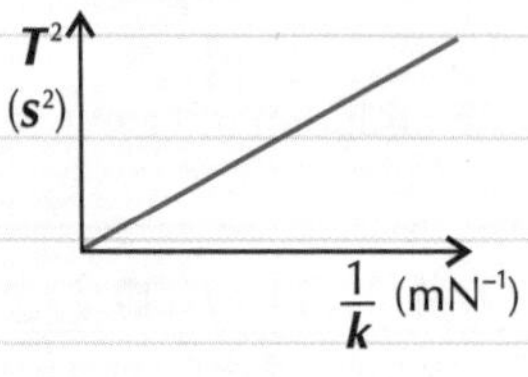

c) T doesn't depend on amplitude, A.

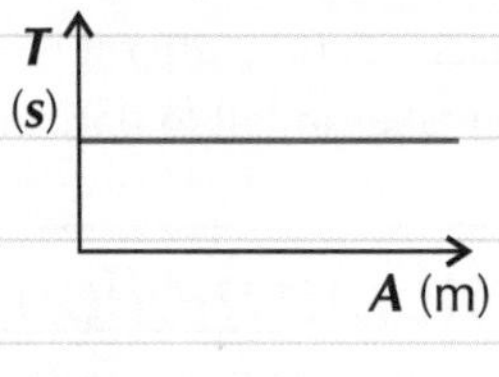

Simple Harmonic Oscillators

The *Simple Pendulum* is the *Classic Example* of an *SHO*

If you set up a simple pendulum attached to an angle sensor and computer like this — then change the length, ***l***, the mass of the bob, ***m***, and the amplitude, ***A***, you get the following results:

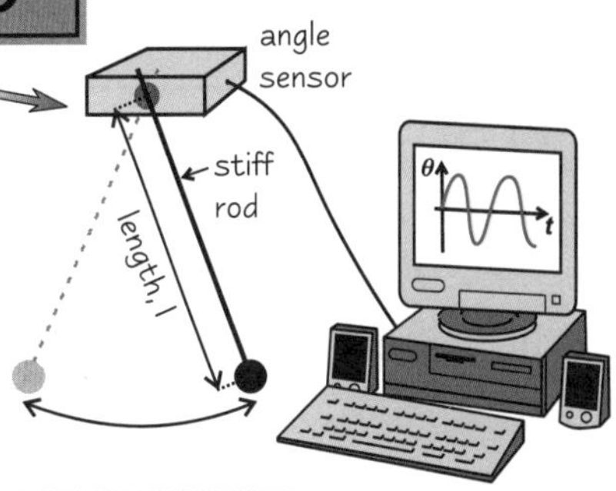

a) $T \propto \sqrt{l}$, so $T^2 \propto l$

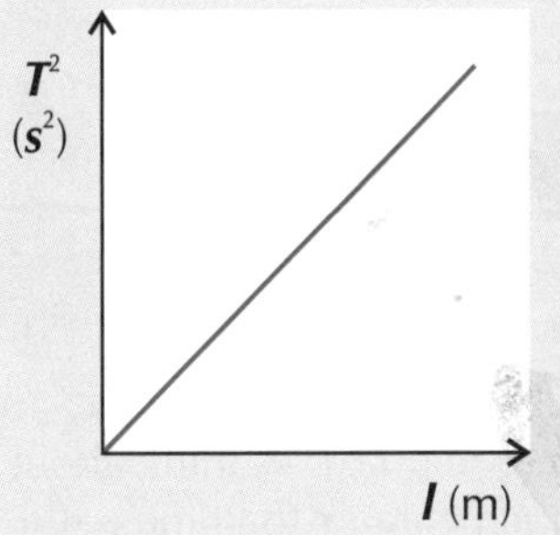

b) ***T*** does not depend on ***m***.

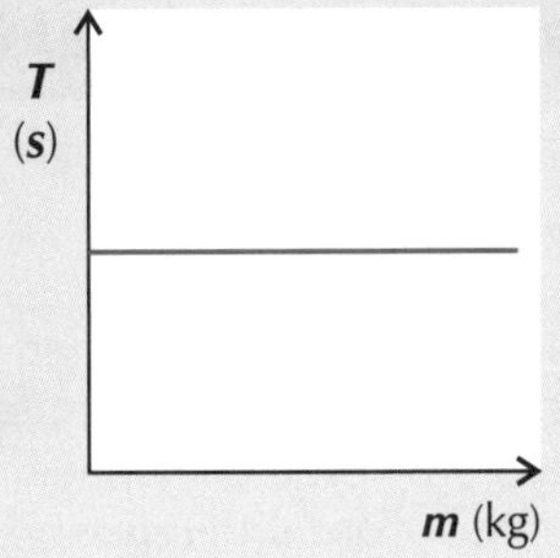

c) ***T*** does not depend on ***A***.

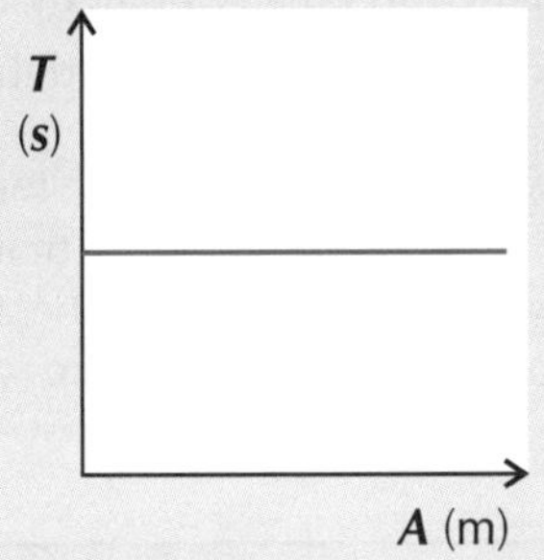

Bob hung around waiting for the experiment to start.

The **formula for the period of a pendulum** is: (The derivation's quite hard, so you don't need to know it.)

This formula only works for small angles of oscillation — up to about 10° from the equilibrium point.

$$T = 2\pi\sqrt{\frac{l}{g}}$$

where ***T*** = period of oscillation in seconds
l = length of pendulum (between pivot and centre of mass of bob) in m
g = gravitational field strength in Nkg^{-1}

Practice Questions

Q1 Write down the formulae for the period of a mass on a spring and the period of a pendulum.

Q2 Describe a method you could use to measure the period of an oscillator.

Q3 For a mass-spring system, what graphs could you plot to find out how the period depends on: a) the mass, b) the spring constant, and c) the amplitude? What would they look like?

Exam Questions

Q1 A spring of original length 0.10 m is suspended from a stand and clamp.
A mass of 0.10 kg is attached to the bottom and the spring extends to a total length of 0.20 m.

(a) Calculate the spring constant of the spring in Nm^{-1}. ($g = 9.81\ Nkg^{-1}$) *The spring isn't moving at this point, so the forces on it must be balanced.* [2 marks]

(b) The mass is pulled down a further 2 cm and then released. Assuming the mass oscillates with simple harmonic motion, calculate the period of the subsequent oscillations. [1 mark]

(c) What mass would be needed to make the period of oscillation twice as long? [2 marks]

Q2 Two pendulums of different lengths were released from rest at the top of their swing.
It took exactly the same time for the shorter pendulum to make five complete oscillations as it took the longer pendulum to make three complete oscillations.
The shorter pendulum had a length of 0.20 m. Show that the length of the longer one was 0.56 m. [3 marks]

Go on — SHO the examiners what you're made of...

The most important things to remember on these pages are those two period equations. You'll be given them in your exam, but you need to know what they mean and be happy using them.

Free and Forced Vibrations

Resonance... hmm... tricky little beast. Remember the Millennium Bridge, that standard-bearer of British engineering? The wibbles and wobbles were caused by resonance. How was it sorted out? By damping, which is coming up too.

Free Vibrations — No Transfer of Energy To or From the Surroundings

1) If you stretch and release a mass on a spring, it oscillates at its **natural frequency**.
2) If **no energy's transferred** to or from the surroundings, it will **keep** oscillating with the **same amplitude forever**.
3) In practice this **never happens**, but a spring vibrating in air is called a **free vibration** anyway.

Forced Vibrations happen when there's an External Driving Force

1) A system can be **forced** to vibrate by a periodic **external force**.
2) The frequency of this force is called the **driving frequency**.

If the **driving frequency** is much **less than** the **natural frequency** then the two are **in phase** — think about a really slow driver and it should make sense. But, if the **driving frequency** is much **greater than** the **natural frequency**, the oscillator won't be able to keep up — you end up with the driver completely **out of phase** with the oscillator. At **resonance** (see below) the **phase difference** between the driver and oscillator is **90°**.

Resonance happens when Driving Frequency = Natural Frequency

When the **driving frequency** approaches the **natural frequency**, the system gains more and more energy from the driving force and so vibrates with a **rapidly increasing amplitude**. When this happens the system is **resonating**.

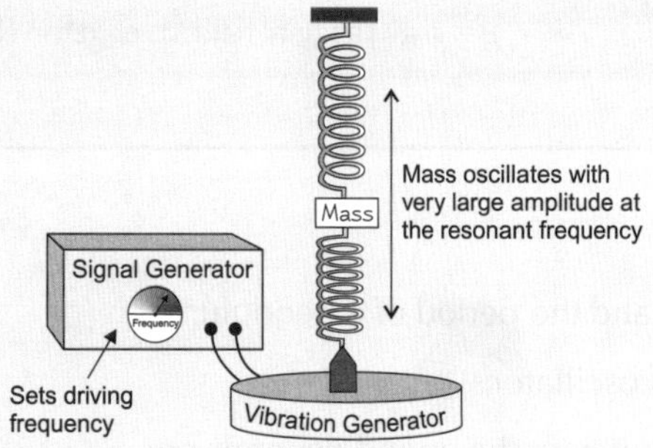

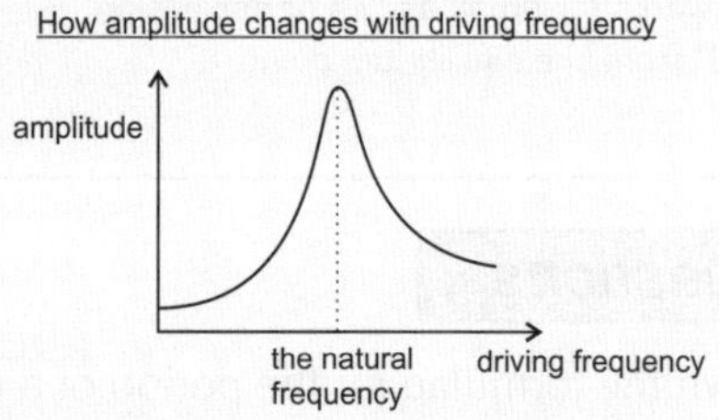

Armies deliberately march 'out of step' when they cross a bridge. This reduces the risk of the bridge resonating and breaking apart.

Examples of resonance:

a) organ pipe

The column of air resonates, driven by the motion of air at the base.

b) swing

A swing resonates if it's driven by someone pushing it at its natural frequency.

c) glass smashing

A glass resonates when driven by a sound wave at the right frequency.

d) radio

A radio is tuned so the electric circuit resonates at the same frequency as the radio station you want to listen to.

Damping happens when Energy is Lost to the Surroundings

1) In practice, **any** oscillating system **loses energy** to its surroundings.
2) This is usually down to **frictional forces** like air resistance.
3) These are called **damping forces**.
4) Systems are often **deliberately damped** to **stop** them oscillating or to **minimise** the effect of **resonance**.

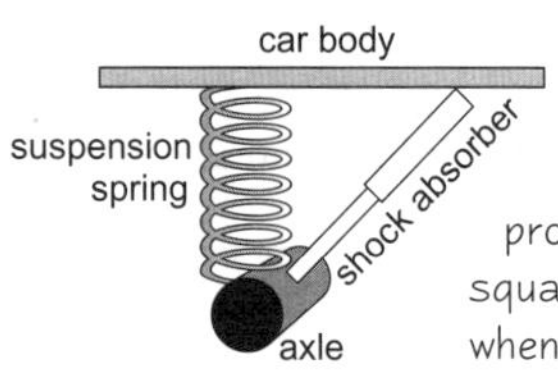

Shock absorbers in a car suspension provide a damping force by squashing oil through a hole when compressed.

Free and Forced Vibrations

Different Amounts of Damping have Different Effects

1) The **degree** of damping can vary from **light** damping (where the damping force is small) to **overdamping**.
2) Damping **reduces** the **amplitude** of the oscillation over time. The **heavier** the damping, the **quicker** the amplitude is reduced to zero.
3) **Critical damping** reduces the amplitude (i.e. stops the system oscillating) in the **shortest possible time**.
4) Car **suspension systems** and moving coil **meters** are critically damped so that they **don't oscillate** but return to equilibrium as quickly as possible.
5) Systems with **even heavier damping** are **overdamped**. They take **longer** to return to equilibrium than a critically damped system.

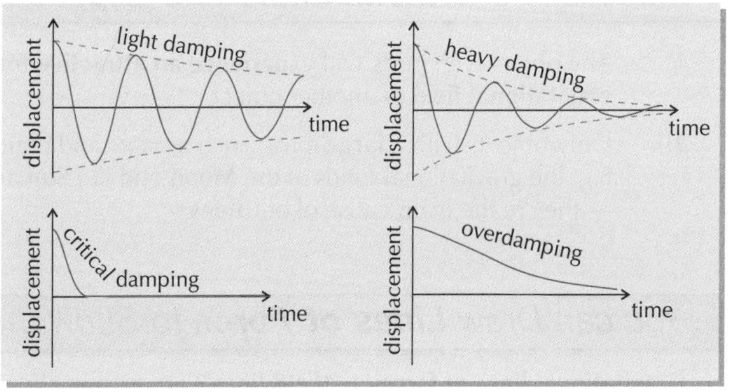

6) **Plastic deformation** of ductile materials **reduces** the **amplitude** of oscillations in the same way as damping. As the material changes shape, it **absorbs energy**, so the oscillation will be smaller.

Damping Affects Resonance too

1) **Lightly damped** systems have a **very sharp** resonance peak. Their amplitude only increases dramatically when the **driving frequency** is **very close** to the **natural frequency**.
2) **Heavily damped** systems have a **flatter response**. Their amplitude doesn't increase very much near the natural frequency and they aren't as **sensitive** to the driving frequency.

Experiment to show how damping affects resonance

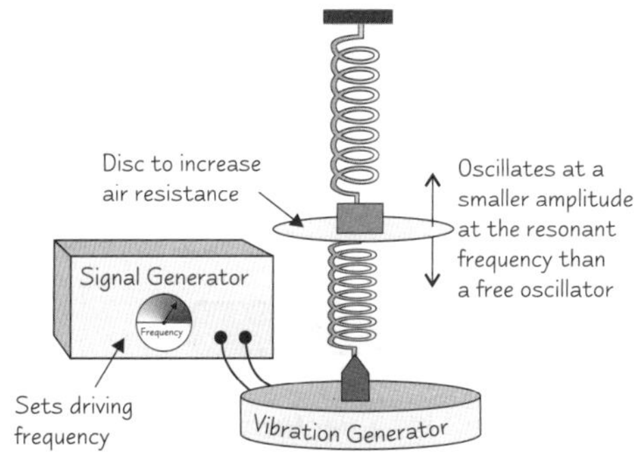

How damping affects resonance

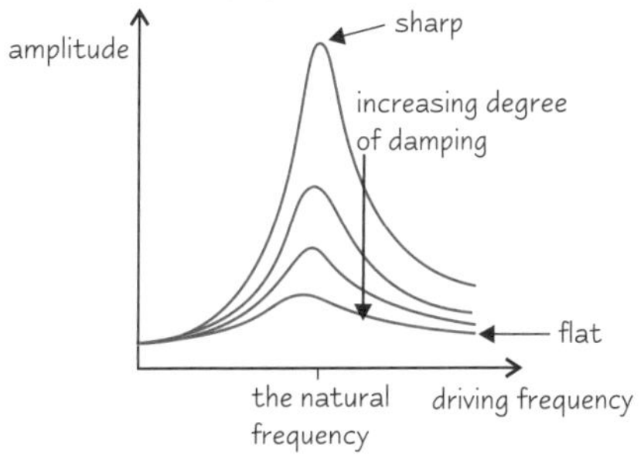

Structures are damped to avoid being damaged by resonance. Loudspeakers are also made to have as flat a response as possible so that they don't 'colour' the sound.

Practice Questions

Q1 What is a free vibration? What is a forced vibration?

Q2 Draw diagrams to show how a damped system oscillates with time when the system is lightly damped and when the system is critically damped.

Exam Questions

Q1 (a) Describe resonance. [2 marks]

(b) Draw a diagram to show how the amplitude of a lightly damped system varies with driving frequency. [2 marks]

(c) On the same diagram, show how the amplitude of the system varies with driving frequency when it is heavily damped. [1 mark]

Q2 (a) Describe critical damping. [1 mark]

(b) State one situation where critical damping is used. [1 mark]

A2 Physics — it can really put a damper on your social life...

Resonance can be really useful (radios, oboes, swings — yay) or very, very bad...

Gravitational Fields

*Gravity's all about masses **attracting** each other. If the Earth didn't have a **gravitational field,** apples wouldn't fall to the ground and you'd probably be floating off into space instead of sitting here reading this page...*

Masses in a **Gravitational Field** Experience a **Force Of Attraction**

1) Any object with mass will **experience an attractive force** if you put it in the **gravitational field** of another object.
2) Only objects with a **large** mass, such as stars and planets, have a significant effect. E.g. the gravitational fields of the **Moon** and the **Sun** are noticeable here on Earth — they're the main cause of our **tides**.

Tides are caused by gravitational fields.

You can Draw **Lines of Force** to Show the **Field** Around an Object

Gravitational lines of force (or "**field lines**") are **arrows** showing the **direction of the force** that masses would feel in a gravitational field.

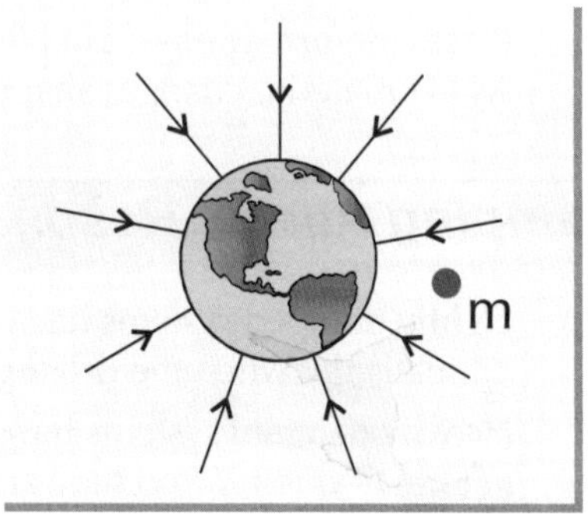

1) If you put a small mass, ***m***, anywhere in the Earth's gravitational field, it will always be attracted **towards** the Earth.
2) The Earth's gravitational field is **radial** — the lines of force meet at the centre of the Earth.
3) If you move mass ***m*** further away from the Earth — where the **lines** of force are **further apart** — the **force** it experiences **decreases**.
4) The small mass, ***m***, has a gravitational field of its own. This doesn't have a noticeable effect on the Earth though, because the Earth is so much **more massive**.

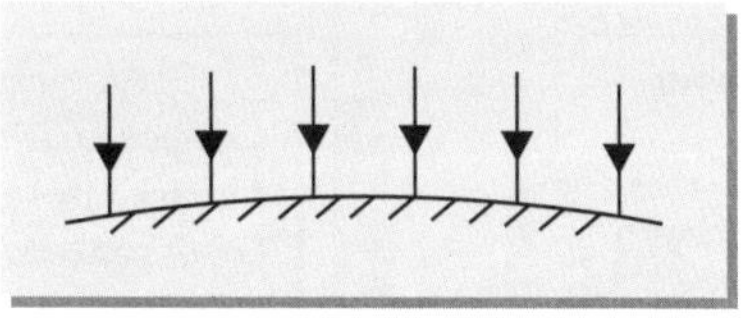

5) Close to the Earth's surface, the field is (almost) uniform — the **field lines** are (almost) **parallel**.

You can **Calculate Forces** Using **Newton's Law of Gravitation**

The **force** experienced by an object in a gravitational field is always **attractive**. It's a **vector** which depends on the **masses** involved and the **distances** between them. It's easy to work this out for **point masses** — or objects which behave as if all their mass is concentrated at the centre, e.g. uniform spheres. You just put the numbers into this equation...

NEWTON'S LAW OF GRAVITATION:

$$F = (-)\frac{GMm}{r^2}$$

There's sometimes a negative sign, to show that the vector F is in the opposite direction to r.

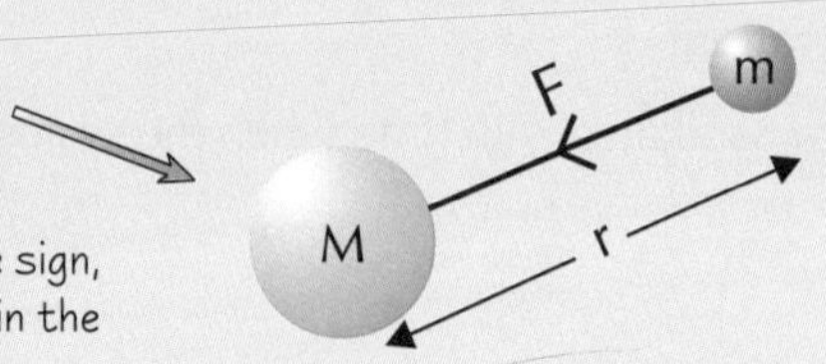

It doesn't matter what you call the masses — M and m, m_1 and m_2, Paul and Larry...

The diagram shows the force acting on ***m*** due to ***M***. (The force on ***M*** due to ***m*** is equal but in the opposite direction.)

M and ***m*** behave as point masses.

G is the **gravitational constant** — 6.67×10^{-11} Nm^{2}kg^{-2}.

r is the distance (in metres) between the centres of the two masses.

The law of gravitation is an **inverse square law** $\left(F \propto \frac{1}{r^2}\right)$ so:

1) if the distance **r** between the masses **increases** then the force **F** will **decrease**.
2) if the **distance doubles** then the **force** will be one **quarter** the strength of the original force.

Gravitational Fields

The **Field Strength** is the **Force per Unit Mass**

Gravitational field strength, ***g***, is the **force per unit mass**. Its value depends on **where you are** in the field. There's a really simple equation for working it out:

$$g = \frac{F}{m}$$

g has units of newtons per kilogram (Nkg^{-1})

1) ***F*** is the force experienced by a mass ***m*** when it's placed in the gravitational field. Divide ***F*** by ***m*** and you get the **force per unit mass**.
2) ***g*** is a **vector** quantity, always pointing towards the centre of the mass whose field you're describing.
3) Since the gravitational field is almost uniform at the Earth's surface, you can assume ***g*** is a constant.
4) ***g*** is just the **acceleration** of a mass in a gravitational field. It's often called the **acceleration due to gravity**.

The **value** of ***g*** at the **Earth's surface** is approximately **9.81** ms^{-2} (or 9.81 Nkg^{-1}).

In a **Radial Field, g** is **Inversely Proportional** to r^2

Point masses have **radial** gravitational fields (see page 104).
The value of ***g*** depends on the distance ***r*** from the point mass ***M***...

$$g = \frac{GM}{r^2}$$

where ***g*** is the **acceleration due to gravity** (ms^{-2}),
G is the **gravitational constant** ($6.67 \times 10^{-11}\ Nm^2kg^{-2}$),
M is a point **mass** and ***r*** is the **distance** from the centre (m).

And it's an **inverse square law** again — as ***r* increases, *g* decreases**.
If you plot a graph of ***g*** against ***r*** for the **Earth**, you get a curve like this.
It shows that ***g*** is greatest at the surface of the Earth, but decreases rapidly as ***r*** increases and you move further away from the centre of the Earth.
The **area** under this curve gives you the **gravitational potential**, ***V*** — see page 106 for more information.

R_E is the **radius** of the Earth.

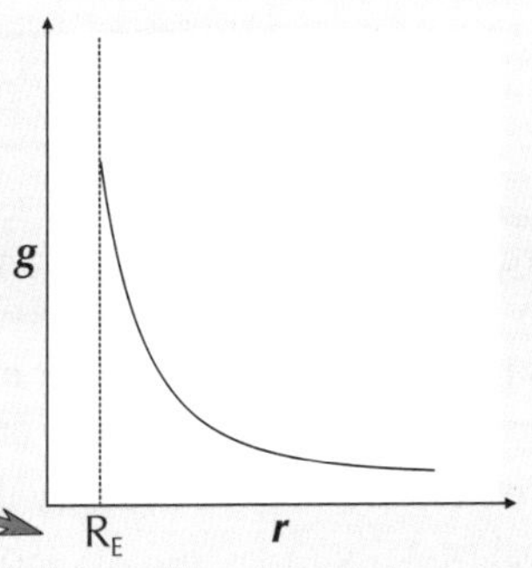

Practice Questions

Q1 Write down Newton's law of gravitation.

Q2 Sketch a graph of distance from mass (*r*) against gravitational field strength (*g*) for a point mass.

Exam Questions

Q1 The Earth's radius is approximately 6400 km. The mass of the Sun is 1.99×10^{30} kg.
The average distance from the Earth to the Sun is 1.5×10^{11} m.

(a) Estimate the mass of the Earth (use g = 9.81 Nkg^{-1} at the Earth's surface). [2 marks]

(b) Estimate the force of gravitational attraction between the Sun and the Earth. [2 marks]

Q2 The Moon has a mass of 7.35×10^{22} kg and a radius of 1740 km.

(a) Calculate the value of ***g*** at the Moon's surface. [1 mark]

(b) Calculate the force acting on a 25 kg mass on the Moon's surface. [1 mark]

If you're really stuck, put 'Inverse Square Law'...

Clever chap, Newton, but famously tetchy. He got into fights with other physicists, mainly over planetary motion and calculus... the usual playground squabbles. Then he spent the rest of his life trying to turn scrap metal into gold. Weird.

Gravitational Fields

Gravitational Potential is Potential Energy per Unit Mass

The **gravitational potential**, ***V***, at a point is the **gravitational potential energy** that a **unit mass** at that point would have. For example, if a **1 kg** mass has **10 J of potential energy** at a point **Z**, then the **gravitational potential at Z is 10 Jkg⁻¹**.

In a **radial field** (like the Earth's), the equation for gravitational potential is:

$$V = (-)\frac{GM}{r}$$

where ***V*** is **gravitational potential** (Jkg^{-1}), ***G*** is the **gravitational constant**, ***M*** is the **mass** of the object causing the gravitational field (kg), and ***r*** is the **distance** from the centre of the object (m).

Gravitational potential is **negative** on the **surface** of the mass and **increases with distance** from the mass. This means that the gravitational potential at an **infinite distance** from the mass will be **zero**. The graph shows how **gravitational potential** varies with **distance**.

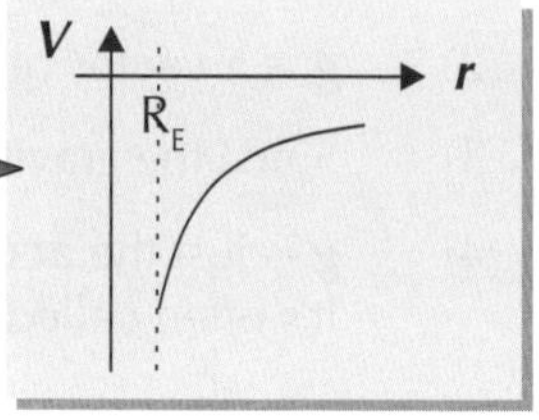

The gradient of a **tangent** gives the value of ***g*** at that point.

If you find the **gradient** of this graph at a particular **point**, you get the value of ***g*** at that point.

In other words:

$$g = -\frac{\Delta V}{\Delta r}$$

g is gravitational field strength (Nkg^{-1}) — see page 105.

Gravitational Potential Difference is the Energy Needed to Move a Unit Mass

Two points at different distances from a mass will have **different** gravitational potentials (because gravitational potential increases with distance) — this means that there is a **gravitational potential difference** between these two points.

When you **move** an object you do **work** against the force of **gravity** — the **amount of energy** you need depends on the **mass** of the object and the **gravitational potential difference** you move it through:

$$\Delta W = m\Delta V$$

where **Δ*W*** is the **work done** (J), ***m*** is the **mass** of the object (kg) and **Δ*V*** is the **gravitational potential difference** (Jkg^{-1}).

Planets and Satellites are Kept in Orbit by Gravitational Forces

1) A **satellite** is just any **smaller mass** which **orbits** a **much larger mass** — the **Moon** is a satellite of the Earth, **planets** are **satellites** of the **Sun**.
2) **Satellites** are kept in **orbit** by the **gravitational 'pull'** of the mass they're orbiting.
3) In our Solar System, the planets have **nearly circular orbits**, so you can use the **equations of circular motion** (see p.96-97) to investigate their **speed** and **orbital period** (see below).

It's either gravitational pull or a giant white rabbit that makes the Earth orbit the Sun. I know which one I believe...

Orbital Period and Speed Depend on Radius

Any object undergoing **circular motion** (e.g. a **satellite**) is kept in its path by a **centripetal force**. What **causes** this force depends on the object — in the case of **satellites** it's the **gravitational attraction** of the mass they're orbiting. This means that the **centripetal** and **gravitational forces** acting on a satellite must be **equal**:

$$\frac{mv^2}{r} = \frac{GMm}{r^2}, \text{ which rearranges to give } v = \sqrt{\frac{GM}{r}}$$

The speed of a satellite is inversely proportional to the square root of the radius of its orbit.

The **time** taken for a satellite to make **one orbit** is called the **orbital period**, ***T***. For circular motion, $T = \frac{2\pi r}{v}$.

The period of an orbit is proportional to the square root of the radius cubed.

Substitute for *v* and rearrange: $T = \sqrt{\frac{4\pi^2 r^3}{GM}}$

The **greater** the **radius** of a satellite's orbit, the **slower** it will travel and the **longer** it will take to complete **one orbit**.

Gravitational Fields

You Can Solve Problems About Orbital Radius, Period and Speed

Example The Moon takes 27.3 days to orbit the Earth. Calculate its distance from the Earth. Take the mass of the Earth to be 5.975×10^{24} kg.

You're trying to find the radius of the orbit, r. Use the formula for period, T:

$$T = \sqrt{\frac{4\pi^2 r^3}{GM}}$$

You've been given the values of T (27.3 days) and M, and you'll be able to look up the value of G on the exam data sheet — $G = 6.67 \times 10^{-11}\ \text{Nm}^2\text{kg}^{-2}$.

and rearrange it for r^3:

$$r^3 = \frac{T^2 GM}{4\pi^2}$$

T = 27.3 days = 2.36×10^6 s
$G = 6.67 \times 10^{-11}\ \text{Nm}^2\text{kg}^{-2}$
$M = 5.975 \times 10^{24}$ kg

and put the numbers in (convert to SI units first):

$$r^3 = \frac{(2.36\times10^6)^2 \times (6.67\times10^{-11}) \times (5.975\times10^{24})}{4\pi^2} = 5.62\times10^{25}$$

$r = 3.83 \times 10^8$ m

= **3.83×10^5 km** (this is the distance between the centre of the Earth and the centre of the Moon)

The Energy of an Orbiting Satellite is Constant

An orbiting **satellite** has **kinetic** and **potential energy** — its **total energy** (i.e. kinetic + potential) is always **constant**.

1) In a **circular orbit**, a satellite's **speed** and **distance** above the mass it's orbiting are **constant**. This means that its **kinetic energy** and **potential energy** are also both **constant**.
2) In an **elliptical orbit**, a satellite will **speed up** as it's **height decreases** (and slow down as its height increases). This means that its **kinetic energy increases** as its **potential energy decreases** (and vice versa), so the **total energy** remains **constant**.

Geosynchronous Satellites Orbit the Earth once in 24 hours

1) Geosynchronous satellites orbit directly over the **equator** and are **always above the same point** on Earth.
2) A geosynchronous satellite travels at the **same angular speed as the Earth** turns below it.
3) These satellites are really useful for sending TV and telephone signals — the satellite is **stationary** relative to a certain point on the **Earth**, so you don't have to alter the angle of your receiver (or transmitter) to keep up.
4) Their orbit takes exactly **one day**.

Practice Questions

Q1 What is gravitational potential? Write an equation for it.

Q2 The International Space Station orbits the Earth with velocity v. If another vehicle docks with it, increasing its mass, what difference, if any, does this make to the speed or radius of the orbit?

Q3 Would a geosynchronous satellite be useful for making observations for weather forecasts? Give reasons.

Exam Questions

(Use $G = 6.67 \times 10^{-11}\ \text{Nm}^2\text{kg}^{-2}$, mass of Earth = 5.98×10^{24} kg, radius of Earth = 6400 km)

Q1 (a) A satellite orbits 200 km above the Earth's surface. Calculate the period of the satellite's orbit. [2 marks]
(b) Calculate the linear speed of the satellite. [1 mark]

Q2 At what height above the Earth's surface would a geosynchronous satellite orbit? [3 marks]

No fluffy bunnies were harmed in the making of these pages...

When I hear the word 'satellite' I just think of man-made ones, e.g. for mobile phones or TV, and tend to forget that planets (including Earth) and the Moon are satellites too — don't make the same mistake. You're probably best off learning all the stuff about satellites and their orbits too — just knowing what they are won't get you all that far in the exam, I'm afraid.

Electric Fields

Electric fields can be attractive or repulsive, so they're different from gravitational ones. It's all to do with ***charge****.*

There is an **Electric Field** *around a* **Charged Object**

Any object with **charge** has an **electric field** around it — the region where it can attract or repel other charges.

1) Electric charge, Q, is measured in **coulombs** (C) and can be either positive or negative.
2) **Oppositely** charged particles **attract** each other. **Like** charges **repel**.
3) If a **charged object** is placed in an electric field, then it will experience a **force**.

You can **Calculate Forces** *using* **Coulomb's Law**

Coulomb's law gives the force of attraction or repulsion between two **point charges** in a **vacuum**:

COULOMB'S LAW:

$$F = \frac{1}{4\pi\varepsilon_0}\frac{Q_1 Q_2}{r^2}$$

ε_0 ("epsilon-nought") is the **permittivity of free space** and is equal to 8.85×10^{-12} Fm^{-1}
Q_1 and Q_2 are the **charges**
r is the **distance** between Q_1 and Q_2

This unit is a 'farad per metre' — see p.112.

1) The force on Q_1 is always **equal** and **opposite** to the force on Q_2 — the **direction** depends on the charges.

If the charges are **opposite** then the force is **attractive**. *F* will be **negative**.

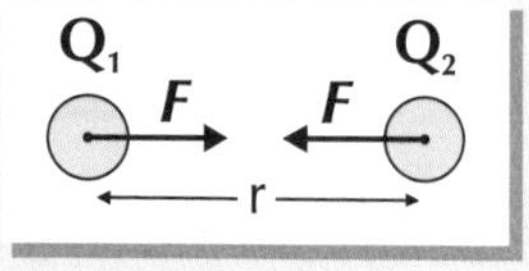

If Q_1 and Q_2 are **alike** then the force is **repulsive**. *F* will be **positive**.

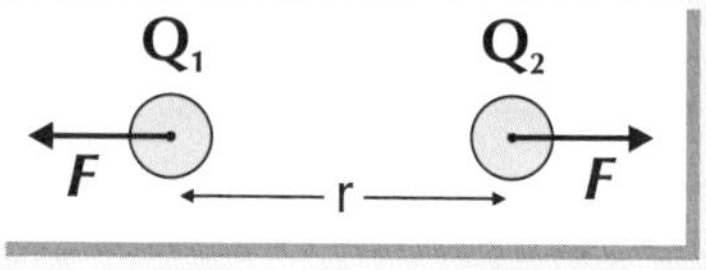

2) Coulomb's law is an **inverse square law**.
The further apart the charges, the weaker the force between them.
3) If the point charges aren't in a vacuum, then the size of the force *F* also depends on the **permittivity**, ε, of the material between them.

Mr Harris liked to explain Coulomb's law using prairie dogs.

Electric Field Strength *is* **Force per Unit Charge**

Electric field strength, ***E***, is defined as the **force per unit positive charge**.
It's the force that a charge of +1 C would experience if it was placed in the electric field.

F is the force on a 'test' charge ***q***.

$$E = \frac{F}{q}$$

where ***E*** is **electric field strength** (NC^{-1}), ***F*** is the **force** (N) and ***q*** is the **charge** (C)

1) ***E*** is a **vector** pointing in the **direction** that a **positive charge** would **move**.
2) The units of ***E*** are **newtons per coulomb** (NC^{-1}).
3) Field strength depends on **where you are** in the field.
4) A **point charge** — or any body that behaves as if all its charge is concentrated at the centre — has a **radial** field.

Electric Fields

In a **Radial Field**, **E** is **Inversely Proportional** to r^2

1) In a **radial field**, the electric field strength, ***E***, depends on the distance ***r*** from the point charge ***Q***:

$$E = \frac{1}{4\pi\varepsilon_0}\frac{Q}{r^2}$$

where ***E*** is the **electric field strength** (NC^{-1}), $\boldsymbol{\varepsilon_0}$ is the **permittivity of free space** ($8.85 \times 10^{-12}\ Fm^{-1}$), ***Q*** is the **point charge** (C) and ***r*** is the **distance** from the point charge (m)

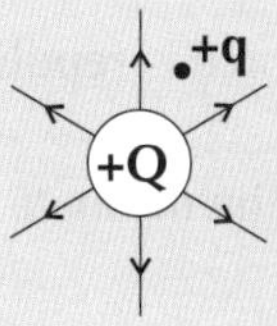

For a **positive *Q***, the small positive 'test' charge ***q*** would be **repelled**, so the field lines point **away** from ***Q***.

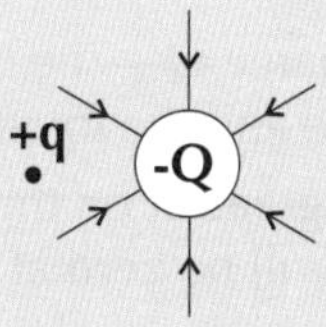

For a **negative *Q***, the small positive charge ***q*** would be **attracted**, so the field lines point **towards *Q***.

2) It's another **inverse square law**: $E \propto \frac{1}{r^2}$

3) Field strength **decreases** as you go **further away** from ***Q*** — on a diagram, the **field lines** get **further apart**.

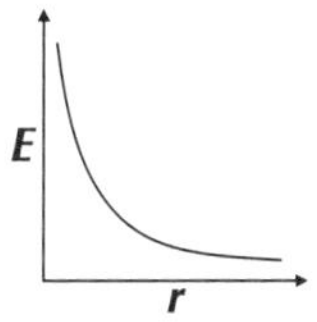

In a **Uniform Field**, **E** is **Inversely Proportional** to **d**

A **uniform field** can be produced by connecting two **parallel plates** to the opposite poles of a battery.

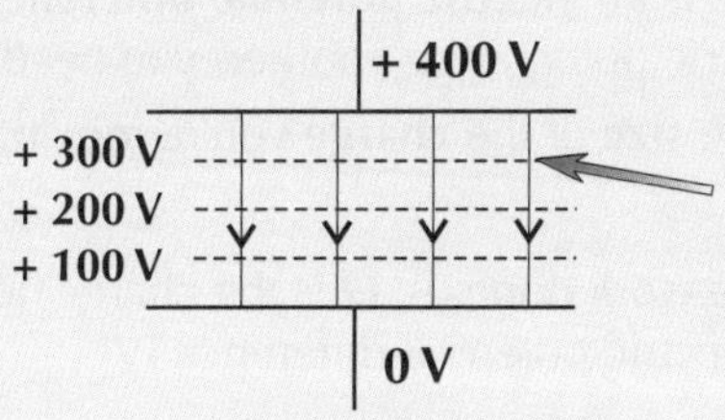

The **electric field lines** are **parallel** to each other.

Field strength ***E*** is the **same** at **all points** between the two plates and is given by:

$$E = \frac{V}{d}$$

where ***E*** is the **electric field strength** (Vm^{-1} or NC^{-1}), ***V*** is the **potential difference** between the plates (V) and ***d*** is the distance between them (m)

Practice Questions

Q1 Write down Coulomb's law.

Q2 Sketch a radial electric field and a uniform electric field. How would you find ***E*** for each?

Exam Questions

(Use $\varepsilon_0 = 8.85 \times 10^{-12}\ Fm^{-1}$ and $e = 1.60 \times 10^{-19}$ C.)

Q1 An alpha particle (charge +2e) was deflected while passing through thin gold foil.
The alpha particle passed within 5×10^{-12} m of a gold nucleus (charge +79e).
What was the magnitude and direction of the electrostatic force experienced by the alpha particle? [4 marks]

Q2 (a) Two parallel plates are connected to a 1500 V dc supply, and separated by an air gap of 4.5 mm.
What is the electric field strength between the plates? State the direction of the field. [3 marks]

(b) The plates are now pulled further apart so that the distance between them is doubled.
The electric field strength remains the same. What is the new voltage between the plates? [2 marks]

Electric fields — one way to roast beef...

At least you get a choice here — uniform or radial, positive or negative, attractive or repulsive, chocolate or strawberry...

Electric Fields

Electric Potential is *Potential Energy per Unit Charge*

All points in an **electric field** have an **electric potential**, ***V***. This is the electric **potential energy** that a **unit positive charge** (+ 1 C) would have at that point. The **electric potential** of a point depends on **how far** it is from the **charge** creating the **electric field** and the **size** of that charge.

In a **radial field**, **electric potential** is given by:

$$V = \frac{1}{4\pi\varepsilon_0}\frac{Q}{r}$$

where ***V*** is **electric potential** (V), ***Q*** is the size of the **charge** (C) and ***r*** is the **distance** from the charge (m)

1) The **sign** of ***V*** depends on the charge ***Q*** — i.e. ***V*** is **positive** when ***Q*** is positive and the force is **repulsive**, and **negative** when ***Q*** is negative and the force is **attractive**.
2) The **absolute magnitude** of ***V*** is **greatest** on the **surface of the charge**, and **decreases** as the **distance** from the charge **increases** — ***V*** will be **zero** at an **infinite distance** from the charge.

Repulsive force
V is initially **positive** and tends to **zero** as *r* increases towards **infinity**.

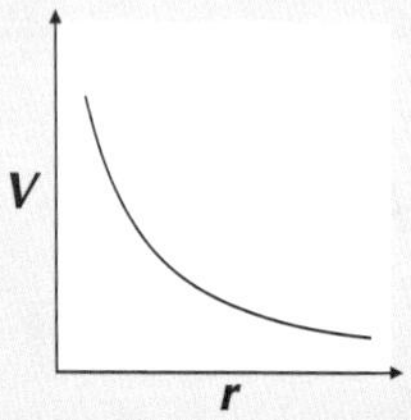

The **gradient** of a **tangent** to either graph gives the **field strength** at that point:

$$E = \frac{\Delta V}{\Delta r}$$

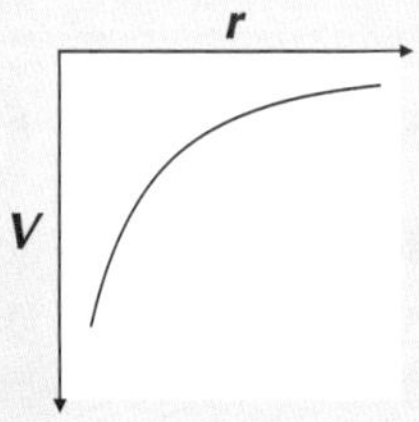

Attractive force
V is initially **negative** and tends to **zero** as *r* increases towards **infinity**.

Electric Potential Difference is the *Energy Needed* to *Move* a *Unit Charge*

If **two points** in an **electric field** have different potential, then there is an **electric potential difference** between them. To **move a charge** across a **potential difference** (i.e. from one electric potential to another) you need to use **energy**.

The **amount of energy** you need (or the **work done**) depends on the **size** of the **charge** you're moving and the size of the **potential difference** you want to move it across:

$$\Delta W = Q\Delta V$$

where ΔW is the work done (J), ***Q*** is the charge being moved (C) and ΔV is the electric potential difference (V).

Deriving the Work Done Formula

There are two parallel plates with a potential difference of ΔV across them, creating a **uniform electric field**.

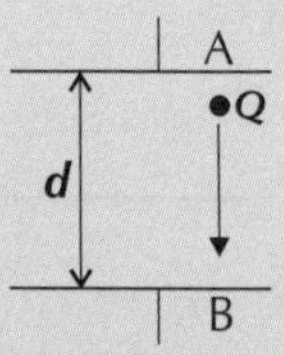

The field strength is $E = \frac{\Delta V}{d} = \frac{F}{Q}$, which rearranges to give $Q\Delta V = Fd$

E in a uniform field

E as force per unit charge

1) To move a charge ***Q*** from A to B, the **work done = force × distance moved** = ***Fd***
2) So the work done in moving a charge ***Q*** against a potential difference ΔV is given by $Q\Delta V$.

Example in a Gravitational Field

At the **Earth's surface** the **gravitational field** is **uniform**.

The field strength is $g = -\frac{\Delta V}{\Delta r} = \frac{F}{m}$, which rearranges to give $m\Delta V = -F\Delta r$

g in a uniform field

g as force per unit mass

F is negative because you're working against gravity

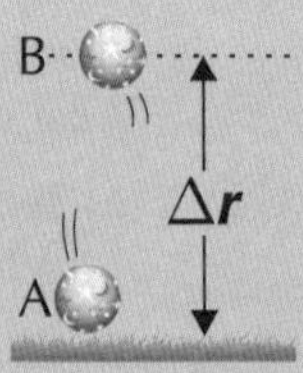

1) To throw a ball ***m*** from A to B, the **work done = force × distance moved** = $m\Delta V$
2) So the energy needed to move a mass ***m*** against a gravitational potential difference is given by $m\Delta V$.

Electric Fields

*There are **Similarities** between **Gravitational** and **Electric Fields...***

If a lot of the stuff on the previous couple of pages sounded strangely familiar it could be because it's very similar to the stuff on gravitational fields (or it could be because you've learnt it before — this is a revision book after all). Anyway, there are **four** big **similarities** between **electric** and **gravitational fields** that you need to know — read on.

1)	Gravitational field strength, g, is **force** per **unit mass**.	Electric field strength, E, is **force** per **unit positive charge**.
2)	Newton's law of gravitation for the **force** between two point masses is an **inverse square law**. $F \propto \frac{1}{r^2}$	Coulomb's law for the electric **force** between two point charges is also an **inverse square law**. $F \propto \frac{1}{r^2}$
3)	The **field lines** for a point mass... 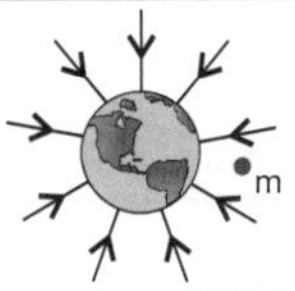	The **field lines** for a **negative** point charge... 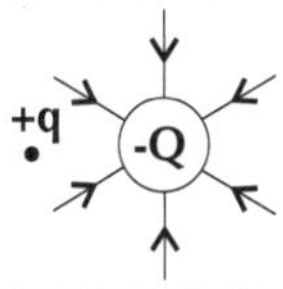
4)	Gravitational potential, V, is **potential energy** per **unit mass** and is **zero** at **infinity**.	Electric potential, V, is **potential energy** per **unit positive charge** and is **zero** at **infinity**.

*... and **Three Differences** too*

Gravitational and electric fields aren't all the same — you need to know the **three main differences**:

1) Gravitational forces are always **attractive**. Electric forces can be either **attractive** or **repulsive**.
2) Objects can be **shielded** from **electric** fields, but not from gravitational fields.
3) The size of an **electric** force depends on the **medium** between the charges, e.g. plastic or air. For gravitational forces, this makes no difference.

Practice Questions

Q1 What is meant by 'electric potential'? How would you find the electric potential in a radial field?

Q2 Sketch a graph of electric potential against distance for an attractive and a repulsive charge.

Q3 What is 'potential difference'? What is the work done when a charge is moved through a potential difference?

Q4 Describe three similarities and three differences between gravitational and electric fields.

Exam Questions

Q1 Compare the magnitude and direction of the gravitational and electric forces between two electrons that are 8×10^{-10} m apart.
(Use $m_e = 9.11 \times 10^{-31}$ kg, $e = 1.60 \times 10^{-19}$ C, $\varepsilon_o = 8.85 \times 10^{-12}$ Fm^{-1} and $G = 6.67 \times 10^{-11}$ Nm^{2}kg^{-2}.) [3 marks]

Q2 A negatively charged oil drop is held stationary between two charged plates. The plates are 3 cm apart vertically, and have an electric potential difference of 5000 V across them.

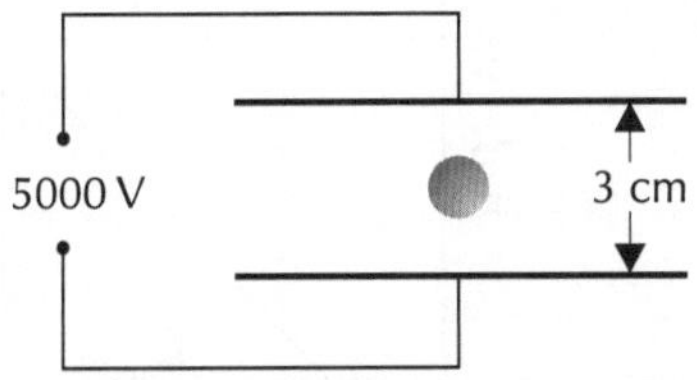

(a) The oil drop has a mass of 1.5×10^{-14} kg. Calculate the size of its charge. [3 marks]

(b) If the polarity of the charged plates was reversed, what would happen to the oil drop? [1 mark]

I prefer gravitational fields — electric fields are repulsive...

Revising fields is a bit like a buy-one-get-one-free sale — you learn all about gravitational fields and they throw electric fields in for free. You just have to remember to change your ms for Qs and your Gs for $1/4\pi\varepsilon_0$s... okay, so it's not quite a BOGOF sale. Maybe more like a buy-one-get-one-half-price sale... anyway, you get the point — go learn some stuff.

Capacitors

Capacitors are things that store electrical charge — like a charge bucket. The capacitance of one of these things tells you how much charge the bucket can hold. Sounds simple enough... ha... ha, ha, ha...

Capacitance is Defined as the Amount of **Charge Stored** per **Volt**

$$C = \frac{Q}{V}$$

where Q is the **charge** in coulombs, V is the **potential difference** in volts and C is the **capacitance** in **farads (F) — 1 farad = 1 C V^{-1}**.

A farad is a **huge** unit so you'll usually see capacitances expressed in terms of:

μF — microfarads (× 10^{-6})
nF — nanofarads (× 10^{-9})
pF — picofarads (× 10^{-12})

You can **Investigate** the **Charge Stored** by a **Capacitor Experimentally**

Investigating the Charge Stored on a Capacitor

1) Set up a **test circuit** to measure current and potential difference:

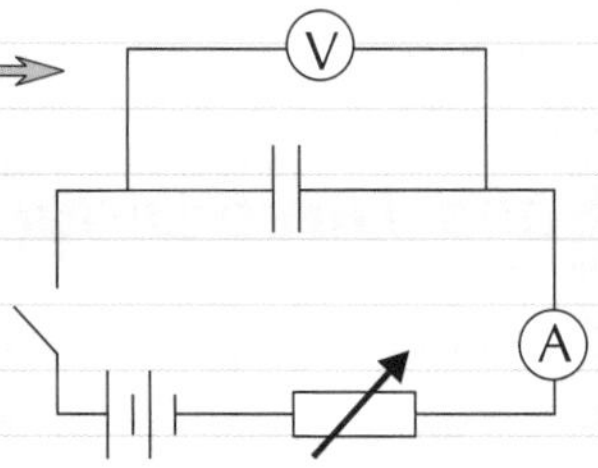

2) Constantly adjust the **variable resistor** to keep the charging current **constant** for as long as you can (it's impossible when the capacitor is nearly fully charged).
3) Record the p.d. at regular intervals until it **equals** the **battery p.d.**
4) From these results, you can plot the following graphs:

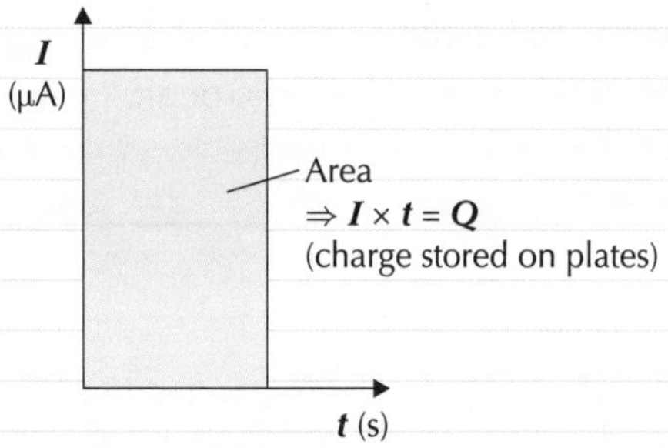

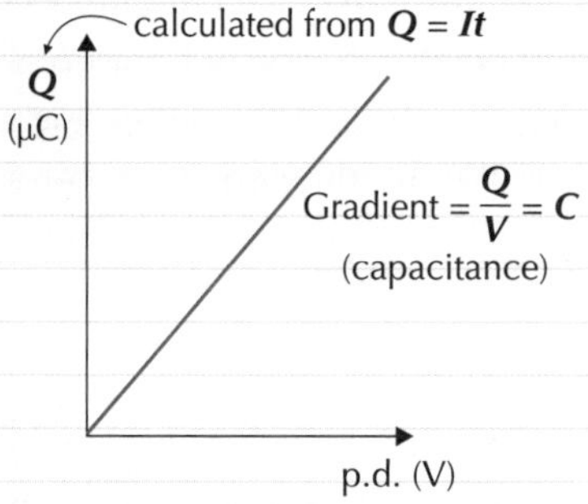

Capacitors **Store Energy**

1) When **charge** builds up on the plates of the **capacitor**, **electrical energy** is **stored** by the capacitor.
2) You can find the **energy stored** in a capacitor from the **area** under a **graph** of **potential difference** against **charge stored** on the capacitor.

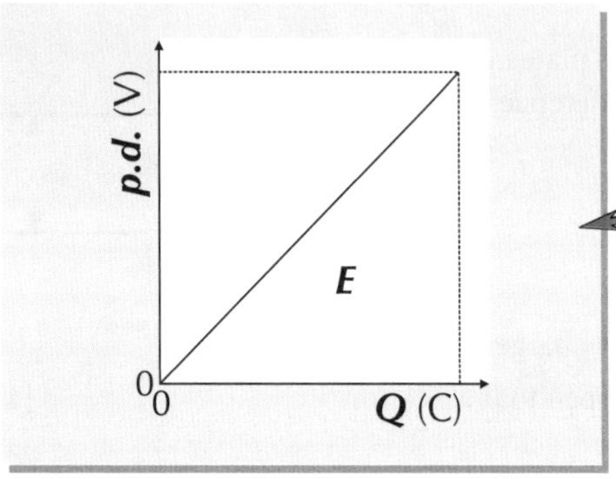

3) The p.d. across the capacitor is **proportional** to the charge stored on it (see above), so the graph will be a **straight line** through the origin. On this graph, the **energy stored** is given by the **yellow triangle**.

Jane had heard there was energy stored on plates.

4) **Area of triangle = ½ × base × height**, so the energy stored by the capacitor is:

$$E = \frac{1}{2}QV$$

You need to remember where this equation comes from.

Capacitors

There are **Three** Expressions for the **Energy Stored** by a Capacitor

1) You know the first one already: $E = \frac{1}{2}QV$

2) $C = \frac{Q}{V}$, so $Q = CV$. Substitute that into the energy equation: $E = \frac{1}{2}CV \times V$. So: $E = \frac{1}{2}CV^2$

3) $V = \frac{Q}{C}$, so $E = \frac{1}{2}Q \times \frac{Q}{C}$. Simplify: $E = \frac{1}{2}\frac{Q^2}{C}$

Example *A 900 μF capacitor is charged up to a potential difference of 240 V. Calculate the energy stored by the capacitor.*

First, choose the best equation to use — you've been given V and C, so you need $E = \frac{1}{2}CV^2$.

Substitute the values in: $E = \frac{1}{2} \times 9 \times 10^{-4} \times 240^2 = 25.92$ J

Practice Questions

Q1 Define capacitance.

Q2 What is the relationship between charge, voltage and capacitance?

Q3 Write the following in standard form: a) 220 μF b) 1000 pF c) 470 nF.

Exam Questions

Q1 The current and potential difference of a test circuit were measured as a capacitor was charged. The graphs below were plotted from the recorded data.

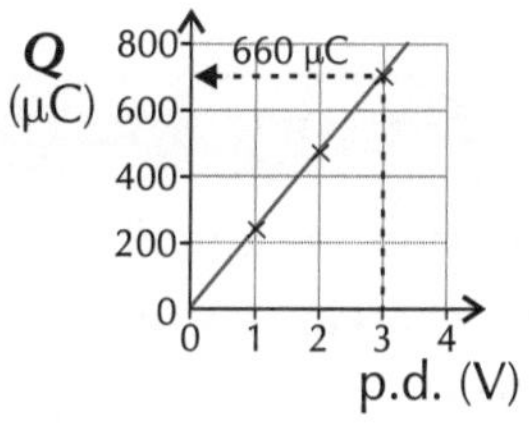

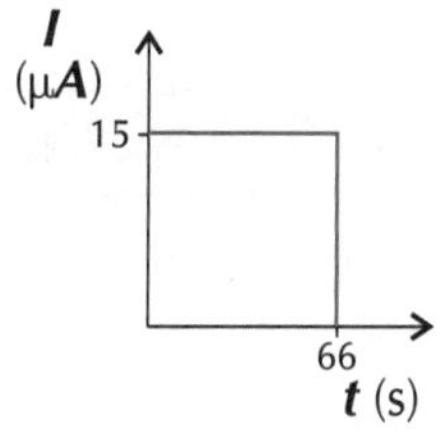

(a) Explain what is meant by the term 'capacitance'. [1 mark]

(b) Calculate: (i) the capacitance of the capacitor. [2 marks]

(ii) the charge stored on the plates of the capacitor. [2 marks]

Q2 (a) Sketch a circuit that could be used to find the energy stored on a capacitor. [2 marks]

(b) Explain how you would use this circuit to find the energy stored. [3 marks]

Q3 A 500 mF capacitor is fully charged up from a 12 V supply.

(a) Calculate the total energy stored by the capacitor. [2 marks]

(b) Calculate the charge stored by the capacitor. [2 marks]

Capacitance — fun, it's not...

Capacitors are really useful in the real world. Pick an appliance, any appliance, and it'll probably have a capacitor or several. If I'm being honest, though, the only saving grace of these pages for me is that they're not especially hard...

Charging and Discharging

Charging and discharging — sounds painful...

You can **Charge** a **Capacitor** by Connecting it to a **Battery**

1) When a capacitor is connected to a **battery**, a **current** flows in the circuit until the capacitor is **fully charged**, then **stops**.
2) The electrons flow onto the plate connected to the **negative terminal** of the battery, so a **negative charge** builds up.
3) This build-up of negative charge **repels** electrons off the plate connected to the **positive terminal** of the battery, making that plate positive. These electrons are attracted to the positive terminal of the battery.

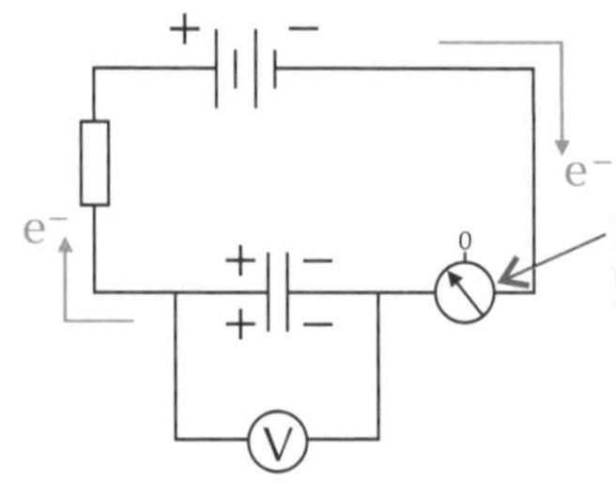

4) An **equal** but **opposite** charge builds up on each plate, causing a **potential difference** between the plates. Remember that **no charge** can flow **between** the plates because they're **separated** by an **insulator** (dielectric).
5) Initially the **current** through the circuit is **high**. But, as **charge** builds up on the plates, **electrostatic repulsion** makes it **harder** and **harder** for more electrons to be deposited. When the p.d. across the **capacitor** is equal to the p.d. across the **battery**, the **current** falls to **zero**. The capacitor is **fully charged**.

an equal but opposite charge

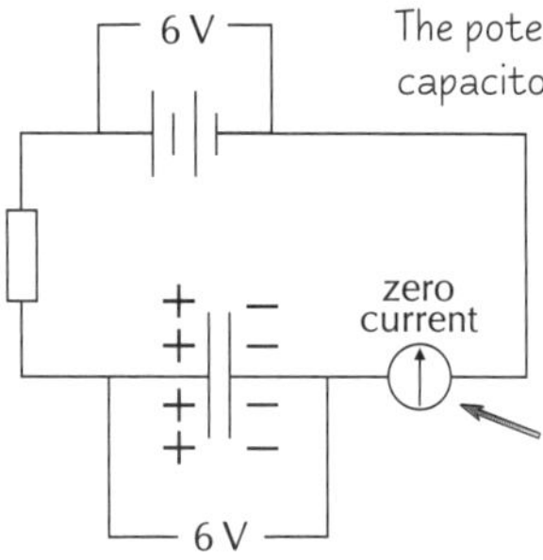

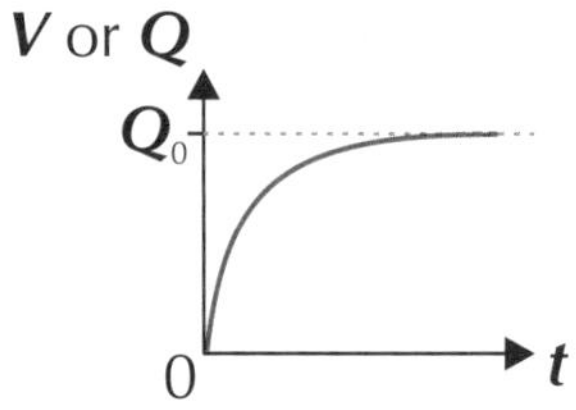

A charging curve for a capacitor

To **Discharge** a Capacitor, **Take Out** the **Battery** and **Reconnect** the **Circuit**

1) When a **charged capacitor** is connected across a **resistor**, the p.d. drives a **current** through the circuit.
2) This current flows in the **opposite direction** from the **charging current**.
3) If you connect a **voltage sensor** attached to a **datalogger** across the **capacitor**, you can plot a **discharge curve** for the capacitor (like the two below). The curve shows how the **voltage changes** as the **capacitor discharges**.
4) The capacitor is **fully discharged** when the **p.d.** across the plates and the **current** in the circuit are both **zero**.

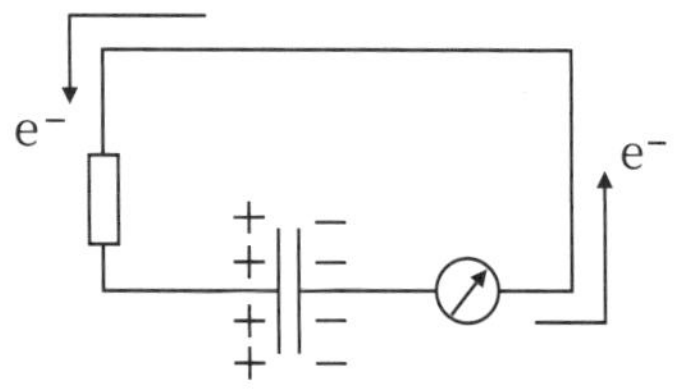

Discharging curves for a capacitor.

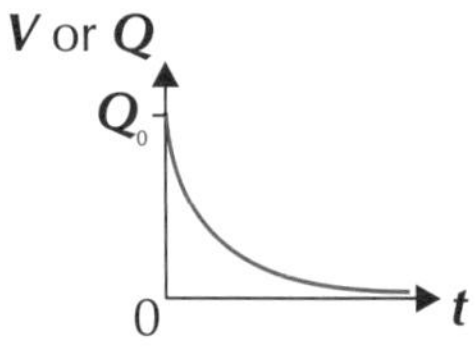

V is proportional to *Q* (see page 112), so you get the same shape graph whether you use *V* or *Q* on the y-axis.

OR

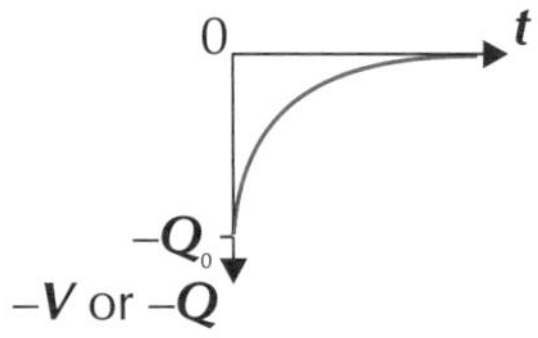

V and *Q* can be written as negative values to show the charge is flowing in the opposite direction to the 'charging current'.

Charging and Discharging

The Time Taken to Charge or Discharge Depends on Two Factors

The **time** it takes to charge or discharge a capacitor depends on:

1) The **capacitance** of the capacitor (***C***). This affects the amount of **charge** that can be transferred at a given **voltage**.
2) The **resistance** of the circuit (***R***). This affects the **current** in the circuit.

The Charge on a Capacitor Decreases Exponentially

1) When a capacitor is **discharging**, the amount of **charge** left on the plates falls **exponentially with time**.
2) That means it always takes the **same length of time** for the charge to **halve**, no matter **how much charge** you start with — like radioactive decay (see p. 128).

   The charge left on the plates of a capacitor discharging from full is given by the equation:

   $$Q = Q_0 e^{-\frac{t}{RC}}$$

   where Q_0 is the charge of the capacitor when it's fully charged.

3) The graphs of ***V*** against ***t*** for charging and discharging are also **exponential**.

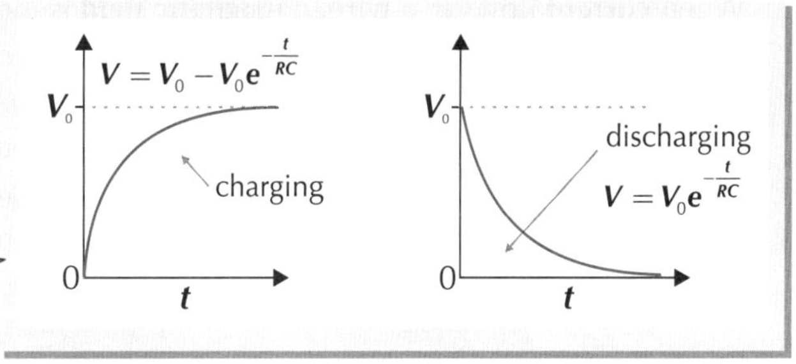

4) ***I*** for a charging or discharging capacitor **always decreases exponentially** — the graph of ***I*** against ***t*** is the **same shape** as the graph of ***V*** against ***t*** for a discharging capacitor.

Time Constant τ = RC

τ is the Greek letter 'tau'

If $t = \tau = RC$ is put into the equation above, then $Q = Q_0 e^{-1}$. So when $t = \tau$: $\frac{Q}{Q_0} = \frac{1}{e}$, where $\frac{1}{e} \approx \frac{1}{2.718} \approx 0.37$.

1) So τ, the **time constant**, is the time taken for the charge on a discharging capacitor (Q) to **fall** to **37%** of Q_0.
2) It's also the time taken for the charge of a charging capacitor to **rise** to **63%** of Q_0.
3) The **larger** the **resistance** in series with the capacitor, the **longer it takes** to charge or discharge.
4) In practice, the time taken for a capacitor to charge or discharge **fully** is taken to be about 5***RC***.

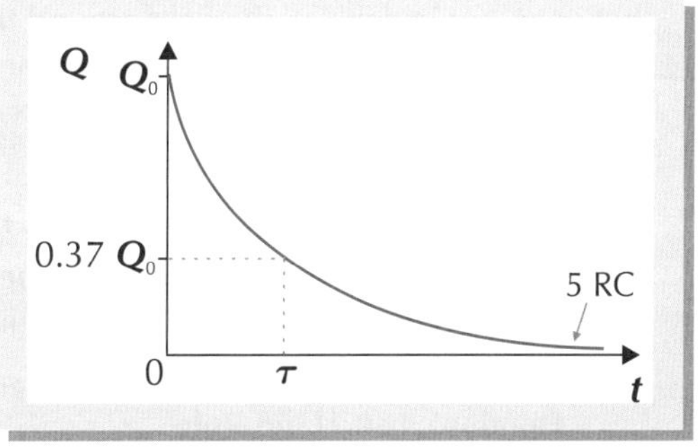

Practice Questions

Q1 Sketch graphs to show the variation of p.d. across the plates of a capacitor with time for:
a) charging a capacitor, b) discharging a capacitor.

Q2 What two factors affect the rate of charge of a capacitor?

Exam Question

Q1 A 250 μF capacitor is fully charged from a 6 V battery and then discharged through a 1 kΩ resistor.

(a) Calculate the time taken for the charge on the capacitor to fall to 37% of its original value. [2 marks]

(b) Calculate the percentage of the total charge remaining on the capacitor after 0.7s. [3 marks]

(c) The charging voltage is increased to 12 V. Explain the effect this has on:

i) the total charge stored ii) the capacitance of the capacitor iii) the time taken to fully charge [3 marks]

An analogy — consider the lowly bike pump...

A good way to think of the charging process is like pumping air into a bike tyre. To start with, the air goes in easily, but as the pressure in the tyre increases, it gets harder and harder to squeeze any more air in. The tyre's 'full' when the pressure of the air in the tyre equals the pressure of the pump. The analogy works just as well for discharging...

Magnetic Fields

Magnetic fields — making pretty patterns with iron filings before spending an age trying to pick them off the magnet.

A **Magnetic Field** is a **Region** Where a **Force** is Exerted on **Magnetic Materials**

1) Magnetic fields can be represented by **field lines**.
2) Field lines go from **north to south**.
3) The **closer** together the lines, the **stronger** the field.

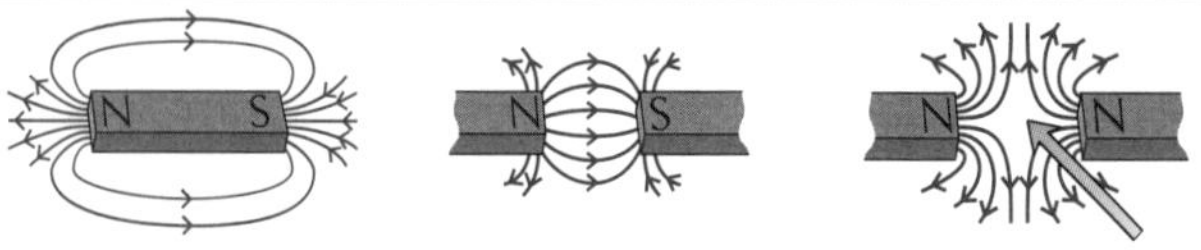

At a neutral point magnetic fields cancel out.

There is a **Magnetic Field** Around a **Wire** Carrying **Electric Current**

When **current** flows in a **wire**, a **magnetic field** is induced around the wire.

1) The **field lines** are **concentric circles** centred on the wire.
2) The **direction** of a magnetic **field** around a current-carrying wire can be worked out with the **right-hand rule**.
3) If you loop the wire into a **coil**, the field is **doughnut shaped**, while lots of coils (a **solenoid**) form a **field** like a **bar magnet**.

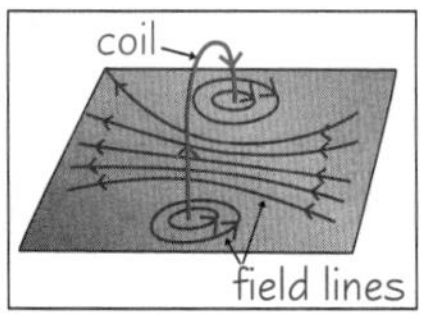

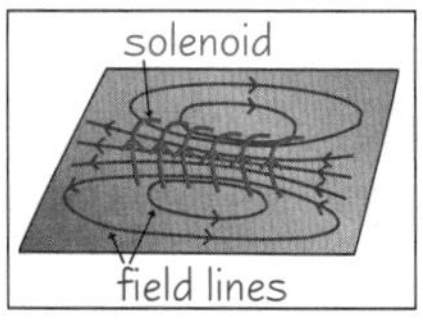

RIGHT-HAND RULE

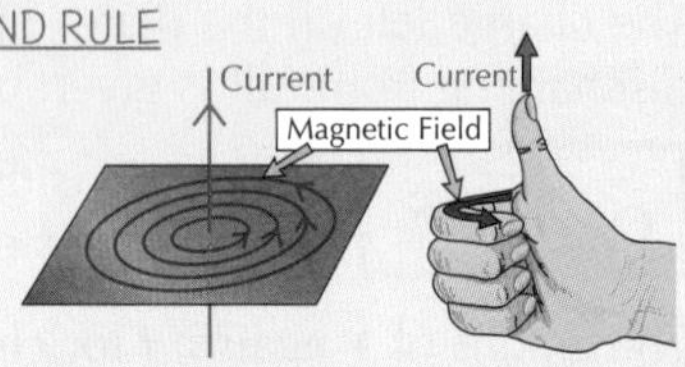

1) Stick your right thumb up, like you're hitching a lift.
2) If your thumb points in the direction of the current...
3) ...your curled fingers point in the direction of the field.

A **Wire** Carrying a **Current** in a **Magnetic Field** will **Experience** a **Force**

1) If you put a **current-carrying wire** into an **external** magnetic field (e.g. between two magnets), the field around the wire and the field from the magnets **interact**. The field lines from the magnet **contract** to form a **'stretched catapult'** effect where the flux lines are closer together.
2) This causes a **force** on the wire.
3) If the current is **parallel** to the flux lines, **no force** acts.
4) The **direction** of the force is always **perpendicular** to both the **current** direction and the **magnetic field** — it's given by **Fleming's left-hand rule**:

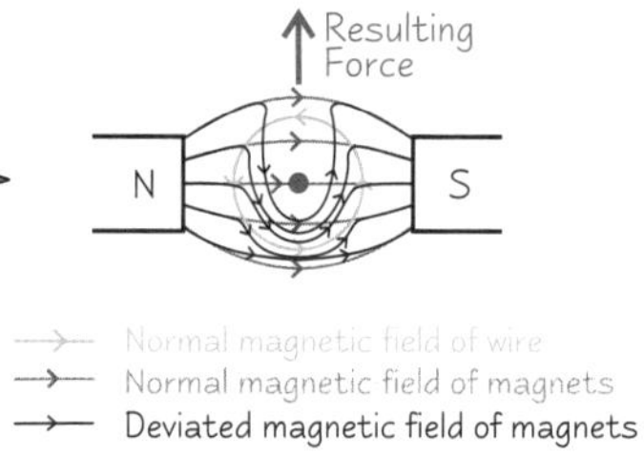

Fleming's Left-Hand Rule

The First finger points in the direction of the uniform magnetic Field, the seCond finger points in the direction of the conventional Current. Then your thuMb points in the direction of the force (in which Motion takes place).

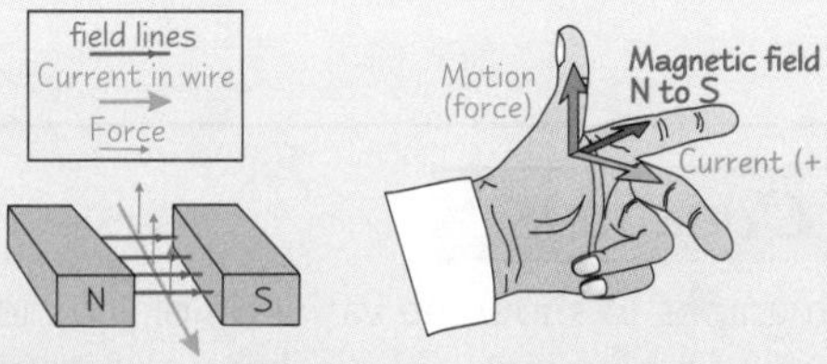

The **Force** on a Wire is **Proportional** to the **Magnetic Field Strength**

1) The size of the **force**, ***F***, on a current-carrying wire at right-angles to a magnetic field is proportional to the **current**, ***I***, the **length of wire** in the field, ***l***, and the **strength of the magnetic field**, ***B***. This gives the equation:

$$F = BIl$$

2) In this equation, the **magnetic field strength**, ***B***, is defined as:

The **force** on **one metre** of wire carrying a **current** of **one amp** at **right angles** to the **magnetic field**.

3) **Magnetic field strength** is also called **flux density** and it's measured in **teslas, T**.

$$1 \text{ tesla} = \frac{\text{Wb}}{\text{m}^2}$$

4) Magnetic field strength is a **vector** quantity with both a **direction** and **magnitude**.

It helps to think of flux density as the number of flux lines (measured in webers (Wb), see p 118) per unit area.

Magnetic Fields

Forces Act on *Charged Particles* in Magnetic Fields

Electric current in a wire is caused by the **flow** of negatively **charged** electrons. These charged particles are affected by **magnetic fields** — so a current-carrying wire experiences a **force** in a magnetic field (see page 116).

1) The **force** on a **current-carrying wire** in a **magnetic field** perpendicular to the current is given by $F = BIl$.
2) Electric **current**, I, is the flow of **charge**, q, per unit **time**, t, i.e. $q = It$.
3) A charged particle which moves a **distance** l in **time** t has a **velocity**, $v = l/t$.
4) Putting all these equations **together** (with a spot of rearranging) gives the **force** acting on a **single charged particle moving through a magnetic field**:

In many exam questions, q is the size of the charge on the electron, which is 1.60×10^{-19} coulomb.

$$F = Bqv$$

where B is the **magnetic field strength** (T), q is the **charge** on the particle (C), and v is its **velocity** (ms^{-1})

Example

What is the force acting on an electron travelling at 2×10^4 ms^{-1} through a uniform magnetic field of strength 2 T? (The magnitude of the charge on an electron is 1.6×10^{-19} C.)

Use $F = Bqv$:
$F = 2 \times 1.6 \times 10^{-19} \times 2 \times 10^4 = \mathbf{6.4 \times 10^{-15}}$ **N**

Charged Particles in a *Magnetic Field* are Deflected in a *Circular Path*

1) By **Fleming's left-hand rule** the force on a **moving charge** in a magnetic field is always **perpendicular** to its **direction of travel**.
2) Mathematically, that is the condition for **circular** motion.
3) This effect is used in **particle accelerators** such as **cyclotrons** and **synchrotrons**, which use **magnetic fields** to accelerate particles to very **high energies** along circular paths.
4) The **radius of curvature** of the **path** of a charged particle moving through a magnetic field gives you information about the particle's **charge** and **mass** — this means you can **identify different particles** by studying how they're **deflected**.

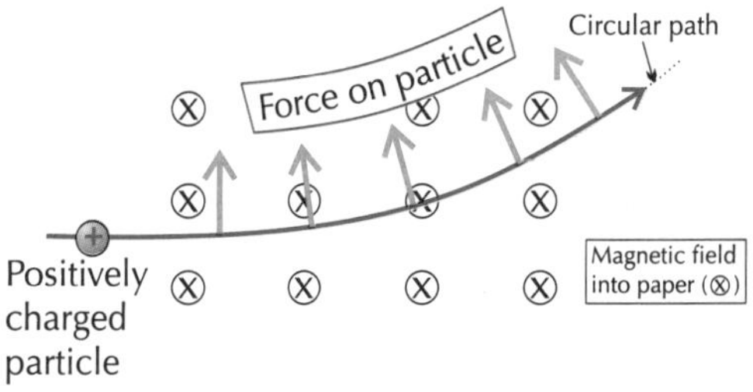

Practice Questions

Q1 Describe why a current-carrying wire at right angles to an external magnetic field will experience a force.

Q2 Write down the equation you would use to find the force on a current-carrying wire.

Q3 Sketch the magnetic fields around a long straight current-carrying wire.
Show the direction of the current and magnetic field.

Q4 A copper bar can roll freely on two copper supports, as shown in the diagram.
When current is applied in the direction shown, which way will the bar roll?

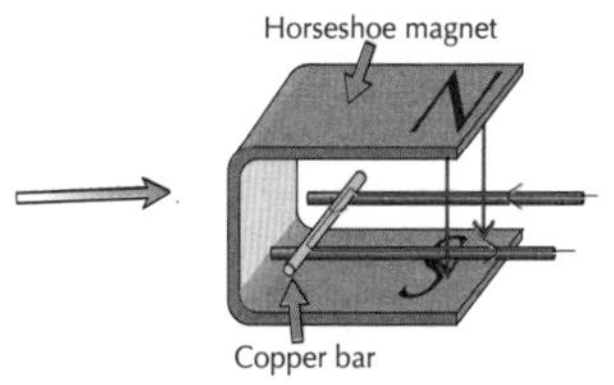

Exam Question

Q1 A wire carrying a current of 3 A runs perpendicular to a magnetic field of strength 2×10^{-5} T.
4 cm of the wire is within the field.

(a) Calculate the magnitude of the force on the wire. [2 marks]

(b) The wire is rotated so that it runs parallel to the magnetic field. What is the force on the wire now? Explain your answer. [2 marks]

I revised the right-hand rule by the A69 and ended up in Newcastle...

Fleming's left-hand rule is the key to this section — so make sure you know how to use it and understand what it all means. Remember that the direction of the magnetic field is from N to S, and that the current is from +ve to –ve — this is as important as using the correct hand. You need to get those right or it'll all go to pot...

Electromagnetic Induction

Think of the **Magnetic Flux** *as the Total* **Number** *of* **Field Lines**

1) **Magnetic flux density**, ***B***, is a measure of the **strength** of the magnetic field **per unit area**.
2) The total **magnetic flux**, ϕ, passing through an **area**, ***A***, perpendicular to a **magnetic field**, ***B***, is defined as:

$$\phi = BA$$

where ϕ is **magnetic flux** (Wb), ***B*** is **magnetic field strength** (T) and ***A*** is **area** (m^2)

3) When you move a **coil** in a magnetic field, the size of the e.m.f. induced depends on the **magnetic flux** passing through the coil, ϕ, and the **number of turns** on the coil. The product of these is called the **flux linkage**, Φ.

For a coil of ***N*** turns perpendicular to B, the flux linkage is given by:

$$\Phi = N\phi = BAN$$

ϕ is the little Greek letter 'phi', and Φ is a capital 'phi'. The unit of both ϕ and Φ is the **weber, Wb**.

If the magnetic flux is **not perpendicular** to ***B***, you can find the **magnetic flux** using this equation:

$$\phi = BA\cos\theta$$

where θ is the **angle** between the **field** and the **normal** to the plane of the coil.

For a **coil** with ***N*** turns the **flux linkage** is:

$$\Phi = BAN\cos\theta$$

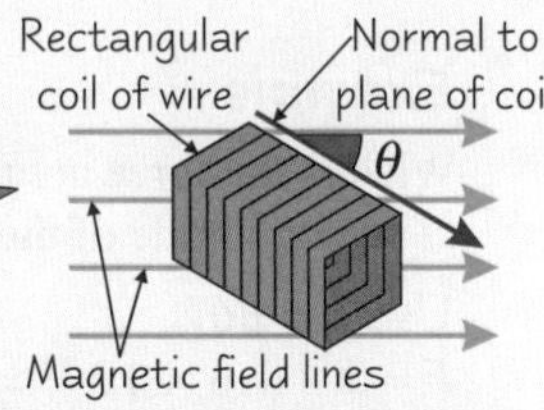

Charges Accumulate *on a* **Conductor** *Moving Through a* **Magnetic Field**

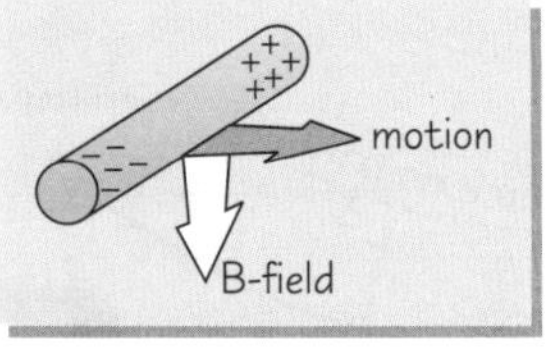

1) If a **conducting rod** moves through a magnetic field its **electrons** will experience a **force** (see p. 116), which means that they will **accumulate** at one end of the rod.
2) This **induces** an **e.m.f.** (**electromotive force**) across the ends of the rod.
3) If the rod is part of a complete **circuit**, then an induced **current** will **flow** through it — this is called **electromagnetic induction**.

Changes *in Magnetic Flux* **Induce** *an* **Electromotive Force**

1) An **electromotive force** (**e.m.f.**) is **induced** when there is **relative motion** between a **conductor** and a **magnet**.
2) The **conductor** can **move** and the **magnetic field** stay **still** or the **other way round** — you get an e.m.f. either way.
3) **Flux cutting** always induces e.m.f. but will only **induce** a **current** if the **circuit** is complete.
4) **Flux linking** is when an e.m.f. is induced by **changing** the **magnitude** or **direction** of the **magnetic flux** (e.g. caused by an **alternating current** electromagnet).

A **change in flux** of **one weber per second** will induce an **electromotive force** of **1 volt** in a loop of wire.

These Results are Summed up by **Faraday's Law...**

FARADAY'S LAW: The **induced e.m.f.** is **directly proportional** to the **rate of change of flux linkage**.

1) **Faraday's law** can be written as: $\text{Induced e.m.f.} = \dfrac{\text{flux change}}{\text{time taken}} = N\dfrac{\Delta\phi}{\Delta t}$
2) The **size** of the e.m.f. is shown by the **gradient** of a graph of Φ against time.
3) The **area under** the graph of e.m.f. against time gives the **flux change**.

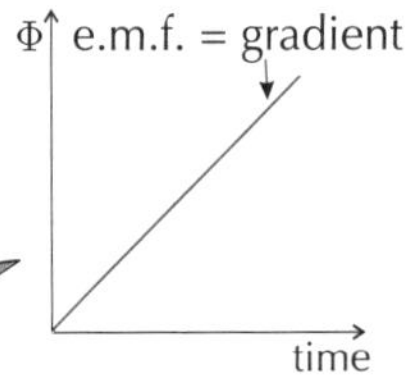

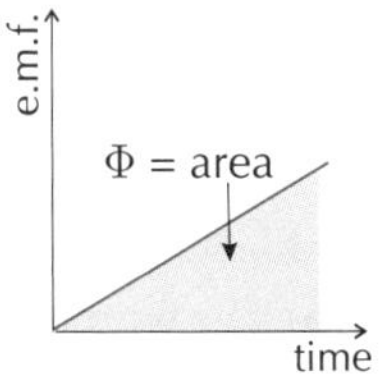

You might be asked to find the e.m.f. induced by the Earth's magnetic field across the wingspan of a plane — just think of it as a moving rod and use the equation as usual.

Example

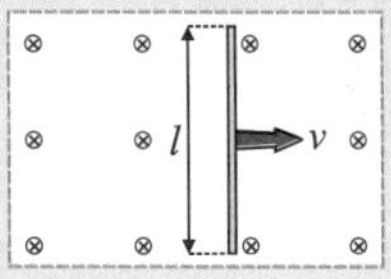

A conducting rod of **length *l*** moves through a perpendicular uniform magnetic field, ***B***, at a constant velocity, ***v***.
Show that the e.m.f. induced in the rod is equal to ***Blv***.

The distance travelled by the rod is $v\Delta t$.
So the area of flux it cuts is $lv\Delta t$, and total magnetic flux is $\phi = Blv\Delta t$.

Faraday's law states that $\text{e.m.f} = \dfrac{\Delta\phi}{\Delta t}$, so $\text{e.m.f} = \dfrac{Blv\Delta t}{\Delta t} = Blv$

Electromagnetic Induction

The **Direction** of the **Induced E.m.f.** and **Current** are given by **Lenz's Law**...

LENZ'S LAW: The **induced e.m.f.** is always in such a **direction** as to **oppose** the **change** that caused it.

1) **Lenz's law** and **Faraday's law** can be **combined** to give one formula that works for both:

$$\text{Induced e.m.f.} = -N\frac{\Delta\phi}{\Delta t}$$

2) The **minus sign** shows the direction of the **induced e.m.f.**
3) The idea that an induced e.m.f. will **oppose** the change that caused it agrees with the principle of the **conservation of energy** — the **energy used** to pull a conductor through a magnetic field, against the **resistance** caused by magnetic **attraction**, is what **produces** the **induced current**.
4) **Lenz's law** can be used to find the **direction** of an **induced e.m.f.** and **current** in a conductor travelling at right angles to a magnetic field.

1) **Lenz's law** says that the **induced e.m.f.** will produce a force that **opposes** the motion of the conductor — in other words a **resistance**.
2) Using **Fleming's left-hand rule** (see p.116), point your thumb in the direction of the force of **resistance** — which is in the **opposite direction** to the motion of the conductor.
3) Your **second finger** will now give you the direction of the **induced e.m.f.**
4) If the conductor is **connected** as part of a **circuit**, a current will be induced in the **same direction** as the induced e.m.f.

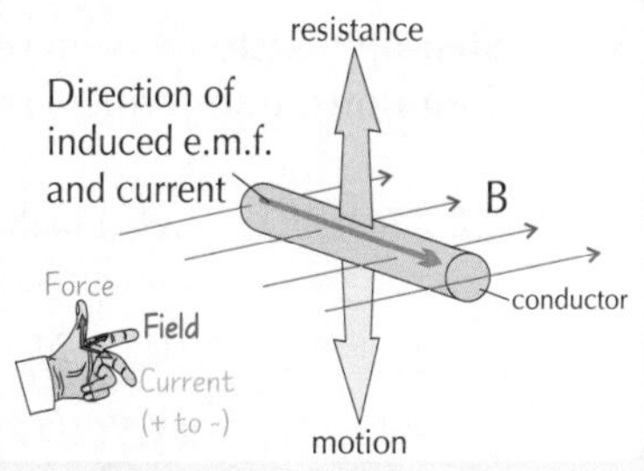

Practice Questions

Q1 What is the difference between magnetic flux density, magnetic flux and magnetic flux linkage?

Q2 A coil consists of N turns, each of area A. What is its flux linkage:
a) if it is perpendicular to a uniform magnetic field, b) if it is at an angle to a uniform magnetic field?

Q3 State Faraday's law.

Q4 State Lenz's law.

Q5 Explain how to find the direction of an induced e.m.f. in a copper bar moving at right angles to a magnetic field.

Exam Questions

Q1 A coil of area 0.23 m^2 is placed at right angles to a magnetic field of 2×10^{-3} T.

(a) Calculate the magnetic flux passing through the coil. [2 marks]

(b) If the coil has 150 turns, what is the magnetic flux linkage in the coil? [2 marks]

(c) Over a period of 2.5 seconds the magnetic field is reduced uniformly to 1.5×10^{-3} T.
Calculate the size of the e.m.f. induced across the ends of the coil. [3 marks]

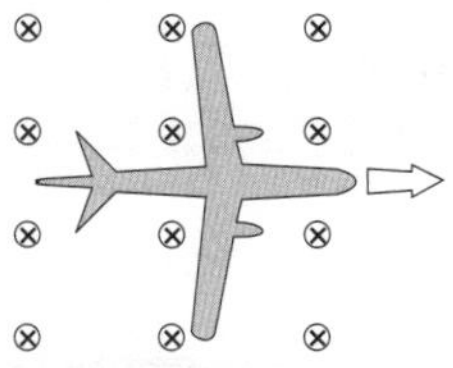

Q2 An aeroplane with a wingspan of 30 m flies at a speed of 100 ms^{-1} perpendicular to the Earth's magnetic field, as shown.
The Earth's magnetic field at the aeroplane's location is 60×10^{-6} T.

(a) Calculate the induced e.m.f. between the wing tips. [2 marks]

(b) Complete the diagram to show the direction of the induced e.m.f. between the wing-tips. [1 mark]

Beware — physics can induce extreme confusion...

Make sure you know the difference between flux and flux linkage, and that you can calculate both. Then all you need to learn is that the induced e.m.f. is proportional to minus the rate of change of flux linkage — and that's it. Remember when you're using Fleming's left-hand rule to work out the direction of the induced e.m.f. that you need to point your thumb in the opposite direction to the direction the conductor is moving... Phew, hope you got all that — I need a cuppa to recover.

Transformers and Alternators

Turns out electromagnetic induction is quite useful in the real world — remember that place?

Transformers *Work by Electromagnetic* ***Induction***

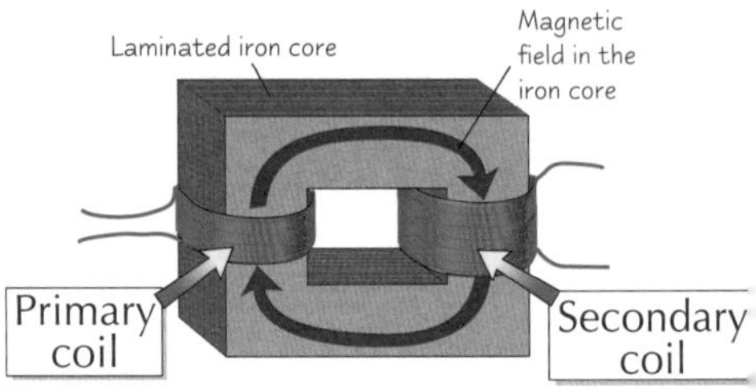

1) **Transformers** are devices that make use of electromagnetic induction to **change** the size of the **voltage** for an **alternating current**.
2) An alternating current flowing in the **primary** (or input) **coil** produces **magnetic flux**.
3) The **magnetic field** is passed through the **iron core** to the **secondary** (or output) coil, where it **induces** an alternating **voltage** of the same frequency.
4) From Faraday's law, the **induced** e.m.f.s in both the **primary** and **secondary** coils can be calculated:

Primary coil: $V_p = N_p \frac{d\Phi}{dt}$

Secondary coil: $V_s = N_s \frac{d\Phi}{dt}$

These can be combined to give the equation for an **ideal transformer**: $\frac{V_p}{V_s} = \frac{N_p}{N_s}$

(where N is the number of turns in a coil)

5) **Step-up** transformers **increase** the **voltage** by having **more turns** on the **secondary** coil than the primary. **Step-down** transformers **reduce** the voltage by having **fewer** turns on the secondary coil.

Example What is the output voltage for a transformer with a primary coil of 100 turns, a secondary coil of 300 turns and an input voltage of 230 V?

$$\frac{V_p}{V_s} = \frac{N_p}{N_s} \Rightarrow \frac{230}{V_s} = \frac{100}{300} \Rightarrow V_s = \frac{230 \times 300}{100} = 690 \text{ V}$$

Transformers are ***Not 100% Efficient***

1) If a transformer was **100% efficient** the **power in** would **equal** the **power out**.
2) This means that for an **ideal transformer**: $V_p I_p = V_s I_s$ or $\frac{V_p}{V_s} = \frac{I_s}{I_p}$
3) However, in practice there will be **small losses** of **power** from the transformer, mostly in the form of **heat**.
4) **Heat** can be produced by **eddy currents** in the transformer's iron **core** — this effect is reduced by **laminating** the core with layers of **insulation**.
5) Heat is also generated by **resistance** in the coils — to minimise this, **thick copper wire** is used, which has a **low resistance**.
6) The **efficiency** of a transformer is simply the **ratio** of **power out** to **power in**, so: efficiency $= \frac{V_s I_s}{V_p I_p}$

You can put the two ideal transformer equations together to give: $\frac{V_p}{V_s} = \frac{N_p}{N_s} = \frac{I_s}{I_p}$

Transformers *are an* ***Important*** *Part of the* ***National Grid...***

1) **Electricity** from power stations is sent round the country in the **national grid** at the **lowest** possible current, because **losses** due to the **resistance** in the cables are proportional to I^2 — so if you double the transmitted current, you quadruple the power lost.
2) Since **power** = **current** × **voltage**, a **low current** means a **high voltage**.
3) **Transformers** allow us to **step up** the voltage to around **400 000 V** for **transmission** through the national grid, and then **reduce** it again to **230 V** for domestic use.

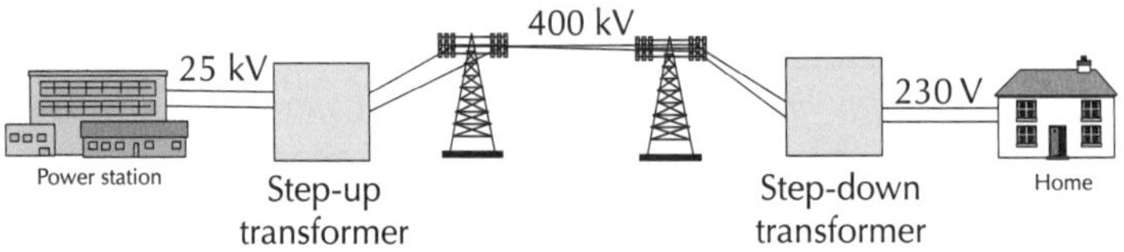

... robots in disguise

Transformers and Alternators

An *Alternator* is a *Generator* of *Alternating Current*

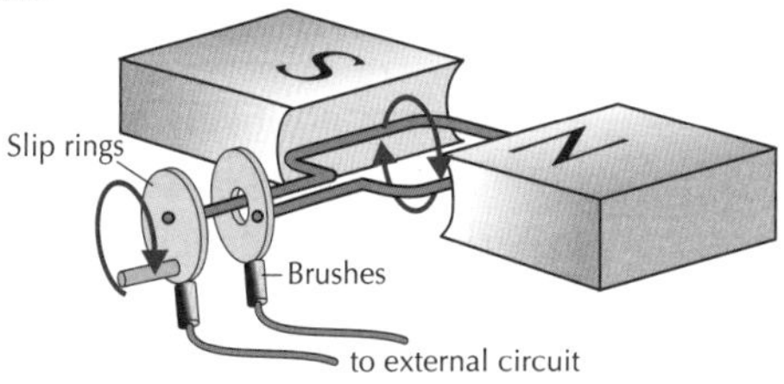

1) **Generators**, or dynamos, **convert** kinetic energy into **electrical energy** — they **induce** an electric **current** by **rotating** a **coil** in a magnetic field.
2) The diagram shows a simple **alternator** — a generator of **AC**. It has **slip rings** and **brushes** to connect the coil to an external circuit.
3) The output **voltage** and **current** change direction with every **half rotation** of the coil, producing **alternating current** (**AC**).

Flux Linkage and *Induced Voltage* are *90° Out of Phase*

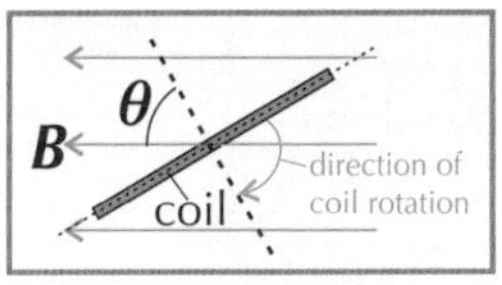

1) The amount of flux cut by the coil (**flux linkage**) is: $\Phi = BAN \cos \theta$ See page 118
(θ is the angle between the normal to the coil and the flux lines)
2) As the coil rotates, θ changes so the **flux linkage** varies **sinusoidally** between $+BAN$ and $-BAN$.

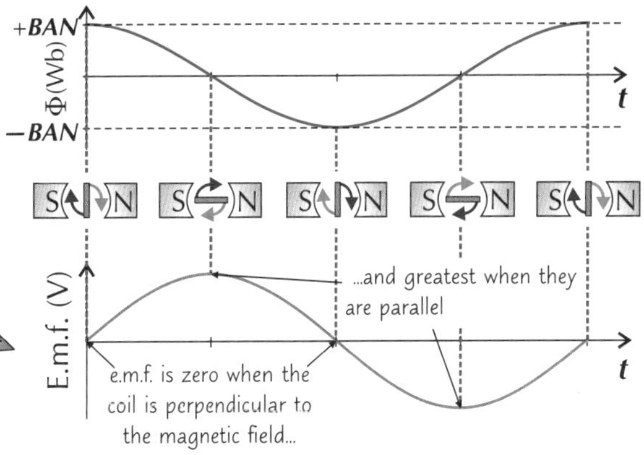

3) How fast θ changes depends on the angular speed, ω, of the coil (see page 96), $\theta = \omega t$. So you can write:

$$\Phi = BAN \cos \omega t$$

4) The **induced e.m.f.**, ε, depends on the **rate of change** of **flux linkage** (Faraday's law), so it also varies **sinusoidally**. The equation for the e.m.f. at time t is:

$$\varepsilon = BAN \omega \sin \omega t$$

Example

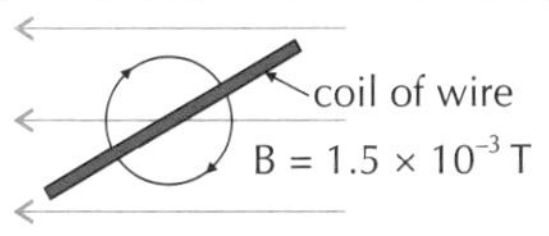

A rectangular coil with 20 turns, each with an area of 0.2 m^2, is rotated at 20 rad s^{-1} in a uniform 1.5 mT magnetic field. Calculate the maximum e.m.f. induced in the coil.

$\varepsilon = BAN\omega \sin \omega t$. So, ε will be greatest when $\sin \omega t = \pm 1$, which gives $\varepsilon = 1.5 \times 10^{-3} \times 0.2 \times 20 \times 20 \times \pm 1 = \pm 0.12$ V

Practice Questions

Q1 Draw a diagram of a simple transformer.

Q2 What is meant by a step-down transformer?

Q3 Show that flux linkage and induced e.m.f. are 90° out of phase.

Exam Questions

Q1 A transformer with 150 turns in the primary coil has an input voltage of 9 V.

(a) Calculate the number of turns needed in the secondary coil to step up the voltage to 45 V. [2 marks]

(b) The input current for the transformer is 1.5 A. Assuming the transformer is ideal, state the output current. [2 marks]

(c) Calculate the actual efficiency of the transformer given that the power output is measured as 10.8 W. [2 marks]

Q2 A 0.01 m^2 coil of 500 turns is rotated on an axis that is perpendicular to a magnetic field of 0.9 T.

(a) Find the flux linkage when the angle between its normal and the magnetic field is 60°. [2 marks]

(b) If the coil is rotated at an angular speed of 40π rad s^{-1}, what is the peak e.m.f. induced? [3 marks]

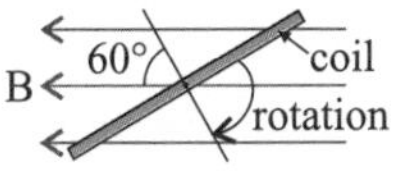

Arrrrrrrrggggggghhhhhhh...

Breathe a sigh of relief, pat yourself on the back and make a brew — well done, you've reached the end of the section. That was pretty nasty stuff, but don't let all of those equations get you down — once you've learnt the main ones and can use them blindfolded, even the trickiest looking exam question will be a walk in the park...

Scattering to Determine Structure

By firing radiation at different materials, you can take a sneaky beaky at their internal structures...

Alpha Particle Scattering Lets Us See Inside the Atom

1) If a beam of **positively charged alpha particles** is directed at a thin gold film, most **pass straight through**. However, if an alpha particle is travelling straight towards, or close by, a nucleus, its path will be **deflected**.
2) **Experimental evidence** shows that some of these alpha particles are deflected by **more than 90°** — in other words they '**bounce**' back.

Rutherford Scattering

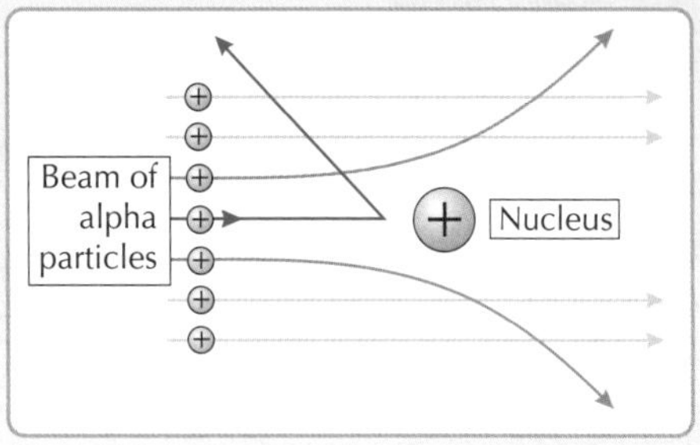

3) This evidence means that inside the atoms there must be **small positively charged nuclei**, which **repel** the passing alpha particles.
4) The nucleus must be **small** since very few alpha particles are deflected by much.
5) It must be **positive** to repel the positively charged alpha particles.

You can Estimate the Closest Approach of a Scattered Particle

1) When you fire an alpha particle at a gold nucleus, you know its **initial kinetic energy**.
2) An alpha particle that 'bounces back' and is deflected through 180° will have stopped a short distance from the nucleus. It does this at the point where its **electrical potential energy** (see p 110) **equals** its **initial kinetic energy**.

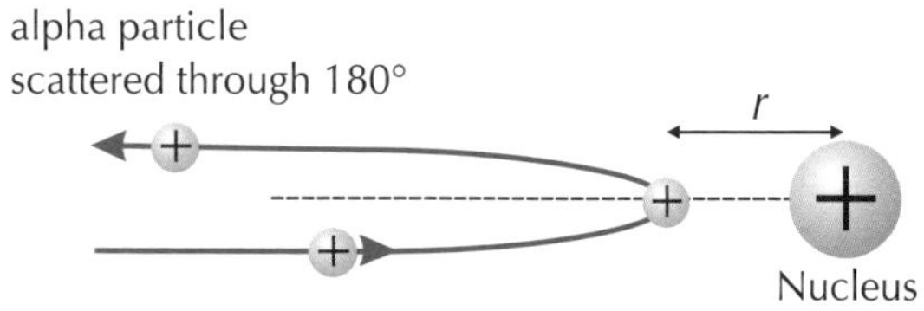

$$\text{Initial K.E.} = E_{elec} = \frac{Q_{gold}q_{alpha}}{4\pi\varepsilon_0 r}$$

ε_0 is the permittivity of free space

3) It's just conservation of energy — and you can use it to find how close the particle can get to the nucleus.
4) To find the charge of a nucleus you need to know the atom's **proton number**, **Z** — that tells you how many protons are in the nucleus (surprisingly).
A proton has a charge of **+e** (where e is the charge on an electron), so the charge of a nucleus must be **+Ze**.

Example An alpha particle with an initial kinetic energy of 6 MeV is fired at a gold nucleus. Find the closest approach of the alpha particle to the nucleus.

Initial particle energy = 6 MeV = 6×10^6 eV

Convert energy into joules: $6\times10^6\times1.6\times10^{-19} = 9.6\times10^{-13}$ J

So, electrical potential energy $= E_{elec} = \frac{Q_{gold}q_{alpha}}{4\pi\varepsilon_0 r} = 9.6\times10^{-13}$ J at closest approach.

Rearrange to get $r = \frac{(+79e)(+2e)}{4\pi\varepsilon_0(9.6\times10^{-13})}$

$$= \frac{2\times79\times(1.6\times10^{-19})^2}{4\pi\times8.9\times10^{-12}\times9.6\times10^{-13}} = \mathbf{3.8\times10^{-14}\ m}$$

Scattering to Determine Structure

You can also use Electron Diffraction to Estimate Nuclear Diameter

1) **Electrons** are a type of particle called a **lepton**. Leptons **don't interact** with the **strong nuclear force** (whereas neutrons and alpha particles do). Because of this, electron diffraction is the **most accurate** method for getting a picture of a crystal's **atomic structure**.
2) Like other particles, electrons show **wave-particle duality** (see p 186) — so **electron beams** can be diffracted.
3) A beam of moving electrons has an associated **de Broglie wavelength**, λ, which at high speeds (where you have to take into account relativistic effects (see p 190)) is approximately: $\lambda \simeq \frac{hc}{E}$
4) If a beam of **high-energy electrons** is directed onto a thin film of material in front of a screen, a **diffraction pattern** will be seen on the screen.
5) As with light diffraction patterns, the first minimum appears where:

$$\sin\theta \simeq \frac{1.22\lambda}{d}$$

d is the diameter of the nucleus it has been scattered by.

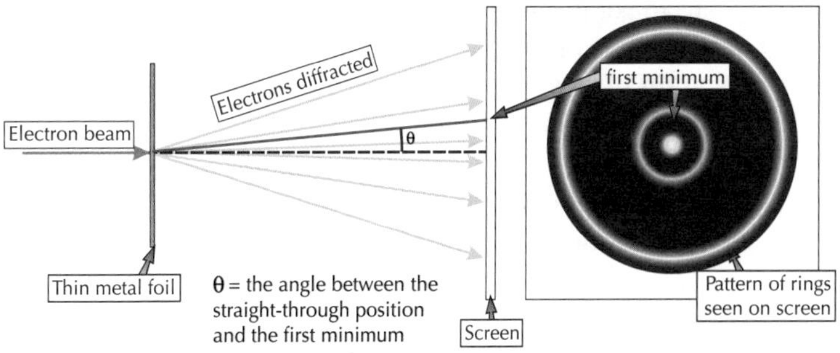

6) Using measurements from this diffraction pattern, the **size and spacing** of the material's atomic **nuclei** can be worked out.

Example A beam of 300 MeV electrons is fired at a piece of thin foil, and produces a diffraction pattern on a fluorescent screen. The first minimum of the diffraction pattern is at an angle of 30° from the straight-through position. Estimate the diameter of the nuclei the electrons were diffracted by.

$$E = 300\text{ MeV} = 3.00\times10^{8}\times1.6\times10^{-19} = 4.8\times10^{-11}\text{J},\ \lambda \simeq \frac{hc}{E} = \frac{6.6\times10^{-34}\times3.0\times10^{8}}{4.8\times10^{-11}} = 4.125\times10^{-15}\text{m}$$

$$\text{So } d \simeq \frac{1.22\lambda}{\sin\theta} = \frac{1.22\times4.125\times10^{-15}}{\sin30^\circ} = 2\times1.22\times4.125\times10^{-15} = \mathbf{1.0\times10^{-14}\ m}$$

Practice Questions

Q1 Explain how alpha particle scattering shows that a nucleus is both small and positively charged.

Q2 Why are X-rays a suitable electromagnetic wave to investigate atomic sizes?

Q3 Describe how you could estimate the distance of closest approach of a scattered particle.

Exam Questions

Q1 A beam of alpha particles is directed onto a very thin gold film.
- (a) Explain why the majority of alpha particles are not scattered. [2 marks]
- (b) Explain how alpha particles are scattered by atomic nuclei. [3 marks]

Q2 Various particles can be used to investigate the structure of matter.
- (a) Explain why particles such as electrons produce diffraction patterns. [2 marks]
- (b) Electron-beam diffraction is currently the most accurate method for finding out about the atomic structure of a crystal. State why this is. [1 mark]
- (c) When electrons are directed at a larger nucleus, the beam suffers less diffraction. Why does this happen? [2 marks]

Alpha scattering — It's positively repulsive...

Scattering and diffraction are the key ideas you need to understand for questions about atomic size and structure. Remember, particles like electrons have wave-like properties, so if you fire them at crystal structures they make a diffraction pattern. This lets you work out size, spacing etc...

Nuclear Radius and Density

The tiny nucleus — such a weird place, but one that you need to become ultra familiar with. Lucky you...

The **Nucleus** is a **Very Small Part** of a Whole **Atom**

1) By **probing atoms** using scattering and diffraction methods, we know that the **diameter of an atom** is about 0.1 nm (1×10^{-10} m) and the diameter of the smallest **nucleus** is about 2 fm (2×10^{-15} m — pronounced "femtometres").
2) So basically, **nuclei** are really, really **tiny** compared with the size of the **whole atom**.
3) To make this **easier to visualise**, try imagining a **large Ferris wheel** (which is pretty darn big) as the size of **an atom**. If you then put a **grain of rice** (which is rather small) in the centre, this would be the size of the atom's **nucleus**.
4) **Molecules** are just a number of **atoms joined together**. As a rough guide, the size of a molecule equals the number of atoms in it multiplied by the size of one atom.

The **Nucleus** is **Made Up** of **Nucleons**

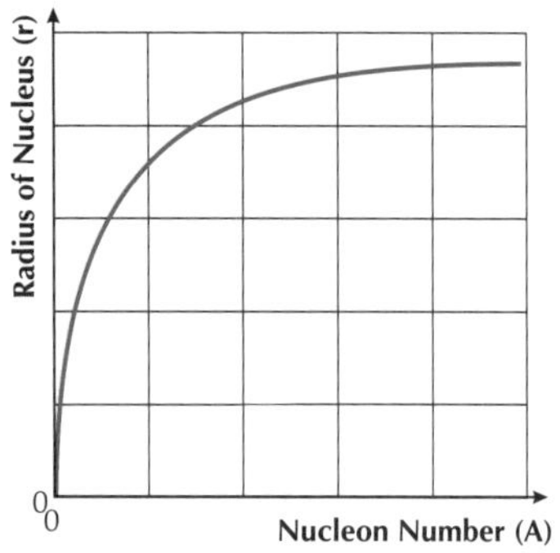

1) The **particles** that make up the nucleus (i.e. **protons** and **neutrons**) are called **nucleons**.
2) The **number of nucleons** in an atom is called the **mass** (or **nucleon**) **number,** ***A***.
3) As **more nucleons** are added to the nucleus, it gets **bigger**.
4) And as we all know by now, you can measure the size of a nucleus by firing particles at it (see p 122-123).

See p. 130 for more on the mass number and how this is used to represent atomic structure in standard notation.

Nuclear Radius is **Proportional** to the **Cube Root** of the **Mass Number**

The **nuclear radius** increases roughly as the cube root of the mass (nucleon) number.

1) This **straight-line graph** shows that the **nuclear radius** (*r*) is **directly proportional** to the cube root of the **nucleon number** (*A*).
2) This relationship can be written as: $r \propto A^{1/3}$.
3) We can make this into an equation by introducing a constant, r_0, giving:

$$r = r_0 A^{1/3}$$

Where r_0 is the value of *r* when *A* = 1, i.e. for a proton (hydrogen nucleus). The value of r_0 is about 1.4 fm.

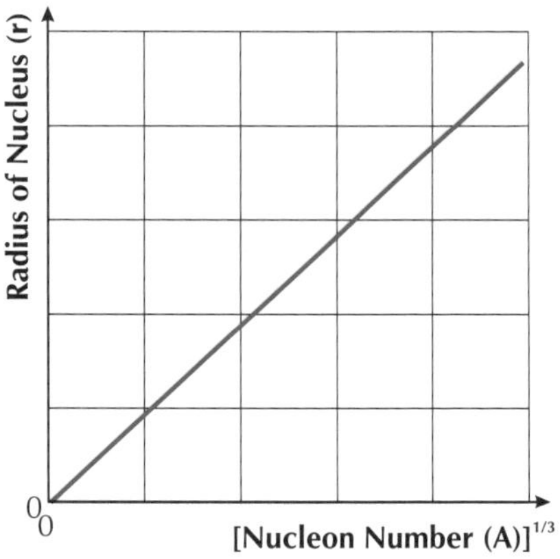

Example Calculate the radius of an oxygen nucleus which has 16 nucleons.

$$r = r_0 A^{1/3} = 1.4 \times 10^{-15} \times (16)^{1/3}$$
$$= 3.5 \times 10^{-15} \text{ m (or 3.5 fm)}$$

Nuclear Radius and Density

The Density of Nuclear Matter is Enormous

1) The **volume** that each nucleon (i.e. a **proton** or a **neutron**) takes up in a nucleus is about the **same**.
2) Because protons and neutrons have nearly the **same mass**, it means that all nuclei have a **similar density** (ρ).
3) But nuclear matter is **no ordinary** stuff. Its density is **enormous**. A **teaspoon** of pure nuclear matter would have a mass of about **five hundred million tonnes**. (Just to make you gasp in awe and wonder, out in space nuclear matter makes up neutron stars, which are several kilometres in diameter.)

The following examples show how **nuclear density** is pretty much the **same**, **regardless of the element**.

Example 1

Work out the density of a carbon nucleus given that its mass is 2.00×10^{-26} kg and $A = 12$.

1) The **radius** (r) of a carbon nucleus $\approx$ **3.2×10^{-15} m**
2) So, the **volume** (v) of the nucleus $= \frac{4}{3}\pi r^3$ $= \mathbf{1.37 \times 10^{-43}\ m^3}$
3) This gives the **density** (ρ) of a carbon nucleus as:

$$\rho = \frac{m}{v} = \frac{2.00 \times 10^{-26}}{1.37 \times 10^{-43}} = 1.46 \times 10^{17}\ \text{kg m}^{-3}$$

Example 2

Work out the density of a gold nucleus given that its mass is 3.27×10^{-25} kg and $A = 197$.

1) The **radius** (r) of a gold nucleus $\approx$ **8.1×10^{-15} m**
2) So, the **volume** (v) of the nucleus $= \frac{4}{3}\pi r^3$ $= \mathbf{2.23 \times 10^{-42}\ m^3}$
3) This gives the **density** (ρ) of a gold nucleus as:

$$\rho = \frac{m}{v} = \frac{3.27 \times 10^{-25}}{2.23 \times 10^{-42}} = 1.47 \times 10^{17}\ \text{kg m}^{-3}$$

Nuclear density is significantly larger than atomic density — this suggests three important facts about the structure of an atom:
a) Most of an atom's mass is in its nucleus.
b) The nucleus is small compared to the atom.
c) An atom must contain a lot of empty space.

Practice Questions

Q1 What is the approximate size of an atom?

Q2 What are nucleons?

Q3 What is the relationship between the nuclear radius and mass number?

Q4 In the formula $r = r_o A^{1/3}$, what does r_o represent?

Q5 Explain why the density of ordinary matter is much less than that of nuclear matter.

Exam Questions

Q1 The radius (r) of a nucleus with A nucleons can be calculated using the equation $r = r_o A^{1/3}$.

(a) If a carbon nucleus containing 12 nucleons has a radius of 3.2×10^{-15} m, show that $r_o = 1.4 \times 10^{-15}$ m. [2 marks]

(b) Calculate the radius of a radium nucleus containing 226 nucleons. [1 mark]

(c) Calculate the density of the radium nucleus if its mass is 3.75×10^{-25} kg. [2 marks]

Q2 A sample of pure gold has a density of 19 300 kg m^{-3}. If the density of a gold nucleus is 1.47×10^{17} kg m^{-3}, discuss what this implies about the structure of a gold atom. [4 marks]

Nuclear and particle physics — heavy stuff...

So basically the nucleus is a tiny part of the atom, but it's incredibly dense. The density doesn't change much from element to element, and the radius depends on the mass number. Learn the theory like your own backyard, but don't worry about remembering equations and values — those friendly examiners have popped them in the exam paper for you. How nice...

Radioactive Emissions

Radiation is all around us... Not quite as catchy as the original, but it is true at least...

Unstable *Atoms are* ***Radioactive***

1) If an atom is **unstable**, it will **break down** to **become** more stable. Its **instability** could be caused by having **too many neutrons, not enough neutrons**, or just **too much energy** in the nucleus.
2) The atom **decays** by **releasing energy** and/or **particles**, until it reaches a **stable form** — this is called **radioactive decay**.
3) An individual radioactive decay is **random** — it can't be predicted.

There are ***Four Types*** *of* ***Nuclear Radiation***

Learn this table.

Radiation	Symbol	Constituent	Relative Charge	Mass (u)
Alpha	α	A helium nucleus — 2 protons & 2 neutrons	+2	4
Beta-minus (Beta)	β or β^-	Electron	-1	(negligible)
Beta-plus	β^+	Positron	+1	(negligible)
Gamma	γ	Short-wavelength, high-frequency electromagnetic wave.	0	0

u stands for atomic mass unit — see p 132.

You met positrons at AS level.

The ***Different Types*** *of Radiation have* ***Different Penetrations***

When a radioactive particle **hits** an **atom** it can **knock off electrons**, creating an **ion** — so, **radioactive emissions** are also known as **ionising radiation**.

Alpha, **beta** and **gamma** radiation can be **fired** at a **variety of objects** with **detectors** placed the **other side** to see whether they **penetrate** the object.

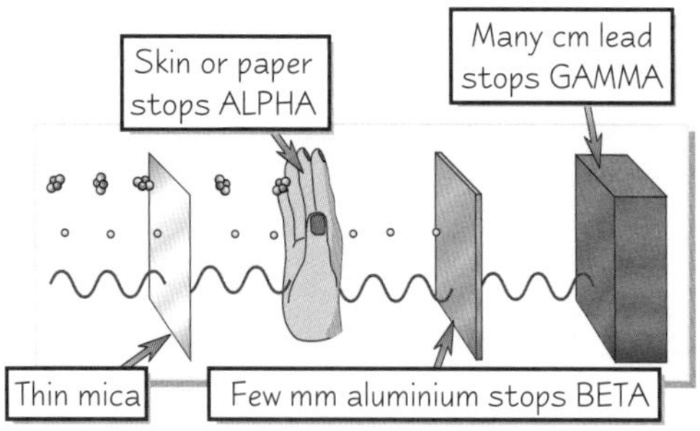

Radiation	Symbol	Ionising	Speed	Penetrating power	Affected by magnetic field
Alpha	α	Strongly	Slow	Absorbed by paper or a few cm of air	Yes
Beta-minus (Beta)	β or β^-	Weakly	Fast	Absorbed by ~3 mm of aluminium	Yes
Beta-plus	β^+	Annihilated by electron — so virtually zero range			
Gamma	γ	Very weakly	Speed of light	Absorbed by many cm of lead, or several m of concrete.	No

Alpha *and* ***Beta*** *Particles have* ***Different Ionising Properties***

What a radioactive source can be used for often depends on its **ionising properties**.

1) **Alpha** particles are **strongly positive** — so they can **easily pull electrons** off atoms.
2) Ionising an atom **transfers** some of the **energy** from the **alpha particle** to the **atom**. The alpha particle **quickly ionises** many atoms (about 10 000 ionisations per alpha particle) and **loses** all its **energy**. This makes alpha-sources suitable for use in **smoke alarms** because they allow **current** to flow, but won't **travel very far**.
3) The **beta**-minus particle has **lower mass** and **charge** than the alpha particle, but a **higher speed**. This means it can still **knock electrons** off atoms. Each **beta** particle will ionise about 100 atoms, **losing energy** at each interaction.
4) This **lower number of interactions** means that beta radiation causes much **less damage** to body tissue.
5) Gamma radiation is even more **weakly ionising** than beta radiation, so will do even **less damage** to body tissue. This means it can be used for **diagnostic techniques** in medicine.

Radioactive Emissions

The Intensity of Gamma Radiation Obeys the Inverse Square Law

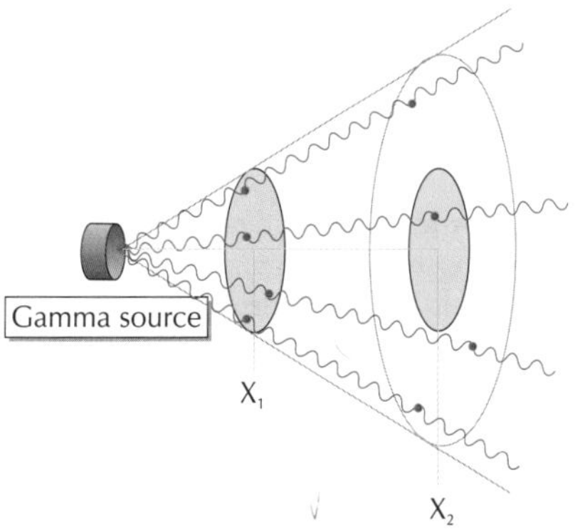

1) A **gamma source** will **emit** gamma **radiation** in **all directions**.
2) This radiation **spreads out** as you get **further away** from the source.
3) However, the amount of **radiation per unit area** (the **intensity**) will **decrease** the further you get from the source.
4) If you took a reading of **intensity**, I, at a **distance**, x, from the source you would find that it **decreases** by the **square of the distance** from the source.
5) This can be written as the equation:

$$I = \frac{kI_0}{x^2}$$

where k is a constant, and I_0 is the intensity at the source.

6) This **relationship** can be **proved** by taking **measurements of intensity** at different distances from a gamma source, using a **Geiger-Müller tube** and **counter**.
7) If the **distance** from the source is **doubled** the **intensity** is found to **fall to a quarter** — which **verifies** the inverse square law.

That's why one of the safety precautions when handling a source is to hold it at arm's length, so you lessen the amount of radiation reaching you.

We're Surrounded by Background Radiation

Put a Geiger-Müller tube **anywhere** and the counter will click — it's detecting **background radiation**.

When you take a **reading** from a radioactive source, you need to **measure** the **background radiation** separately and **subtract** it from your **measurement**.

There are many **sources** of background radiation:

1) **The air:** Radioactive **radon gas** is released from **rocks**. It emits alpha radiation. The concentration of this gas in the atmosphere varies a lot from place to place, but it's usually the largest contributor to the background radiation.
2) **The ground and buildings: All rock** contains radioactive isotopes.
3) **Cosmic radiation:** Cosmic rays are particles (mostly high-energy protons) from **space**. When they collide with particles in the upper atmosphere, they produce nuclear radiation.
4) **Living things:** All plants and animals contain **carbon**, and some of this will be radioactive **carbon-14**.
5) **Man-made radiation:** In most areas, radiation from **medical** or **industrial** sources makes up a tiny, tiny fraction of the background radiation.

Practice Questions

Q1 What makes an atom radioactive?

Q2 Name three types of nuclear radiation and give three properties of each.

Q3 Give three sources of background radiation.

Exam Questions

Q1 Briefly describe an absorption experiment to distinguish between alpha, beta and gamma radiation. You may wish to include a sketch in your answer. [4 marks]

Q2 The count rate detected by a G-M tube, 10 cm from a gamma source, is 240 counts per second. What would you expect the count rate to be at 40 cm from the source? [3 marks]

Radioactive emissions — as easy as α, β, γ...

You need to learn the different types of radiation and their properties. Remember that alpha particles are by far the most ionising and so cause more damage if they get inside your body than the same dose of any other radiation — which is one reason we don't use alpha sources as medical tracers. Learn this all really well, then go and have a brew and a bickie...

Exponential Law of Decay

Oooh look — some maths. Good.

It could be you.

Every Isotope Decays at a Different Rate

1) **Radioactive decay** is completely **random**. You **can't predict which** atom will decay **when**.
2) Although you can't predict the decay of an **individual atom**, if you take a **very large number of atoms**, their **overall behaviour** shows a **pattern**.
3) Any sample of a particular **isotope** has the **same rate of decay**, i.e. the same **proportion** of atoms will **decay** in a **given time**.

Isotopes of an element have the same number of protons, but different numbers of neutrons in their nuclei.

The Rate of Decay is Measured by the Decay Constant

The **activity** of a sample — the **number** of atoms that **decay each second** — is **proportional** to the **size of the sample**. For a **given isotope**, a sample **twice** as big would give **twice** the **number of decays** per second.

The **decay constant** (λ) measures how **quickly** an isotope will **decay** — the **bigger** the value of λ, the faster the rate of decay. Its unit is s^{-1}.

activity = decay constant × number of atoms   Or in symbols: $A = \lambda N$

Don't get λ confused with wavelength.

Activity is measured in **becquerels** (Bq): 1 Bq = 1 decay per second (s^{-1})

You Need to Learn the Definition of Half-Life

The **half-life** ($T_{1/2}$) of an **isotope** is the **average time** it takes for the **number of undecayed atoms** to **halve**.

Measuring the **number of undecayed atoms** isn't the easiest job in the world. **In practice**, half-life isn't measured by counting atoms, but by measuring the **time it takes** the **activity** to **halve**.

The **longer** the **half-life** of an isotope, the **longer** it stays **radioactive**.

The Number of Undecayed Particles Decreases Exponentially

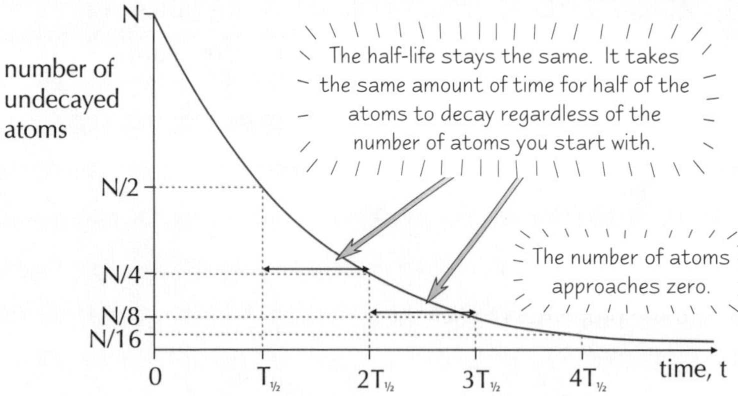

You'd be **more likely** to actually meet a **count rate-time graph** or an **activity-time graph**. They're both **exactly the same shape** as the graph above, but with different **y-axes**.

Plotting the natural log (ln) of the number of radioactive atoms (or the activity) against time gives a straight-line graph (see p 192).

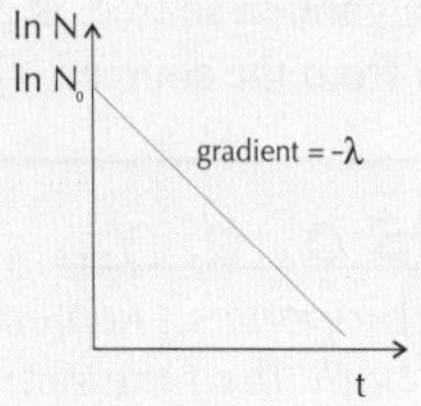

When you're **measuring** the **activity** and **half-life** of a **source** you've got to **remember background radiation**. The **background radiation** needs to be **subtracted** from the **activity readings** to give the **source activity**.

How to find the half-life of an isotope

STEP 1: Read off the value of count rate, particles or activity when t = 0.

STEP 2: Go to half the original value.

STEP 3: Draw a horizontal line to the curve, then a vertical line down to the x-axis.

STEP 4: Read off the half-life where the line crosses the x-axis.

STEP 5: Check the units carefully.

STEP 6: It's always a good idea to check your answer. Repeat steps 1-4 for a quarter the original value. Divide your answer by two. That will also give you the half-life. Check that you get the same answer both ways.

Exponential Law of Decay

You Need to Know the Equations for Half-Life and Decay...

1) The number of radioactive nuclei decaying per second (**activity**) is proportional to the number of nuclei remaining.
2) The **half-life** can be **calculated** using the equation: $T_{\frac{1}{2}} = \frac{\ln 2}{\lambda} \simeq \frac{0.693}{\lambda}$
(where ln is the natural log)
3) The **number of radioactive atoms** remaining, N, depends on the **number originally** present, N_o. The **number remaining** can be calculated using the equation: $N = N_0 e^{-\lambda t}$
Here t = time, measured in seconds.
4) As a sample decays, its activity goes down — there's an equation for that too: $A = A_0 e^{-\lambda t}$

Example:
A sample of the radioactive isotope ^{13}N contains 5×10^6 atoms. The decay constant for this isotope is $1.16 \times 10^{-3}\ s^{-1}$.

a) What is the half-life for this isotope?

$$T_{\frac{1}{2}} = \frac{\ln 2}{1.16 \times 10^{-3}} = 598\ s$$

b) How many atoms of ^{13}N will remain after 800 seconds?

$$N = N_0 e^{-\lambda t} = 5 \times 10^6 e^{-(1.16\times10^{-3})(800)} = 1.98 \times 10^6 \text{ atoms}$$

Radioactive Isotopes Have Many Uses

Radioactive substances are extremely useful. You can use them for all sorts — to **date** organic material, diagnose **medical problems**, **sterilise** food, and in **smoke alarms**.

Radiocarbon Dating
The radioactive isotope **carbon-14** is used in **radiocarbon dating**. Living plants take in carbon dioxide from the atmosphere as part of **photosynthesis**, including the **radioactive isotope carbon-14**. When they die, the **activity** of carbon-14 in the plant starts to **fall**, with a **half-life** of around **5730 years**. Archaeological finds made from once living material (like wood) can be tested to find the **current amount** of carbon-14 in them, and date them.

Medical Diagnosis
Technetium-99m is widely used in medical tracers — **radioactive substances** that are used to show tissue or **organ function**. The tracer is **injected** into or **swallowed** by the patient and then **moves** through the **body** to the region of interest. The **radiation emitted** is **recorded** and an **image** of inside the patient produced. Technetium-99m is suitable for this use because it emits **γ-radiation**, has a **half-life of 6 hours** (long enough for data to be recorded, but short enough to limit the radiation to an acceptable level) and **decays** to a **much more stable isotope**.

Practice Questions

Q1 Define radioactive activity. What units is it measured in?

Q2 Sketch a general radioactive decay graph showing the number of undecayed particles against time.

Q3 What is meant by the term 'half-life'?

Q4 Describe how radiocarbon dating works.

Exam Questions

Q1 Explain what is meant by the random nature of radioactive decay. [1 mark]

Q2 You take a reading of 750 Bq from a pure radioactive source. The radioactive source initially contains 50 000 atoms, and background activity in your lab is measured as 50 Bq.
(a) Calculate the decay constant for your sample. [3 marks]
(b) Determine the half-life of this sample. [2 marks]
(c) Approximately how many atoms of the radioactive source will there be after 300 seconds? [2 marks]

Radioactivity is a random process — just like revision shouldn't be...

Remember the shape of that graph — whether it's count rate, activity or number of atoms plotted against time, the shape's always the same. This is all pretty straightforward mathsy-type stuff: plugging values in equations, reading off graphs, etc. Not very interesting, though. Ah well, once you get onto relativity you'll be longing for a bit of boredom.

Nuclear Decay

The stuff on these pages covers the most important facts about nuclear decay that you're just going to have to make sure you know inside out. I'd be very surprised if you didn't get a question about it in your exam...

Atomic Structure** can be Represented Using **Standard Notation

STANDARD NOTATION:

$$^{12}_{6}\mathrm{C}$$

The **proton number** or **atomic number** (Z) — there are six protons in a carbon atom.

The **nucleon number** or **mass number** (A) — there are a total of 12 protons and neutrons in a carbon-12 atom.

The symbol for the element carbon.

Atoms with the **same number of protons** but **different numbers of neutrons** are called **isotopes**. The following examples are all isotopes of carbon: $^{12}_{6}\mathrm{C}$, $^{13}_{6}\mathrm{C}$, $^{14}_{6}\mathrm{C}$

*Some Nuclei are **More Stable** than Others*

The nucleus is under the **influence** of the **strong nuclear force holding** it **together** and the **electromagnetic force pushing** the **protons apart**. It's a very **delicate balance**, and it's easy for a nucleus to become **unstable**. You can get a stability graph by plotting **Z** (atomic number) against **N** (number of neutrons).

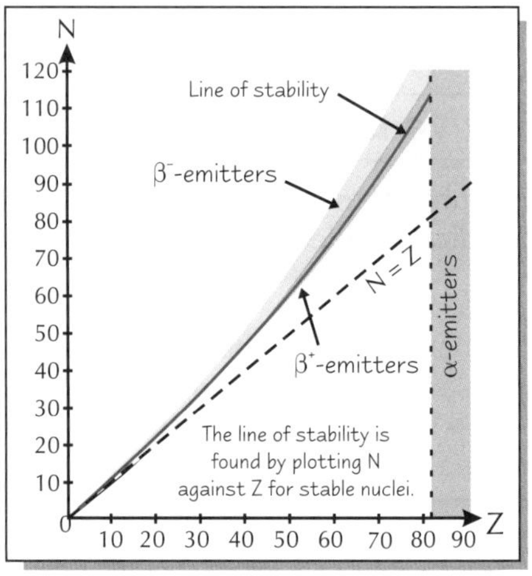

A nucleus will be **unstable** if it has:

1) **too many neutrons**
2) **too few neutrons**
3) **too many nucleons** altogether, i.e. it's **too heavy**
4) **too much energy**

*α **Emission** Happens in **Heavy Nuclei***

When an alpha particle is **emitted**:

The **proton number decreases** by **two**, and the **nucleon number decreases** by **four**.

$$^{238}_{92}\mathrm{U} \longrightarrow {}^{234}_{90}\mathrm{Th} + {}^{4}_{2}\alpha$$

nucleon number decreases by 4

proton number decreases by 2

1) **Alpha emission** only happens in **very heavy** atoms (with more than 82 protons), like **uranium** and **radium**.
2) The **nuclei** of these atoms are **too massive** to be stable.

*β⁻ **Emission** Happens in **Neutron Rich** Nuclei*

1) **Beta-minus** (usually just called beta) decay is the emission of an **electron** from the **nucleus** along with an **antineutrino**.
2) Beta decay happens in isotopes that are **"neutron rich"** (i.e. have many more **neutrons** than **protons** in their nucleus).
3) When a nucleus ejects a beta particle, one of the **neutrons** in the nucleus is **changed** into a **proton**.

When a **beta-minus** particle is **emitted**:

The **proton number increases** by **one**, and the **nucleon number stays the same**.

$$^{188}_{75}\mathrm{Re} \longrightarrow {}^{188}_{76}\mathrm{Os} + {}^{0}_{-1}\beta + {}^{0}_{0}\bar{\nu}_e$$

nucleon number stays the same

proton number increases by 1

In **beta-plus emission**, a **proton** gets **changed** into a **neutron**.
The **proton number decreases** by **one**, and the **nucleon number stays the same**.

Nuclear Decay

γ *Radiation* is *Emitted from* **Nuclei** *with* **Too Much Energy**

1) **After alpha** or **beta** decay, the **nucleus** often has **excess energy** — it's **excited**. This energy is **lost** by emitting a **gamma ray**.
2) **Another way** that gamma radiation is produced is when a nucleus **captures** one of its own orbiting **electrons**.

$$p + e^- \rightarrow n + \nu_e$$

3) **Electron capture** causes a **proton** to **change** into a **neutron**. This makes the **nucleus unstable** and it **emits** gamma radiation.

The artificial isotope technetium-99ᵐ is formed in an excited state from the decay of another element. It is used as a tracer in medical imaging (see p 129).

During **gamma emission**, there is **no change** to the nuclear **constituents** — the nucleus just **loses excess energy**.

There are **Conservation Rules** *in* **Nuclear Reactions**

In every nuclear reaction **energy**, **momentum**, **proton number / charge** and **nucleon number** must be conserved.

238 = 234 + 4 — nucleon numbers balance

$$^{238}_{92}U \longrightarrow {}^{234}_{90}Th + {}^{4}_{2}\alpha$$

92 = 90 + 2 — proton numbers balance

Mass *is* **Not Conserved**

1) The **mass** of the **alpha particle** is less than the **individual masses** of **two protons** and **two neutrons**. The difference is called the **mass defect**.
2) Mass **doesn't** have to be **conserved** because of **Einstein's equation**: $E = mc^2$
3) This says that **mass and energy** are **equivalent**.
The **energy released** when the nucleons **bonded together** accounts for the missing mass — so the **energy released** is the same as the **mass defect** × c^2.

Practice Questions

Q1 What makes a nucleus unstable? Describe the changes that happen in the nucleus during alpha, beta and gamma decay.

Q2 Explain the circumstances in which gamma radiation may be emitted.

Q3 Define the mass defect.

Exam Questions

Q1 (a) Radium-226 undergoes alpha decay to radon. Complete the balanced nuclear equation for this reaction.

$^{226}_{88}Ra \rightarrow \quad Rn +$ [3 marks]

(b) Potassium-40 ($Z = 19$, $A = 40$) undergoes beta decay to calcium.
Write a balanced nuclear equation for this reaction. [3 marks]

Q2 Calculate the energy released during the formation of an alpha particle, given that the total mass of two protons and two neutrons is 6.695×10^{-27} kg, the mass of an alpha particle is 6.645×10^{-27} kg and the speed of light, c, is 3.00×10^{8} ms^{-1}. [3 marks]

Nuclear decay — it can be enough to make you unstable...

$E = mc^2$ is an important equation that says mass and energy are equivalent. Remember it well, 'cos you're going to come across it a lot in questions about mass defect and the energy released in nuclear reactions over the next few pages...

Binding Energy

Turn off the radio and close the door, 'cos you're going to need to concentrate hard on this stuff about binding energy...

*The **Mass Defect** is **Equivalent** to the **Binding Energy***

1) The **mass** of a **nucleus** is **less than** the mass of its **constituent parts** — the difference is called the **mass defect** (see p. 131).
2) Einstein's equation, $\boldsymbol{E = mc^2}$, says that mass and energy are **equivalent**.
3) So, as nucleons join together, the total mass **decreases** — this '**lost**' mass is **converted** into energy and **released**.
4) The amount of **energy released** is **equivalent** to the **mass defect**.
5) If you **pulled** the nucleus completely **apart**, the **energy** you'd have to use to do it would be the **same** as the energy **released** when the nucleus formed.

> The energy needed to **separate** all of the nucleons in a nucleus is called the **binding energy** (measured in **MeV**), and it is **equivalent** to the **mass defect**.

Atomic mass is usually given in atomic mass units (u), where $1\text{ u} = 1.66 \times 10^{-27}$ kg.

Example Calculate the binding energy of the nucleus of a lithium atom, ${}^{6}_{3}\text{Li}$, given that its mass defect is 0.0343 u.

1) Convert the mass defect into kg.

   Mass defect $= 0.0343 \times 1.66 \times 10^{-27} = 5.70 \times 10^{-29}$ kg

2) Use $\boldsymbol{E = mc^2}$ to calculate the binding energy.

   $E = 5.70 \times 10^{-29} \times (3 \times 10^8)^2 = 5.13 \times 10^{-12}\text{ J} = 32\text{ MeV}$

$1\text{ MeV} = 1.6 \times 10^{-13}$ J

6) The **binding energy per unit of mass defect** can be calculated (using the example above):

$$\frac{\text{binding energy}}{\text{mass defect}} = \frac{32\text{ MeV}}{0.0343\text{ u}} \approx 931.3\text{ MeV u}^{-1}$$

7) This means that a mass defect of **1 u** is equivalent to about **931.3 MeV** of binding energy.

Captain Skip didn't believe in ghosts, marmalade and that things could be bound without rope.

*The **Binding Energy Per Nucleon** is at a **Maximum** around **N = 50***

A useful way of **comparing** the binding energies of different nuclei is to look at the **binding energy per nucleon**.

$$\text{Binding energy per nucleon (in MeV)} = \frac{\text{Binding energy (B)}}{\text{Nucleon number (A)}}$$

So, the binding energy per nucleon for ${}^{6}_{3}\text{Li}$ (in the example above) is $32 \div 6 = 5.3$ MeV.

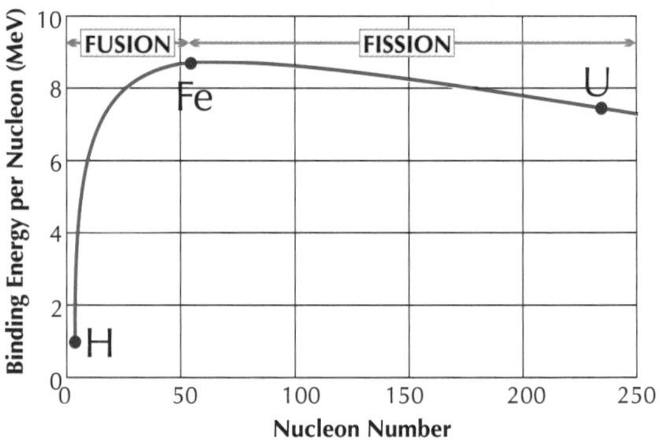

1) A **graph** of **binding energy per nucleon** against **nucleon number**, for all elements, shows a **curve**.
2) **High** binding energy per nucleon means that **more energy** is needed to **remove** nucleons from the nucleus.
3) In other words the **most stable** nuclei occur around the **maximum point** on the graph — which is at **nucleon number 56** (i.e. **iron**, Fe).
4) **Combining small nuclei** is called nuclear **fusion** (see p. 135) — this **increases** the **binding energy per nucleon** dramatically, which means a lot of **energy is released** during nuclear fusion.
5) **Fission** is when **large nuclei** are **split in two** (see p. 134) — the **nucleon numbers** of the two **new nuclei** are **smaller** than the original nucleus, which means there is an **increase** in the binding energy per nucleon. So, energy is also **released** during nuclear fission (but not as much energy per nucleon as in nuclear fusion).

Binding Energy

*The **Change** in **Binding Energy** Gives the **Energy Released...***

The **binding energy per nucleon graph** can be used to **estimate** the **energy released** from nuclear reactions.

Energy released in nuclear fusion

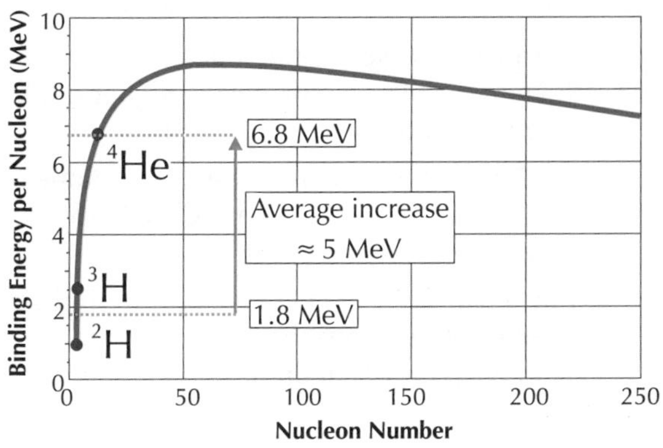

1) If ^{2}**H** and ^{3}**H** nuclei were **fused** together to form ^{4}He (and a neutron), the **average increase** in binding energy per ^{4}He nucleon would be about **5 MeV**.
2) There are **4 nucleons** in ^{4}He, so we can **estimate** the **energy released** as $4 \times 5 =$ **20 MeV**.

Energy released in nuclear fission

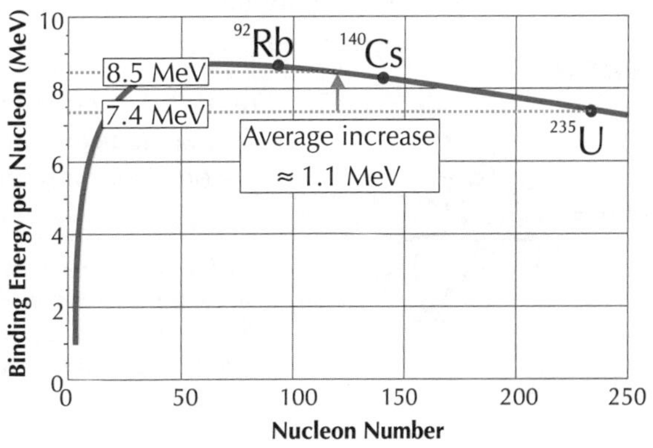

1) If a ^{235}**U** nucleus **splits** into ^{92}**Rb** and ^{140}**Cs** (plus a few neutrons) during nuclear **fission**, the **average increase** in **binding energy per nucleon** would be about 1.1 MeV.
2) There are **235 nucleons** in ^{235}U to begin with, so we can **estimate** the energy **released** as $235 \times 1.1 \approx$ **260 MeV**.

Practice Questions

Q1 What is the binding energy of a nucleus?

Q2 How can we calculate the binding energy for a particular nucleus?

Q3 What is the binding energy per nucleon?

Q4 Which element has the highest value of binding energy per nucleon?

Q5 Do nuclear fusion or fission reactions release more energy per nucleon?

Exam Questions

Q1 The mass of a $^{14}_{6}$**C** nucleus is 13.999948 u. The mass of a proton is 1.007276 u, and a neutron is 1.008665 u.

(a) Calculate the mass defect of a $^{14}_{6}$**C** nucleus (given that $1\text{ u} = 1.66 \times 10^{-27}$ kg). [3 marks]

(b) Use $E = mc^2$ to calculate the binding energy of the nucleus in MeV (given that $c = 3 \times 10^8$ ms^{-1} and $1\text{ MeV} = 1.6 \times 10^{-13}$ J). [2 marks]

Q2 The following equation represents a nuclear reaction that takes place in the Sun:

$$^1_1p + ^1_1p \rightarrow ^2_1H + ^{0}_{+1}\beta + \text{energy released}$$

where p is a proton and β is a positron (opposite of an electron)

(a) State the type of nuclear reaction shown. [1 mark]

(b) Given that the binding energy per nucleon for a proton is 0 MeV and for a ^{2}H nucleus it is approximately 0.86 MeV, estimate the energy released by this reaction. [2 marks]

A mass defect of 1 u is equivalent to a binding energy of 931.3 MeV...

Remember this useful little fact, and it'll save loads of time in the exam — because you won't have to fiddle around with converting atomic mass from u → kg and binding energy from J → MeV. What more could you possibly want...

Nuclear Fission and Fusion

What did the nuclear scientist have for his tea? Fission chips... hohoho.

Fission Means Splitting Up into Smaller Parts

1) **Large nuclei**, with at least 83 protons (e.g. uranium), are **unstable** and some can randomly **split** into two **smaller** nuclei — this is called **nuclear fission**.
2) This process is called **spontaneous** if it just happens **by itself**, or **induced** if we **encourage** it to happen.

Example

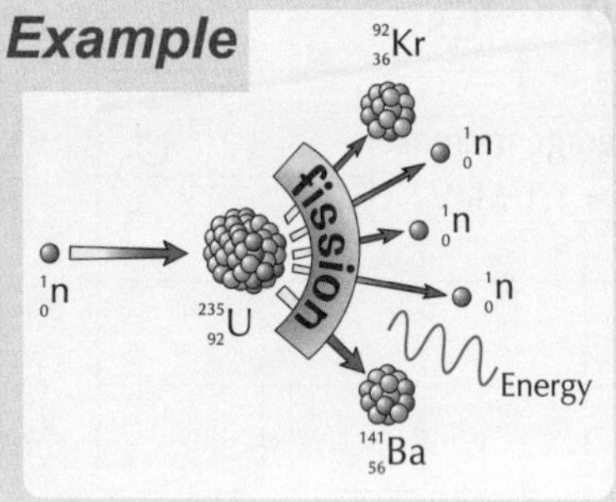

Fission can be induced by making a neutron enter a ^{235}U nucleus, causing it to become very unstable.

Only low energy neutrons can be captured in this way. A low energy neutron is called a **thermal neutron**.

3) **Energy is released** during nuclear fission because the new, smaller nuclei have a **higher binding energy per nucleon** (see p. 132).
4) The **larger** the nucleus, the more **unstable** it will be — so large nuclei are **more likely** to **spontaneously fission**.
5) This means that spontaneous fission **limits** the **number of nucleons** that a nucleus can contain — in other words, it **limits** the number of **possible elements**.

Controlled Nuclear Reactors Produce Useful Power

We can **harness** the **energy** released during nuclear **fission reactions** in a **nuclear reactor**, but it's important that these reactions are very **carefully controlled**.

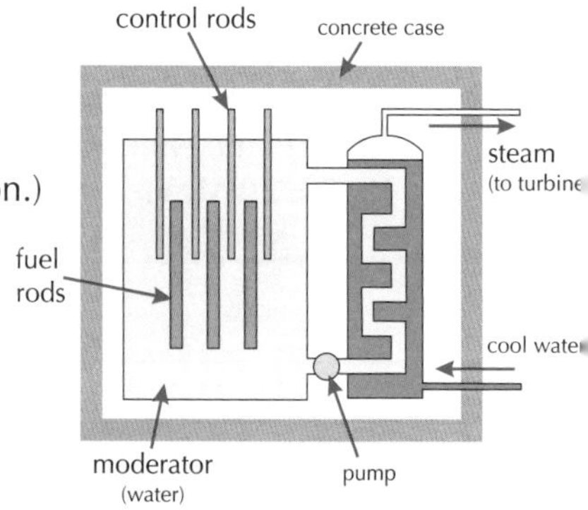

1) Nuclear reactors use **rods of uranium** that are rich in ^{235}**U** as '**fuel**' for fission reactions. (The rods also contain a lot of ^{238}**U**, but that doesn't undergo fission.)
2) These **fission** reactions produce more **neutrons** which then **induce** other nuclei to fission — this is called a **chain reaction**.
3) The **neutrons** will only cause a chain reaction if they are **slowed down**, which allows them to be **captured** by the uranium nuclei — these slowed down neutrons are called **thermal neutrons**.
4) ^{235}U **fuel rods** need to be placed in a **moderator** (for example, **water**) to **slow down** and/or absorb **neutrons**. You need to choose a moderator that will slow down some neutrons enough so they can cause **further fission**, keeping the reaction going at a steady rate. Choosing a moderator that absorbs **more neutrons the higher the temperature** will **decrease** the chance of **meltdown** if the reactor **overheats** — as it will naturally **slow down** the reaction.
5) You want the chain reaction to continue on its own at a **steady rate**, where **one** fission follows another. The amount of 'fuel' you need to do this is called the **critical mass** — any less than the critical mass (**sub-critical mass**) and the reaction will just peter out. Nuclear reactors use a **supercritical** mass of fuel (where several new fissions normally follow each fission) and **control the rate of fission** using **control rods**.
6) Control rods control the **chain reaction** by **limiting** the number of **neutrons** in the reactor. These **absorb neutrons** so that the **rate of fission** is controlled. **Control rods** are made up of a material that **absorbs neutrons** (e.g. boron), and they can be inserted by varying amounts to control the reaction rate.
In an **emergency**, the reactor will be **shut down** automatically by the **release of the control rods** into the reactor, which will stop the reaction as quickly as possible.
7) **Coolant** is sent around the reactor to **remove heat** produced in the fission — often the coolant is the **same water** that is being used in the reactor as a **moderator**. The **heat** from the reactor can then be used to make **steam** for powering **electricity-generating turbines**.

If the chain reaction in a nuclear reactor is **left to continue unchecked**, large amounts of **energy** are **released** in a very **short time**.
Many new fissions will follow each fission, causing a **runaway reaction** which could lead to an **explosion**.
This is what happens in a **fission (atomic) bomb**.

8) The nuclear reactor is surrounded by a thick **concrete case**, which acts as **shielding**. This prevents **radiation escaping** and reaching the people working in the power station.

Nuclear Fission and Fusion

Waste Products of Fission Must be Disposed of Carefully

1) The **waste products** of **nuclear fission** usually have a **larger proportion of neutrons** than nuclei of a similar atomic number — this makes them **unstable** and **radioactive**.
2) The products can be used for **practical applications** such as **tracers** in medical diagnosis (see p129).
3) However, they may be **highly radioactive** and so their **handling** and **disposal** needs **great care**.
4) When material is removed from the reactor, it is initially **very hot**, so is placed in **cooling ponds** until the **temperature falls** to a safe level.
5) The radioactive waste is then **stored** underground in **sealed containers** until its **activity has fallen** sufficiently.

Fusion Means Joining Nuclei Together

1) **Two light nuclei** can **combine** to create a larger nucleus — this is called **nuclear fusion**.
2) A lot of **energy** is released during nuclear fusion because the new, heavier nuclei have a **much higher binding energy per nucleon** (see p. 132).

Example

In the Sun, **hydrogen nuclei** fuse in a series of reactions to form **helium**.

$$^{2}_{1}H + ^{1}_{1}H \rightarrow ^{3}_{2}He + \text{energy}$$

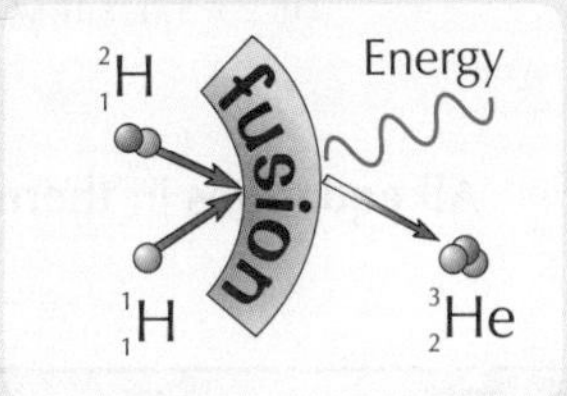

Nuclei Need Lots of Energy to Fuse

1) All nuclei are **positively charged** — so there will be an **electrostatic** (or Coulomb) **force** of **repulsion** between them.
2) Nuclei can only **fuse** if they **overcome** this electrostatic force and get **close** enough for the attractive force of the **strong interaction** to hold them both together.
3) About **1 MeV** of kinetic energy is **needed** to make nuclei fuse together — and that's **a lot of energy**.

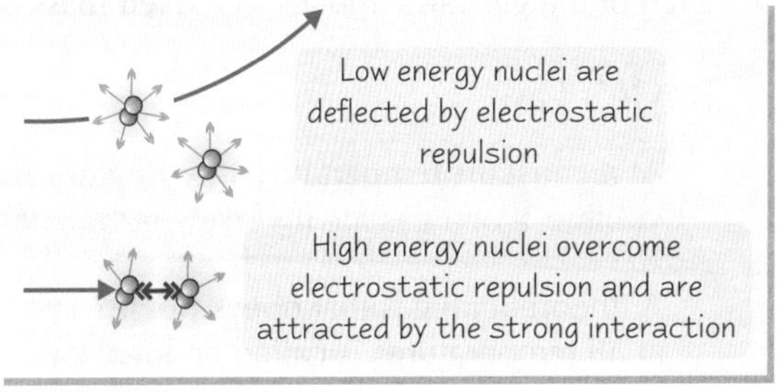

Practice Questions

Q1 What is spontaneous fission?

Q2 How can fission be induced in ^{235}U?

Q3 Why must the waste products of nuclear fission be disposed of very carefully?

Q4 Why is a lot of energy required for nuclear fusion to occur?

Exam Questions

Q1 Nuclear reactors use carefully controlled chain reactions to produce energy.

(a) Explain what is meant by the expression 'chain reaction' in terms of nuclear fission. [2 marks]

(b) Describe and explain one feature of a nuclear reactor whose role is to control the rate of fission. Include an example of a suitable material for the feature you have chosen. [3 marks]

(c) Explain what happens in a nuclear reactor during an emergency shut-down. [2 marks]

Q2 Discuss two advantages and two disadvantages of using nuclear fission to produce electricity. [4 marks]

If anyone asks, I've gone fission... that joke never gets old...

So, controlled nuclear fission reactions can provide a shedload of energy to generate electricity. There are pros and cons to using fission reactors... But then, you already knew that — now you need to learn all the grisly details.

Ideal Gases

*Aaahh... great... another one of those 'our equation doesn't work properly with **real gases**, so we'll invent an **ideal** gas that it **does work** for and they'll think we're dead clever' situations. Hmm. Physicists, eh...*

*There's an **Absolute Scale** of **Temperature***

There is a **lowest possible temperature** called **absolute zero***. Absolute zero is given a value of **zero kelvin**, written **0 K**, on the absolute temperature scale.

At **0 K** all particles have the **minimum** possible **kinetic energy** — everything pretty much stops — at higher temperatures, particles have more energy. In fact, with the **Kelvin scale**, a particle's **energy** is **proportional** to its **temperature** (see page 140).

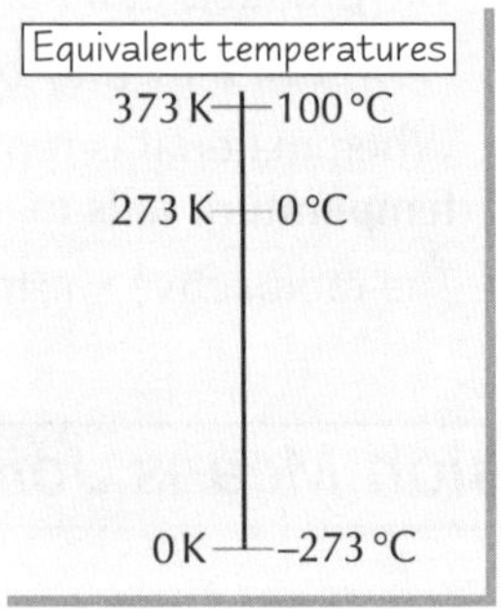

1) The Kelvin scale is named after Lord Kelvin who first suggested it.
2) A change of **1 K** equals a change of **1 °C**.
3) To change from degrees Celsius into kelvin you **add 273** (or 273.15 if you need to be really precise).

$$K = C + 273$$

All equations in **thermal physics** use temperatures measured in kelvin.

**It's true. –273.15 °C is the lowest temperature theoretically possible. Weird, huh. You'd kinda think there wouldn't be a minimum, but there is.*

*There are **Three Gas Laws***

The three gas laws were each worked out **independently** by **careful experiment**. Each of the gas laws applies to a **fixed mass** of gas.

$p \propto \frac{1}{V}$

Boyle's Law

At a **constant temperature** the **pressure *p*** and **volume *V*** of a gas are **inversely proportional**.

A (theoretical) gas that obeys Boyle's law at all temperatures is called an **ideal gas**.

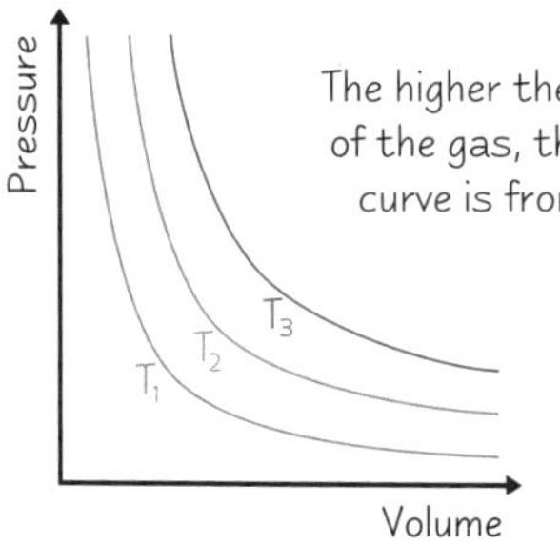

The higher the temperature of the gas, the further the curve is from the origin.

$$pV = \text{constant}$$

Charles' Law

At constant **pressure**, the **volume *V*** of a gas is **directly proportional** to its **absolute temperature *T***.

Ideal gases obey this law and the pressure law as well.

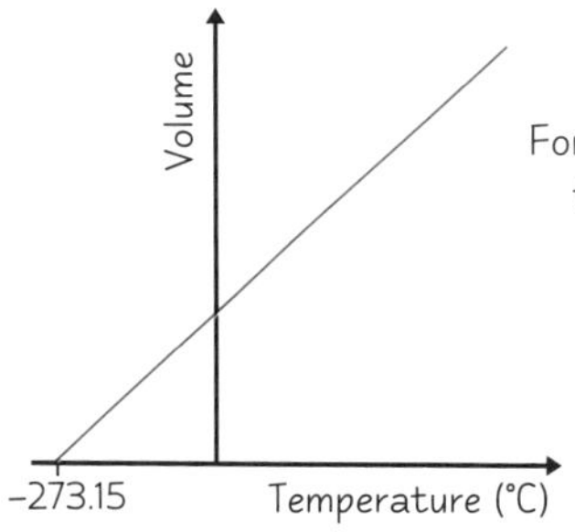

For any ideal gas, the line meets the temperature axis at –273.15 °C — that is, absolute zero.

$$V/T = \text{constant}$$

'Ello, 'ello...

If you'd plotted these graphs in kelvin, they'd both have gone through the origin.

The Pressure Law

At constant **volume**, the **pressure *p*** of a gas is **directly proportional** to its **absolute temperature *T***.

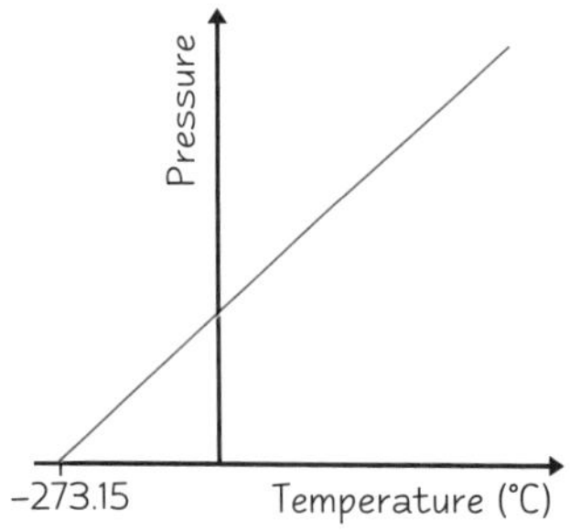

$$p/T = \text{constant}$$

Ideal Gases

If you Combine All Three you get the Ideal Gas Equation

Combining all three gas laws gives the equation: $\frac{pV}{T} = \textbf{constant}$

1) The constant in the equation depends on the amount of gas used. The amount of **gas** can be **measured** in **moles**, ***n***. ← (Pretty obvious... if you have more gas it takes up more space.)
2) The constant then becomes ***nR***, where ***R*** is called the **molar gas constant**. Its value is 8.31 J mol^{-1} K^{-1}.
3) Plugging this into the equation gives: $\frac{pV}{T} = nR$ or rearranging, $pV = nRT$ — ***the ideal gas equation***

This equation works well (i.e., a real gas approximates to an ideal gas) for gases at **low pressure** and fairly **high temperatures**.

Boltzmann's Constant k is like a Gas Constant for One Particle of Gas

One mole of any **gas** contains the same number of particles.
This number is called **Avogadro's constant** and has the symbol N_A. The value of N_A is **6.02 × 10^{23} particles per mole**.

1) The **number of particles** in a **mass of gas** is given by the **number of moles**, ***n***, multiplied by **Avogadro's constant**. So the number of particles, $N = nN_A$.
2) **Boltzmann's constant**, ***k***, is equivalent to R/N_A — you can think of Boltzmann's constant as the **gas constant** for **one particle of gas**, while ***R*** is the gas constant for **one mole of gas**.
3) The value of Boltzmann's constant is **1.38 × 10^{-23} JK^{-1}**.
4) If you combine $N = nN_A$ and $k = R/N_A$ you'll see that $Nk = nR$ — which can be substituted into the ideal gas equation: → $pV = NkT$ — ***the equation of state***

   The equation $pV = NkT$ is called the equation of state of an ideal gas.

Practice Questions

Q1 State Boyle's law, Charles' law and the pressure law.

Q2 What is the ideal gas equation?

Q3 The pressure of a gas is 100 000 Pa and its temperature is 27 °C. The gas is heated — its volume stays fixed but the pressure rises to 150 000 Pa. Show that its new temperature is 177 °C.

Q4 What is the equation of state of an ideal gas?

Exam Questions

Q1 The mass of one mole of nitrogen gas is 0.028 kg. R = 8.31 J mol^{-1} K^{-1}.

(a) A flask contains 0.014 kg of nitrogen gas. State the number of:

i) moles of nitrogen gas in the flask. [1 mark]

ii) nitrogen molecules in the flask. [1 mark]

(b) The flask has a volume of 0.01 m^3 and is at a temperature of 27 °C. Calculate the pressure inside it. [2 marks]

(c) Explain what would happen to the pressure if the number of molecules of nitrogen in the flask was halved. [2 marks]

Q2 A large helium balloon has a volume of 10 m^3 at ground level.
The temperature of the gas in the balloon is 293 K and the pressure is 1 × 10^5 Pa.
The balloon is released and rises to a height where its volume becomes 25 m^3 and its temperature is 260 K.
Calculate the pressure inside the balloon at its new height. [3 marks]

Ideal revision equation — marks = (pages read × questions answered)^2...

All this might sound a bit theoretical, but most gases you'll meet in the everyday world come fairly close to being 'ideal'. They only stop obeying these laws when the pressure's too high or they're getting close to their boiling point.

The Pressure of an Ideal Gas

Kinetic theory *tries to* ***explain*** *the* ***gas laws***. *It basically models a gas as a series of hard balls that obey Newton's laws.*

You Need to be Able to **Derive** the **Pressure** of an **Ideal Gas**

Start by **Deriving** the **Pressure** on **One Wall** of a Box — in the x direction

Imagine a cubic box with sides of length l containing N particles each of mass m.

This isn't an easy page. Work through it properly and make sure you understand it.

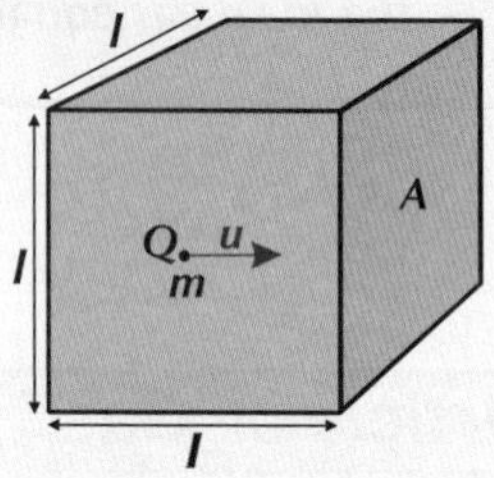

1) Say particle **Q** moves directly towards **wall A** with velocity u. Its **momentum** approaching the wall is mu. It strikes wall **A**. Assuming the **collisions** are perfectly **elastic**, it rebounds and heads back in the opposite direction with momentum $-mu$. So the **change in momentum** is $mu - (-mu) = 2mu$.

2) Assuming **Q** suffers no collisions with other particles, the **time between collisions** of **Q** and wall **A** is $2l/u$. The number of **collisions per second** is therefore $u/2l$.

3) This gives the **rate of change of momentum** as $2mu \times u/2l$.

4) Force equals the rate of change of momentum (Newton's second law), so the **force exerted on the wall** by this one particle = $2mu^2/2l = mu^2/l$.

5) Particle **Q** is only one of many in the cube. Each particle will have a different velocity u_1, u_2 etc. towards **A**. The total force, **F**, of all these particles on wall **A** is:

$$F = \frac{m(u_1^2 + u_2^2 + etc.)}{l}$$

6) You can define a quantity called the **mean square speed**, $\overline{u^2}$ as:

$$\overline{u^2} = \frac{u_1^2 + u_2^2 + etc.}{N}$$

7) If you put that into the equation above, you get:

$$F = \frac{Nm\overline{u^2}}{l}$$

8) So, the pressure of the gas on end **A** is given by:
where V = volume of the cube

$$pressure, p = \frac{force}{area} = \frac{Nm\overline{u^2}/l}{l^2} = \frac{Nm\overline{u^2}}{V}$$

...Then for the **General Equation** you need to think about **All 3 Directions** — *x*, *y* and *z*

A gas particle can move in **three dimensions** (i.e. the x, y and z directions).

1) You can calculate its **speed**, c, from Pythagoras' theorem:

$c^2 = u^2 + v^2 + w^2$ where u, v and w are the components of the particle's velocity in the x, y and z directions.

2) If you treat all N particles in the same way, this gives an **overall** mean square speed of: $\overline{c^2} = \overline{u^2} + \overline{v^2} + \overline{w^2}$

3) Since the particles move **randomly**: $\overline{u^2} = \overline{v^2} = \overline{w^2}$ and so $\overline{c^2} = 3\overline{u^2}$

4) You can substitute this into the equation for pressure that you derived above to give:

$$pV = \frac{1}{3}Nm\overline{c^2}$$

The Pressure of an Ideal Gas

A Useful Quantity is the **Root Mean Square Speed** $\sqrt{\overline{c^2}}$

As you saw on the previous page, it often helps to think about the motion of a **typical particle** in kinetic theory.

1) $\overline{c^2}$ is the **mean square speed** and has **units m^2s^{-2}**.
2) $\overline{c^2}$ is the **square** of the **speed** of an **average particle**, so the square root of it gives you the typical speed.
3) This is called the **root mean square speed** or, usually, the **r.m.s. speed**.
   The **unit** is the same as any speed — **ms^{-1}**.

$$\text{r.m.s. speed} = \sqrt{\text{mean square speed}} = \sqrt{\overline{c^2}}$$

Lots of **Simplifying Assumptions** are Used in **Kinetic Theory**

In **kinetic theory**, physicists picture gas particles moving at **high speed** in **random directions**.
To get **equations** like the one you just derived though, some **simplifying assumptions** are needed:

1) The gas contains a **large number of particles**.
2) The particles **move rapidly** and **randomly**.
3) The motion of the particles follows **Newton's laws**.
4) **Collisions** between particles themselves or at the walls of a container are **perfectly elastic**.
5) There are **no attractive forces** between particles.
6) Any **forces** that act during collisions are **instantaneous**.
7) Particles have a **negligible volume** compared with the volume of the container.

A **gas obeying** these **assumptions** is called an **ideal** gas. Real gases behave like ideal gases as long as the **pressure isn't too big** and the **temperature** is **reasonably high** (compared with their boiling point), so they're useful assumptions.

Practice Questions

Q1 What is the change in momentum when a gas particle hits a wall of its container head-on?

Q2 What is the force exerted on the wall by this one particle? What is the total force exerted on the wall?

Q3 What is the pressure exerted on this wall? What is the total pressure on the container?

Q4 What is 'root mean square speed'? How would you find it?

Q5 What are the seven assumptions made about ideal gas behaviour?

Exam Question

Q1 Some helium gas is contained in a flask of volume 7×10^{-5} m^3. Each helium atom has a mass of 6.6×10^{-27} kg, and there are 2×10^{22} atoms present. The pressure of the gas is 1×10^5 Pa.

(a) Calculate the mean square speed of the atoms. [2 marks]

(b) State the r.m.s. speed of a typical atom. [1 mark]

(c) If the absolute temperature of the gas is doubled, what will the r.m.s. speed of an atom become? [2 marks]

Help — these pages are de-riving me crazy...

Make sure you know the derivation inside out and back to front — it's not easy, so you might want to go through it a few times, but it is worth it. Oh, and remember — mean square speed is the average of the squared speeds — i.e. square all the speeds, then find the average. Don't make the mistake of finding the average speed first and then squaring. No, no, no...

Energy and Temperature

The energy of a particle depends on its temperature on the ***thermodynamic scale*** *(that's Kelvin to you and me).*

The ***Speed Distribution*** *of* ***Gas Particles*** *Depends on* ***Temperature***

The **particles** in a **gas don't** all **travel** at the **same speed**.
Some particles will be moving fast but others much more slowly.
Most will travel around the average speed. The shape of the **speed distribution** depends on the **temperature** of the gas.

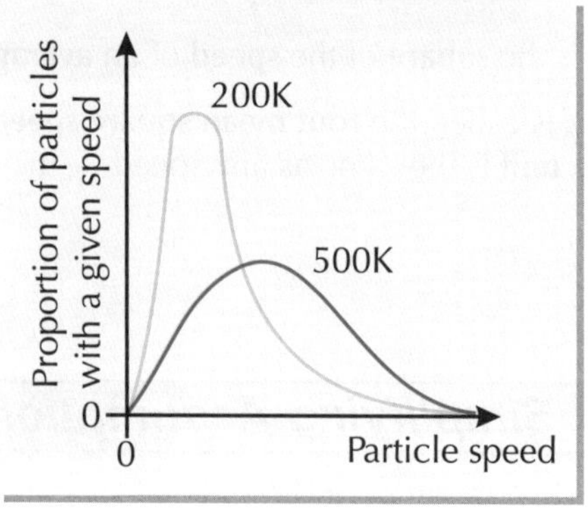

As the temperature of the gas increases:

1) the **average** particle speed increases.
2) the **maximum** particle speed increases.
3) the distribution curve becomes more **spread out.**

Energy Changes *Happen Between Particles*

The particles of a gas **collide** with each other **all the time**. Some of these collisions will be '**head-on**' (particles moving in **opposite directions**) while others will be '**shunts from behind**' (particles moving in the **same direction**).

1) As a result of the collisions, **energy** will be **transferred** between particles.
2) Some particles will **gain speed** in a collision and others will **slow down**.
3) **Between collisions**, the particles will travel at **constant speed**.
4) The energy of individual particles changes at each collision, but the **total energy** of the **system** doesn't change.
5) So, the **average** speed of the particles will stay the same provided the **temperature** of the gas **stays the same**.

Average Kinetic Energy *is* ***Proportional*** *to* ***Absolute Temperature***

There are **two equations** for the **product *pV*** of a gas — the ideal gas equation (page 137), and the equation involving the mean square speed of the particles (page 138). You can **equate these** to get an expression for the **average kinetic energy**.

N is the number of molecules in the gas.

c is the velocity of a particle. $\overline{c^2}$ is the average of the squared speeds of the particles, called the mean square speed.

m is the mass of one molecule in the gas.

1) The **ideal gas equation**: $pV = nRT$
2) The **pressure** of an **ideal gas** given by kinetic theory: $pV = \frac{1}{3}Nm\overline{c^2}$
3) **Equating** these two gives: $\frac{1}{3}Nm\overline{c^2} = nRT$
4) **Multiplying** by **3/2** gives: $\frac{3}{2} \times \frac{1}{3}Nm\overline{c^2} = \frac{3nRT}{2}$, so: $$\frac{1}{2}m\overline{c^2} = \frac{3}{2}\frac{nRT}{N}$$
5) $\frac{1}{2}m\overline{c^2}$ is the **average kinetic energy** of a **particle**.
6) You can substitute ***Nk*** for ***nR***, where ***k*** is the **Boltzmann constant** (see page 137) to show that the **average kinetic energy** of a particle is **directly proportional** to ***T*** (**absolute temperature**). $$\frac{1}{2}m\overline{c^2} = \frac{3}{2}kT$$
You can use ***kT*** as an **approximation** for the **average kinetic energy** of the molecules in **any substance**.
7) Finally, the Boltzmann constant is equivalent to R/N_A (see page 137), so you can substitute this for ***k*** in the equation above, to get: $$\frac{1}{2}m\overline{c^2} = \frac{3RT}{2N_A}$$

Energy and Temperature

Specific Heat Capacity is how much Energy it Takes to Heat Something

When you heat something, its particles get more **kinetic energy** and its **temperature** rises.

The **specific heat capacity** (*c*) of a substance is the amount of **energy** needed to **raise** the **temperature** of **1 kg** of the substance by **1 K** (or 1°C).

or put another way: **energy change = mass × specific heat capacity × change in temperature**

in symbols: $\Delta Q = mc\Delta\theta$ ← ΔT is sometimes used instead of $\Delta\theta$ for the change in temperature.

ΔQ is the energy change in J, m is the mass in kg and $\Delta\theta$ is the temperature change in K or °C. Units of **c** are J kg^{-1} K^{-1} or J kg^{-1} $°C^{-1}$.

Specific Latent Heat is the Energy Needed to Change State

To **melt** a **solid** or **boil or evaporate a liquid**, you need **energy** to **break the bonds** that hold the particles in place. The **energy** needed for this is called **latent heat**. The **larger** the **mass** of the substance, the **more energy** it takes to **change** its **state**. That's why the **specific latent heat** is defined per kg:

The **specific latent heat** (*l*) of **fusion** or **vaporisation** is the quantity of **thermal energy** required to **change the state** of **1 kg** of a substance.

which gives: **energy change = specific latent heat × mass of substance changed**

or in symbols: $\Delta Q = ml$ ← You'll usually see the latent heat of vaporisation written l_v and the latent heat of fusion written l_f.

Where ΔQ is the energy change in J and m is the mass in kg. The units of l are J kg^{-1}.

Practice Questions

Q1 Describe the changes in the distribution of gas particle speeds as the temperature of a gas increases.

Q2 What happens to the average kinetic energy of a particle if the temperature of a gas doubles?

Q3 Define specific heat capacity and specific latent heat.

Q4 Show that the thermal energy needed to heat 2 kg of water from 20 °C to 50 °C is ~250 kJ (c_{water} = 4180 J$kg^{-1}K^{-1}$).

Exam Questions

Q1 The mass of one mole of nitrogen molecules is 2.8×10^{-2} kg. There are 6.02×10^{23} molecules in one mole.

(a) Calculate the typical speed of a nitrogen molecule at 300 K. [4 marks]

(b) Explain why all the nitrogen molecules will not be moving at this speed. [2 marks]

Q2 Some air freshener is sprayed at one end of a room. The room is 8.0 m long and the temperature is 20 °C.

(a) Assuming the average freshener molecule moves at 400 ms^{-1}, how long will it take for a particle to travel directly to the other end of the room? [1 mark]

(b) The perfume from the air freshener only slowly diffuses from one end of the room to the other. Explain why this takes much longer than suggested by your answer to part (a). [2 marks]

Q3 A 2 kg metal cylinder is heated uniformly from 4.5 °C to 12.7 °C in 3 minutes.
The electric heater used is rated at 12 V, 7.5 A.
Assuming that heat losses were negligible, calculate the specific heat capacity of the metal. [3 marks]

Q4 A 3 kW electric kettle contains 0.5 kg of water already at its boiling point.
Neglecting heat losses, determine how long it will take to boil dry. (l_v (water) = 2.26×10^6 J kg^{-1}) [3 marks]

My specific eat capacity — 24 pies...

*Phew... there's a lot to take in here. Go back over it, step by step, and make sure you're comfortable using those equations. Interesting(ish) fact — it's the **huge** difference in SHC between the land and the sea that causes the monsoon in Asia.*

Optical Telescopes

Some optical telescopes use lenses (no, really), so first, here's a bit of lens theory...

Converging Lenses Bring Light Rays **Together**

1) **Lenses** change the **direction** of light rays by **refraction**.
2) Rays **parallel** to the **principal axis** of the lens converge onto a point called the **principal focus**. Parallel rays that **aren't** parallel to the principal axis converge somewhere else on the **focal plane** (see diagram).
3) The **focal length**, ***f***, is the distance between the **lens axis** and the **principal focus**.

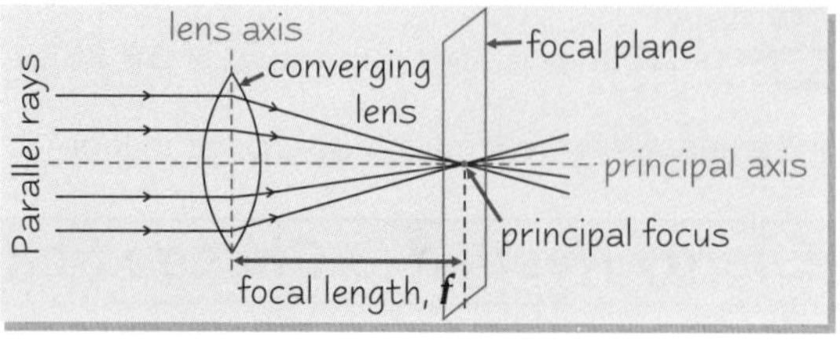

Images can be **Real** or **Virtual**

1) A **real image** is formed when light rays from an object are made to **pass through** another point in space. The light rays are **actually there**, and the image can be **captured** on a **screen**.
2) A **virtual image** is formed when light rays from an object **appear** to have come from another point in space. The light rays **aren't really where the image appears to be**, so the image **can't** be captured on a screen.
3) Converging lenses can form both **real** and **virtual** images, depending on where the object is. If the object is **further** than the **focal length** away from the lens, the image is **real**. If the object's **closer**, the image is **virtual**.
4) To work out where an image will appear, you can draw a **ray diagram**. You only need to draw **two rays** on a ray diagram: one **parallel** to the principal axis that passes through the **principal focus**, and one passing through the **centre** of the lens that **doesn't get bent**.

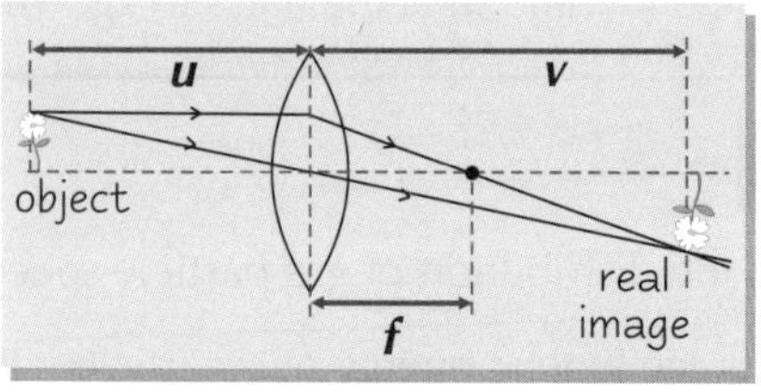

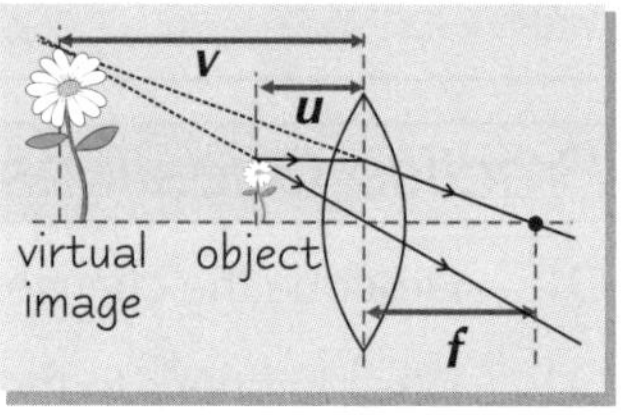

In the diagram, u = distance between object and lens axis, v = distance between image and lens axis (**positive** if image is **real**, **negative** if image is **virtual**), f = focal length.

5) The values u, v and f are related by the **lens equation**:

$$\frac{1}{f} = \frac{1}{u} + \frac{1}{v}$$

A **Refracting Telescope** uses **Two** Converging Lenses

1) The **objective lens** converges the rays from the object to form a **real image**.
2) The **eye lens** acts as a **magnifying glass** on this real image to form a **magnified virtual image**.

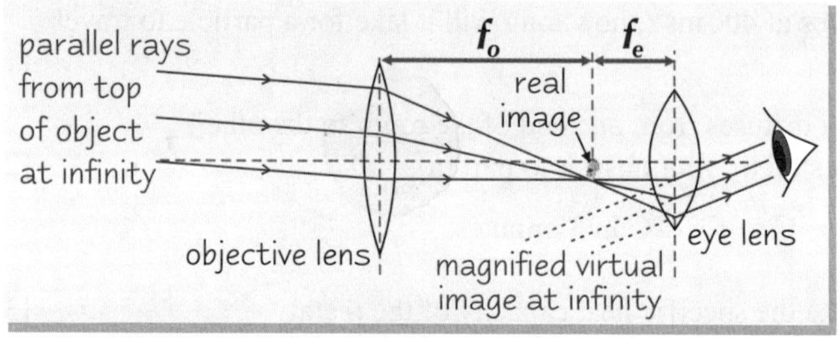

3) If you assume the object is at infinity, then the rays from it are **parallel**, and the real image is formed on the **focal plane**.
4) A **telescope** (in normal adjustment) is set up so that the **principal focus** of the **objective** lens is in the **same position** as the principal focus of the **eye** lens, so the **final magnified image** appears to be at **infinity**.
5) The **magnification**, M, of the telescope can be calculated in terms of angles, or the focal length. The **angular magnification** is the **angle** subtended by the **image** θ_i over the **angle** subtended by the **object** θ_o at the eye:

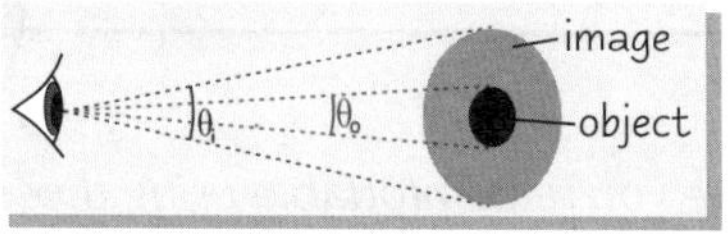

$$M = \frac{\theta_i}{\theta_o}$$

or in terms of **focal length** (with the telescope in normal adjustment as shown above):

$$M = \frac{f_o}{f_e}$$

Optical Telescopes

A **Reflecting Telescope** uses a **Concave Mirror** and a **Converging Lens**

1) A **parabolic concave mirror** (the **primary mirror**) converges parallel rays from an object, forming a **real image**.
2) An **eye lens magnifies** the image as before.
3) The focal point of the mirror (where the image is formed) is **in front** of the mirror, so an arrangement needs to be devised where the observer doesn't **block out** the light. A set-up called the **Cassegrain arrangement**, which uses a **convex secondary mirror**, is a common solution to this problem.

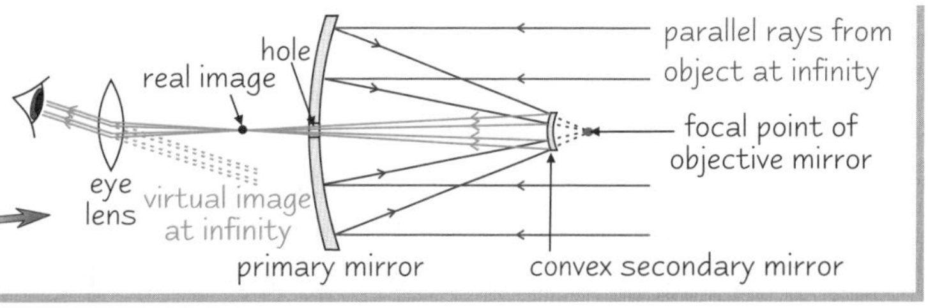

The **Resolving Power** of a Telescope — how much **Detail** you can See

1) The **resolving power** of an instrument is the **smallest angle** at which it can **distinguish** two points.

About half of the stars that we see in the night sky are actually collections of two or more stars. Our eyes see them as a single star since the angle between them is too small to resolve.

2) Resolution is limited by diffraction. If a beam of light passes through a circular **aperture**, then a **diffraction pattern** is formed. The central circle is called the **Airy disc** (see p.186 for an example of the pattern).
3) **Two** light sources can **just** be distinguished if the **centre** of the **Airy disc** from one source is **at least as far away** as the **first minimum** of the other source. This led to the **Rayleigh criterion**:

$$\theta \approx \frac{\lambda}{D}$$

where θ is the **minimum angle** that can be resolved in **radians**, λ is the **wavelength** of the light in **metres** and D is the **diameter** of the **aperture** in **metres**.

4) For **telescopes**, D is the diameter of the **objective lens** or the **objective mirror**. So **very large** lenses or mirrors are needed to see **fine detail**.

There are **Big Problems** with **Refracting Telescopes**

1) Glass refracts **different colours** of light by **different amounts** and so the image for each colour is in a slightly **different position**. This **blurs** the image and is called **chromatic aberration**.

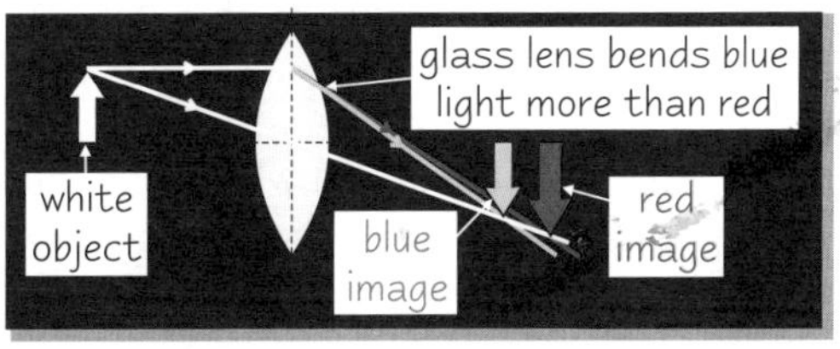

2) Any **bubbles** and **impurities** in the glass **absorb** some of the light, which means that **very faint** objects **aren't seen**. Building large lenses that are of a **sufficiently good quality** is **difficult** and **expensive**.
3) **Large lenses** are very **heavy** and can only be **supported** from their **edges**, so their **shape** can become **distorted**.
4) For a **large magnification**, the **objective lens** needs to have a **very long focal length**. This means that refracting telescopes have to be **very long**, leading to very **large** and **expensive buildings** needed to house them.

Reflecting Telescopes are **Better** than Refractors but they have **Problems** too

1) **Large mirrors** of **good quality** are much **cheaper** to build than large lenses. They can also be **supported** from **underneath** so they don't **distort** as much as lenses.
2) Mirrors don't suffer from **chromatic aberration** (see above) but can have **spherical aberration**:

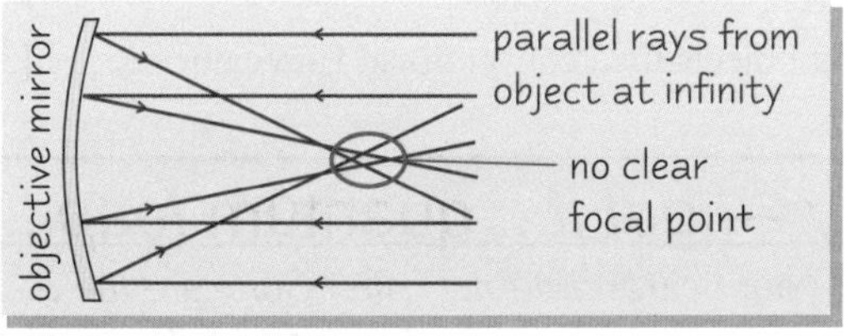

If the **shape** of the mirror isn't quite **parabolic**, parallel rays reflecting off different parts of the mirror do not all **converge** onto the same point.

When the **Hubble Space Telescope** was first launched it suffered from **spherical aberration**. They had to find a way round the problem before it could be used.

Optical Telescopes

Charge-Coupled Devices (CCDs) are Very Sensitive Light Detectors

1) CCDs are **silicon chips** about the size of a postage stamp, divided up into a grid of millions of **identical pixels**.
2) Silicon is a **semiconductor** so it doesn't usually have many **free electrons**. When **light** shines on a pixel, **electrons** are released from the silicon, with the **number** of electrons released being proportional to the **brightness/intensity** of the light.
3) **Underneath** each **pixel** is a **potential well** (a kind of controllable electrical bucket), which traps the electrons.
4) Once a picture has been taken, the electrons are **shunted** from **one potential well** to **another** so that they all come out **in sequence** from **one corner** of the CCD. (This is called an 'electrical bucket brigade', to use the lingo.)
5) This sequence can be converted into a **digital signal** and sent to computers **anywhere in the world**.

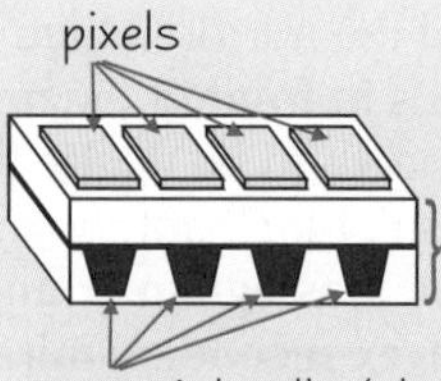

CCDs use Quantum Effects

1) **Quantum physics** tells us that EM radiation is formed in **discrete packets** of energy called **photons**.
2) The incoming photons release electrons in the silicon due to the **photoelectric effect**. (See p. 184.)
3) Electrons are released by **more than 70%** of the photons that hit a pixel, so the **quantum efficiency** of a CCD is greater than 70%. On average, a cell in the **eye** needs about **100 photons** before it responds and so has a quantum efficiency of about **1%**. The quantum efficiency of a **photographic emulsion** is about **4%**.

Practice Questions

Q1 Define the focal length of a converging lens.

Q2 Draw ray diagrams to show how an image is formed in a refracting and a reflecting (Cassegrain) telescope.

Q3 Explain resolving power and state the Rayleigh criterion.

Q4 What does CCD stand for?

Q5 What does quantum efficiency mean and what is the quantum efficiency of a CCD?

Exam Questions

Q1 (a) Define the *principal focus* and the *focal length* of a converging lens. [2 marks]

(b) An object was placed 0.20 m in front of a converging lens with a focal length of 0.15 m. Calculate how far behind the lens the image was formed. [3 marks]

(c) The object was placed 0.10 m in front of the same lens. Determine where the image was formed. [2 marks]

Q2 An objective lens with a focal length of 5.0 m and an eye lens with a focal length of 0.10 m are used in a refracting telescope.

(a) How far apart should the lenses be placed for the telescope to be in normal adjustment? [1 mark]

(b) Define angular magnification and calculate the angular magnification of this telescope. [2 marks]

Q3 (a) Describe the basic function of a CCD. [3 marks]

(b) Explain why CCDs are much better at taking pictures of faint objects than conventional film cameras. [3 marks]

CCDs were a quantum leap for astronomy — get it... quantum leap... *sigh*

With CCDs, you can get all the images you want from the comfort of your nearest internet café. Gone are the days of standing on a hill with a telescope and a flask hoping the sky clears before your nose turns black and falls off. Shame.

Non-Optical Telescopes

Some telescopes don't use visible light — they use radio waves, IR, UV or X-rays instead — read on to learn more...

Radio Telescopes are Similar to Optical Telescopes in Some Ways

1) The most obvious feature of a radio telescope is its **parabolic dish**.
This works in exactly the same way as the **objective mirror** of an **optical reflecting** telescope.
2) An **antenna** is used as a detector at the **focal point** instead of an eye or camera in an optical telescope, but there is **no equivalent** to the **eye lens**.

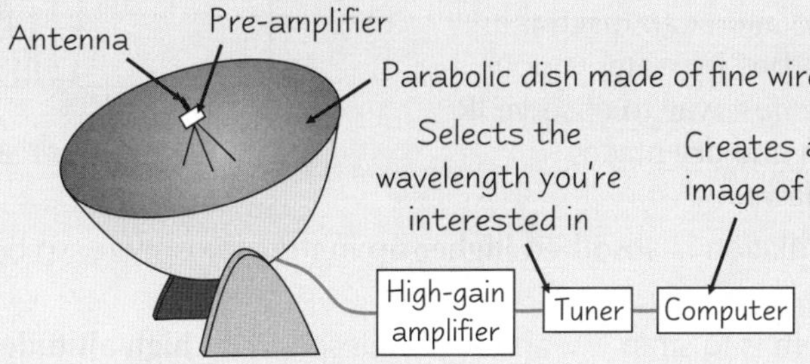

3) Most radio telescopes are **manoeuvrable**, allowing the source of the waves to be **tracked** (in the same way as optical telescopes). The telescope moves with the source, stopping it 'slipping out of view' as the Earth rotates.

Radio Waves have a Much Longer Wavelength than Light...

1) The **wavelengths** of **radio waves** are about a **million times longer** than the wavelengths of **light**.
2) The **Rayleigh criterion** (see p. 143) gives the **resolving power** of a telescope as $\theta \approx \lambda/D$.
3) So for a radio telescope to have the **same resolving power** as an optical telescope, its dish would need to be a **million times bigger** (about the size of the UK for a decent one). The **resolving power** of a radio telescope is **worse** than the **unaided eye**.

Radio astronomers get around this by **linking** lots of telescopes together.
Using some nifty computer programming, their data can be combined to form a **single image**. This is equivalent to one **huge dish** the size of the **separation** of the telescopes.
Resolutions **thousands** of times better than optical telescopes can be achieved this way.

...so Radio Telescopes aren't as Fiddly to Make as Optical Reflectors

1) Instead of a **polished mirror**, a **wire mesh** can be used since the long wavelength radio waves don't notice the gaps. This makes their **construction** much **easier** and **cheaper** than optical reflectors.
2) The **shape** of the dish has to have a **precision** of about **λ/20** to avoid **spherical aberration** (see page 143).
So the dish does not have to be **anywhere near as perfect** as a mirror.
3) Unlike an optical telescope, a radio telescope has to **scan across** the radio source to **build up** the **image**.

Non-Optical Telescopes

The **Atmosphere Blocks** Certain **EM Wavelengths**

1) One of the big problems with doing astronomy on Earth is trying to look through the atmosphere.
2) Our atmosphere only lets **certain wavelengths** of **electromagnetic radiation** through and is **opaque** to all the others. The graph shows how the **transparency** of the atmosphere varies with **wavelength**.

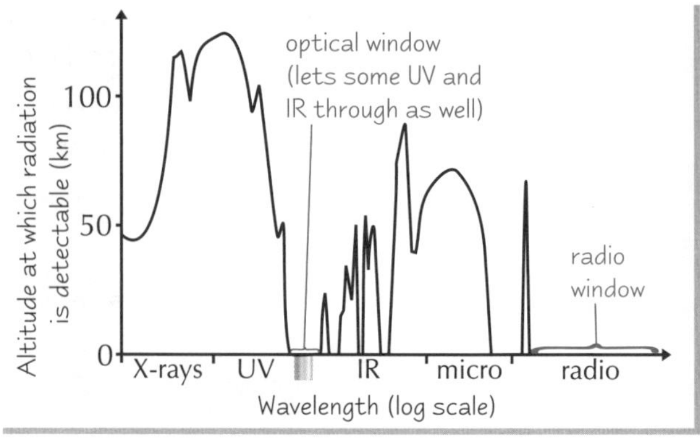

3) We can use **optical** and **radio** telescopes on the surface of the Earth because the atmosphere is **transparent** to these wavelengths. Observing other wavelengths can be a bit more tricky.
4) A few wavelengths of **infrared** radiation can reach the Earth's surface, but most are absorbed by water vapour in the atmosphere. On Earth, the best way to observe IR radiation is to set up shop in **high** and **dry** places, like the Mauna Kea volcano in Hawaii.
5) But most **ultraviolet** and **X-ray** radiation is absorbed **higher up** in the atmosphere, so being on a mountain doesn't help.
6) One way to get round this problem is to strap UV and X-ray telescopes to **high altitude weather balloons** or **aeroplanes**. They can take the telescope high enough into the atmosphere to detect the radiation.
7) The ideal situation is to get your telescope **above the atmosphere** altogether, by launching it into **space** and setting it in orbit around the Earth.

IR and **UV** Telescopes have a **Very Similar Structure** to **Optical** Telescopes

1) Infrared and ultraviolet telescopes are very similar to optical reflecting telescopes. They use the same **parabolic mirror** set-up to focus the radiation onto a detector.
2) In both cases, **CCDs** (see p 144) or **special photographic paper** are used as the radiation detectors, just as in optical telescopes.
3) The **longer** the **wavelength** of the radiation, the **less** it's affected by imperfections in the mirror (see previous page). So the mirrors in **infrared** telescopes **don't** need to be as perfectly shaped as in optical telescopes. But the mirrors in **UV** telescopes have to be even **more** precisely made.

IR telescopes have the added problem that they produce their **own** infrared radiation due to their **temperature**. They need to be **cooled** to very low temperatures using liquid helium, or refrigeration units.

X-ray Telescopes have a **Different Structure** from Other Telescopes

1) X-rays don't reflect off surfaces in the same way as most other EM radiation. Usually X-ray radiation is either **absorbed** by a material or it **passes straight through** it.

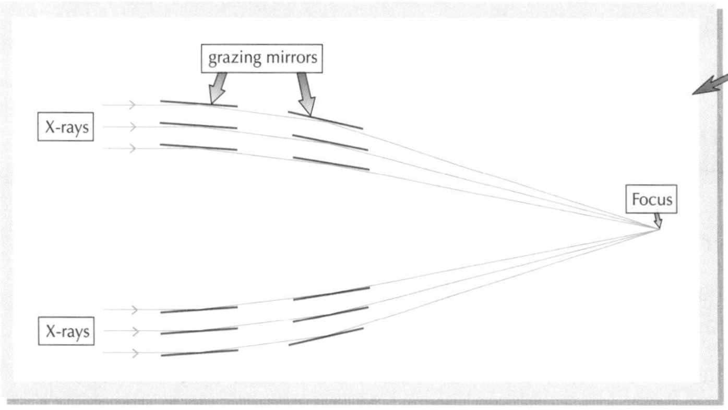

2) X-rays **do** reflect if they just **graze** a mirror's surface though. By having a series of **nested mirrors**, you can gradually alter the direction of X-rays enough to bring them to a **focus** on a detector. This type of telescope is called a **grazing telescope**.
3) The X-rays can be detected using a modified **Geiger counter** or a **fine wire mesh**. Modern X-ray telescopes such as the XMM-Newton telescope use highly sensitive X-ray **CCD** cameras.

Non-Optical Telescopes

Different Telescopes have Different **Resolving** and **Collecting Powers**

The **RESOLVING POWER** of a telescope is limited by two main factors:

1) The **Rayleigh criterion** (see page 143):
This depends on the **wavelength** of the radiation and the **diameter** of the objective mirror or dish.
So, for the **same size** of dish, a UV telescope has a much better resolving power than a radio telescope.
2) The quality of the **detector**:
Just like in digital cameras, the resolving power of a telescope is limited by the resolving power of the detector. That can be how many **pixels** there are on a CCD, or for a wire mesh X-ray detector, how **fine** the wire mesh is.

The **COLLECTING POWER** of a telescope is proportional to its **collecting area**.

1) A **bigger dish** or **mirror** collects **more energy** from an object in a given time.
This gives a **more intense image**, so the telescope can observe **fainter** objects.
2) The **collecting power** (energy collected per second) is proportional to the area:

$$\textbf{Power} \propto \textbf{Diameter}^2$$

3) For a **radio**, **optical**, **UV** or **IR** telescope, this is the area of the objective mirror or dish.
4) For **X-ray** telescopes, it's the size of the **opening** through which X-rays can enter the telescope.
In general, X-ray telescopes have a much **smaller collecting power** than other types of telescope.

The bigger the dish, the greater the collecting power. Mmm....

Practice Questions

Q1 Why do radio telescopes tend to have poor resolving powers?

Q2 Why is it easier to make a parabolic dish for a radio telescope than it is to make a mirror for an optical telescope?

Q3 Why don't astronomers install UV and X-ray telescopes on the top of mountains?

Exam Questions

Q1 Describe and explain the differences in resolving and collecting powers between radio and UV telescopes. [4 marks]

Q2 In 1983, the IRAS satellite observed the entire sky in infrared wavelengths. The satellite was kept at a temperature of 2 K by a reservoir of liquid helium which cooled the satellite by evaporation.

(a) Why did the satellite need to be kept at such a low temperature? [2 marks]

(b) Some infrared telescopes are on the surface of the Earth. What is special about their location? [1 mark]

Q3 (a) Many X-ray and UV telescopes are housed on satellites that orbit high above the Earth's atmosphere. Where else are X-ray and UV telescopes positioned? Explain why this is necessary. [2 marks]

(b) Describe and explain the major differences between the mirrors in X-ray and UV telescopes. [3 marks]

Q4 (a) How is the collecting power of a telescope related to its objective diameter? [1 mark]

(b) The Arecibo radio telescope has a dish diameter of 300 m. The Lovell radio telescope has a dish diameter of 76 m. Calculate the ratio of their collecting powers. [2 marks]

Power is proportional to diameter^2? Bring on the cakes...

If you can't observe the radiation you want to from Earth, just strap your telescope to a rocket and blast it into space. Sounds easy enough till you remember it's going to be reeeally hard to repair if anything goes wrong.

Distances and Magnitude

There are a couple of ways to classify stars — the first is by luminosity, using the magnitude scale.

*The **Luminosity** of a Star is the **Total Energy** Emitted **per Second***

1) Stars can be **classified** according to their luminosity — that is, the **total** amount of energy emitted in the form of electromagnetic radiation **each second** (see p.150).
2) The **Sun's** luminosity is about 4×10^{26} W (luminosity is measured in watts, since it's a sort of power). The **most luminous** stars have a luminosity about a **million** times that of the Sun.
3) The **intensity**, I, of an object that we observe is the power **received** from it per unit area **at Earth**. This is the effective **brightness** of an object.

Apparent Magnitude, m,** is based on how **Bright** things **Appear** from **Earth

1) The **brightness** of a star in the night sky depends on **two** things — its **luminosity** and its **distance from us** (if you ignore weather and light pollution, etc.). So the **brightest** stars will be **close** to us and have a **high luminosity**.
2) The ancient Greeks invented a system where the very **brightest** stars were given an **apparent magnitude** of **1** and the **dimmest** stars an apparent magnitude of **6**, with other levels catering for the stars in between.
3) In the 19th century, the scale was redefined using a strict **logarithmic** scale:

   > A **magnitude 1** star has an **intensity 100 times** greater than a **magnitude 6** star.

   This means a difference of **one magnitude** corresponds to a difference in **intensity** of $\mathbf{100^{1/5}}$ **times**. So a magnitude 1 star is about **2.5 times brighter** than a magnitude 2 star.
4) At the same time, the range was **extended** in **both directions** with the very brightest objects in the sky having **negative apparent magnitude**.

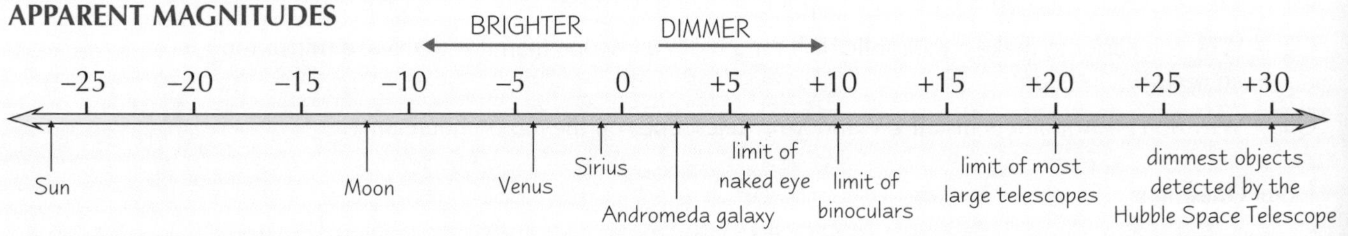

5) The **apparent magnitude**, m, is related to the **intensity**, I, by the following formula:

$$m = -2.5 \log I + \text{constant}$$

Note that log is $\log_{10}$, not the natural logarithm ln. And that 2.5 is exactly 2.5 — not $100^{1/5}$.

*The **Distance** to **Nearby Stars** can be Measured in **Parsecs***

1) Imagine you're in a **moving car**. You see that (stationary) objects in the **foreground** seem to be **moving faster** than objects in the **distance**. This **apparent change in position** is called **parallax**.
2) Parallax is measured in terms of the **angle of parallax**. The **greater** the **angle**, the **nearer** the object is to you.
3) The distance to **nearby stars** can be calculated by observing how they **move relative** to **very distant stars** when the Earth is in **different parts** of its **orbit**. This gives a **unit** of distance called a **parsec** (**pc**).

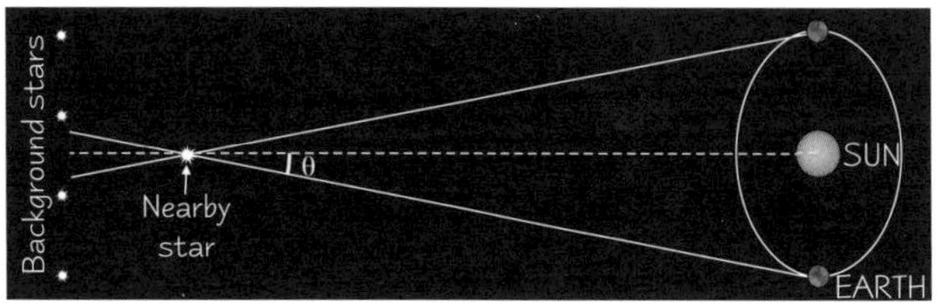

> A star is exactly **one parsec (pc)** away from Earth if the **angle of parallax**,
>
> $$\theta = 1 \text{ arcsecond} = \left(\frac{1}{3600}\right)^{\circ}$$

Distances and Magnitude

*Absolute Magnitude, M, is based only on the **Luminosity** of the Star*

1) The **absolute magnitude** of a star or galaxy, M, does not depend on its distance from Earth. It is defined as what its apparent magnitude **would be** if it were **10 parsecs** away from Earth.
2) The relationship between M and m is given by the following formula:

$$m - M = 5\log\left(\frac{d}{10}\right)$$ where d is the distance in parsecs

If you know the absolute magnitude of a star, you can use this equation to calculate its **distance** from Earth. This is really handy, since the distance to most stars is **too big** to measure using parallax (see previous page).

This method uses objects like **supernovae** that are known as **standard candles**. Standard candles are objects that you can calculate the luminosity of **directly**. So, if you find a supernova within a galaxy, you can work out how far that galaxy is from us. This is how the **Hubble constant** was worked out (see p. 158).

Distances** in the Solar System are Often Measured in **Astronomical Units (AU)

The **parsec** is only one measurement used in **astrophysics** — luckily the others you need to know are much **simpler**.

1) From **Copernicus** onwards, astronomers were able to work out the **distance** the **planets** are from the Sun **relative** to the Earth, using **astronomical units** (AU). But they could not work out the **actual distances**.

   One **astronomical unit** (AU) is defined as the **mean distance** between the **Earth** and the **Sun**.

2) The **size** of the AU (1.50×10^{11} m) wasn't known until 1769 — when it was carefully **measured** during a **transit of Venus** (when Venus passed between the Earth and the Sun).

*Another Measure of Distance is the **Light-Year (ly)***

1) All **electromagnetic waves** travel at the **speed of light**, c, in a vacuum ($c = 3.00 \times 10^8$ ms^{-1}).

   The **distance** that electromagnetic waves travel through a vacuum in **one year** is called a **light-year** (ly).

2) If we see the light from a star that is, say, **10 light-years away** then we are actually seeing it as it was **10 years ago**. The further away the object is, the further **back in time** we are actually seeing it.
3) **1 ly** is equivalent to about **63 000 AU**.

Practice Questions

Q1 What is the relationship between apparent magnitude and intensity?

Q2 What is the equation that links apparent magnitude, absolute magnitude and distance?

Q3 Give three units of distance used in astrophysics. Explain the meaning of each one.

Exam Questions

Q1 Define the *absolute magnitude* of a star. [2 marks]

Q2 Calculate the absolute magnitude of the Sun given that the Sun's apparent magnitude is –27.
[1 pc = 2×10^5 AU] [4 marks]

Q3 The star Sirius has an apparent magnitude of –1.46 and an absolute magnitude of +1.4.
The star Canopus has an apparent magnitude of –0.72 and an absolute magnitude of –5.5.

(a) State which of the two stars appears brighter from Earth. [1 mark]

(b) Calculate the distance of Canopus from Earth. [3 marks]

*Logs — the cheap and easy alternative pet...**

The magnitude scale is a pretty weird system, but like with a lot of astronomy, the old ways have stuck. Remember — the lower the number, the brighter the object. The definition of absolute magnitude is a bit random as well — I mean, why ten parsecs? Ours not to reason why, ours but to... erm... learn it. (Doesn't have quite the same ring does it.)

* Too young to remember Twin Peaks...? Ah well, never mind.

Stars as Black Bodies

Now they're telling us the Sun's black. Who writes this stuff?

A **Black Body** is a **Perfect Absorber** and **Emitter**

1) Objects emit **electromagnetic radiation** due to their **temperature**. At everyday temperatures this radiation lies mostly in the **infrared** part of the spectrum (which we can't see) — but heat something up enough and it will start to **glow**.
2) **Pure black** surfaces emit radiation **strongly** and in a **well-defined way**. We call it **black body radiation**.
3) A black body is defined as:

> A body that **absorbs all wavelengths** of electromagnetic radiation (that's why it's called a **black** body) and can **emit all wavelengths** of electromagnetic radiation.

4) To a reasonably good approximation **stars** behave as **black bodies** and their black body radiation produces their **continuous spectrum**.
5) The graph of **intensity** against **wavelength** for black body radiation varies with **temperature**, as shown in the graph:

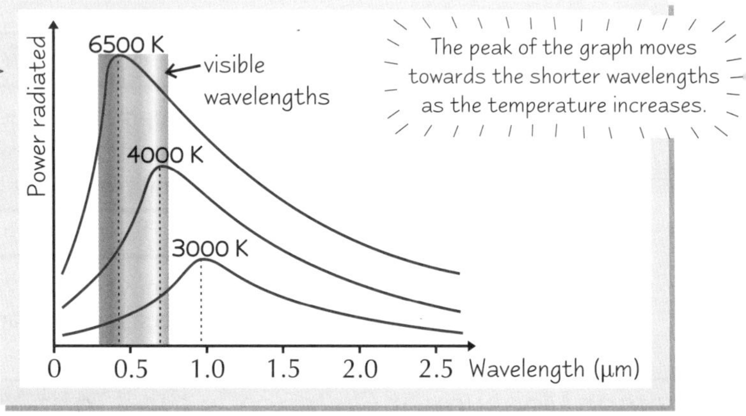

The **Peak Wavelength** gives the **Temperature**

1) For each temperature, there is a **peak** in the black body curve at a wavelength called the **peak wavelength**, λ_{max}.
2) λ_{max} is related to the **temperature** by **Wien's displacement law**:

$$\lambda_{max} T = 0.0029 \text{ m·K}$$

where T is the temperature in kelvin and m·K is a metre-kelvin.

The **Luminosity** of a Star Depends on its **Temperature** and **Surface Area**

1) The **luminosity** of a star is the **total energy** it emits **per second** and is related to the **temperature** of the star and its **surface area**.
2) The luminosity is proportional to the **fourth power** of the star's **temperature** and is **directly proportional** to the **surface area**. This is **Stefan's law**:

$$L = \sigma A T^4$$

where L is the luminosity of the star (in W), A is its surface area (in m^2), T is its surface temperature (in K) and σ (a little Greek "sigma") is Stefan's constant.

3) Measurements give Stefan's constant as $\sigma = 5.67 \times 10^{-8}$ **Wm^{-2}K^{-4}**.
4) From **Earth**, we can measure the **intensity** of the star. The intensity is the **power** of radiation **per square metre**, so as the radiation spreads out and becomes **diluted**, the intensity **decreases**. If the energy has been emitted from a **point** or a **sphere** (like a star, for example) then it obeys the **inverse square law**:

$$I = \frac{L}{4\pi d^2}$$

where L is the luminosity of the star (in W), and d is the distance from the star.

Stars as Black Bodies

You Can Put the **Equations Together** to **Solve Problems**

Example The star Sirius B has a surface area of 4.1×10^{13} m^2 and produces a black body spectrum with a peak wavelength of 115 nm. The intensity of the light from Sirius B when it reaches Earth is 1.12×10^{-11} Wm^{-2}. How long does the light from Sirius B take to reach Earth? *($\sigma = 5.67 \times 10^{-8}$ Wm^{-2}K^{-4}, $c = 3.0 \times 10^8$ ms^{-1})*

First, find the temperature of Sirius B:

$$\lambda_{max}T = 0.0029 \text{ m·K, so } T = 0.0029 \div \lambda_{max} = 0.0029 \div 115 \times 10^{-9} = 25\,217 \text{ K.}$$

Now, you can use Stefan's law to find the luminosity:

$$L = \sigma AT^4 = (5.67 \times 10^{-8}) \times (4.1 \times 10^{13}) \times 25\,217^4 = 9.4 \times 10^{23} \text{ W}$$

Then use $I = \dfrac{L}{4\pi d^2}$ to find the distance of Sirius B from Earth:

$$d = \sqrt{\frac{L}{4\pi I}} = \sqrt{\frac{9.4\times10^{23}}{4\times\pi\times1.12\times10^{-11}}} = \sqrt{6.68\times10^{23}} = 8.2\times10^{16}\text{m}$$

Finally, use $c = d \div t$ to find the time taken $t = d \div c = 8.2 \times 10^{16} \div 3 \times 10^8 = 273\,333\,333$ s $\approx$ **8.7 years**

It's Hard to get **Accurate** Measurements

1) **Wien's displacement law**, **Stefan's law** and the **inverse square law** can all be used to work out various **properties** of stars. This needs very **careful measurements**, but our **atmosphere** mucks up the results.
2) It only lets through **certain wavelengths** of **electromagnetic radiation** — **visible** light, most **radio** waves, **very near infrared** and a bit of **UV**. It's **opaque** to the rest.
3) And then there are things like **dust** and **man-made light pollution** to contend with. Observatories are placed at **high altitudes**, well away from **cities**, and in **low-humidity** climates to minimise the problem. The best solution, though, is to send up **satellites** that can take measurements **above** the atmosphere.
4) Our **detectors** don't do us any favours either. The **measuring devices** that astronomers use aren't perfect since their **sensitivity** depends on the **wavelength**.
5) For example, **glass absorbs UV** light but is **transparent** to **visible light**, so any instruments that use glass affect UV readings straight off.
6) All you can do about this is choose the best materials for what you want to measure, and then **calibrate** your instruments really carefully.

Practice Questions

Q1 What is Wien's displacement law and what is it used for?

Q2 What is the relationship between luminosity, surface area and temperature?

Q3 Why are accurate measurements of black body radiation difficult on the Earth's surface?

Exam Questions

Q1 A star has a surface temperature of 4000 K and the same luminosity as the Sun (3.9×10^{26} W).

(a) Which radiation curve represents this star — X, Y or Z? Explain your answer. [2 marks]

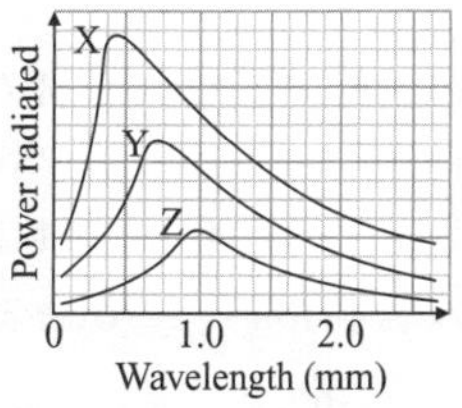

(b) Calculate the star's surface area. [2 marks]

Q2 The star Procyon A, which has a luminosity of 2.3×10^{27} W, produces a black body spectrum with a peak wavelength of 436 nm.

Calculate the surface area of Procyon A. [4 marks]

Astronomy — theories, a bit of guesswork and a whole load of calibration...

Astronomy isn't the most exact of sciences, I'm afraid. The Hubble Space Telescope's improved things a lot, but try and get a look at some actual observational data. Then look at the error bars — they'll generally be about the size of a house.

Spectral Classes and the H-R Diagram

As well as classifying stars by luminosity (the magnitude scale, p.58), they can be classified by colour.

The **Visible** Part of **Hydrogen's Spectrum** is called the **Balmer Series**

1) The lines in **emission** and **absorption spectra** occur because electrons in an atom can only exist at certain well-defined **energy levels**.
2) In **atomic hydrogen**, the electron is usually in the **ground state** (**n** = 1), but there are lots of energy levels (**n** = 2 to **n** = ∞ — called excitation levels) that the electron **could** exist in if it was given more energy.

> The wavelengths corresponding to the **visible bit** of hydrogen's spectrum are caused by electrons moving from **higher energy levels** to the **first excitation level** (**n = 2**).
> This leads to a series of **lines** called the **Balmer series**.

The **Strengths** of the **Spectral Lines** Show the **Temperature** of a Star

1) For a **hydrogen absorption line** to occur in the **visible** part of a star's spectrum, electrons in the hydrogen atoms already need to be in the **n = 2** state.
2) This happens at **high temperatures**, where **collisions** between the atoms give the electrons extra energy.
3) If the temperature is **too high**, though, the majority of the electrons will reach the **n = 3** level (or above) instead, which means there won't be so many Balmer transitions.
4) So the **intensity** of the Balmer lines depends on the **temperature** of the star.
5) For a particular intensity of the Balmer lines, **two temperatures** are possible. Astronomers get around this by looking at the **absorption lines** of **other atoms** and **molecules** as well.

The **Relative Strength** of Absorption Lines gives the **Spectral Class**

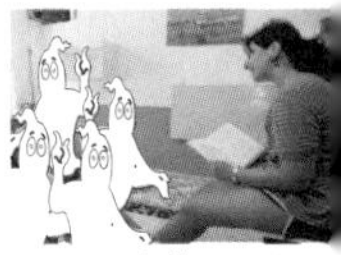

Well... quite

1) For historical reasons the stars are classified into:

   spectral classes: O (hottest), **B, A, F, G, K** and **M**

   Use a **mnemonic** to remember the order. The standard one is the rather non-PC '**Oh Be A Fine Girl, Kiss Me**'.
2) The graph shows how the **intensity** of the visible spectral lines changes with **temperature**:

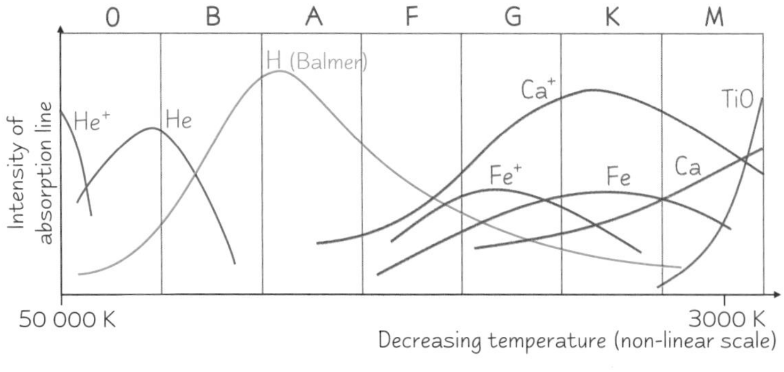

A quick note on the temperature axis: These diagrams, and the H-R diagram (next page), tend to be drawn with spectral class along the horizontal axis. The relationship between spectral class and temperature isn't linear or logarithmic.

> **THE VISIBLE SPECTRAL CHARACTERISTICS OF SPECTRAL CLASSES**
>
> **O** — **Blue** stars: temperature **25 000 – 50 000 K**. The strongest spectral lines are **helium ion** and **helium atom** absorptions, since these need a really high temperature. They have weak **hydrogen Balmer** lines too.
>
> **B** — **Blue** stars: **11 000 – 25 000 K**. These spectra show strong **helium atom** and **hydrogen** absorptions.
>
> **A** — **Blue-white** stars: **7500 – 11 000 K**. Visible spectra are governed by very strong Balmer **hydrogen** lines, but there are also some **metal ion** absorptions.
>
> **F** — **White** stars: **6000 – 7500 K**. These spectra have strong **metal ion** absorptions.
>
> **G** — **Yellow-white** stars: **5000 – 6000 K**. These have both **metal ion** and **metal atom** absorptions.
>
> **K** — **Orange** stars: **3500 – 5000 K**. At this temperature, spectral lines are mostly from neutral **metal atoms**.
>
> **M** — **Red** stars: **< 3500 K**. **Molecular band** absorptions from compounds like **titanium oxide** are present in the spectra of these stars, since they're cool enough for molecules to form.

Spectral Classes and the H-R Diagram

Absolute Magnitude vs Temperature/Spectral Class — the H-R diagram

1) Independently, Hertzsprung and Russell noticed that a plot of **absolute magnitude** (see p. 149) against **temperature** (or **spectral class**) didn't just throw up a random collection of stars but showed **distinct areas**.

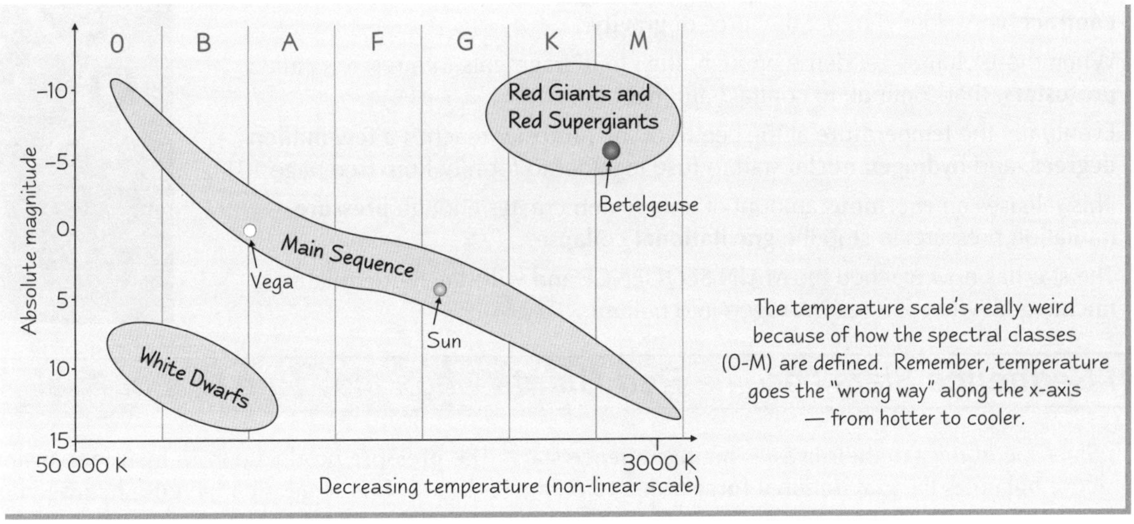

2) The **long, diagonal band** is called the **main sequence**. Main sequence stars are in their long-lived **stable phase** where they are fusing **hydrogen** into **helium**. The Sun is a main sequence star.

3) Stars that have a **high luminosity** and a relatively **low surface temperature** must have a **huge** surface area because of Stefan's law (page 150). These stars are called **red giants** and are found in the **top-right** corner of the H-R diagram. Red giants are stars that have **moved off** the main sequence and fusion reactions other than hydrogen to helium are also happening in them.

4) Stars that have a **low luminosity** but a **high temperature** must be very **small**, again because of Stefan's law. These stars are called **white dwarfs** and are about the size of the Earth. They lie in the **bottom-left** corner of the H-R diagram. White dwarfs are stars at the **end** of their lives, where all of their fusion reactions have stopped and they are just **slowly cooling down**.

Practice Questions

Q1 Why does hydrogen have to be at a particular temperature before Balmer absorption lines are seen?

Q2 List the spectral classes in order of decreasing temperature and outline their spectral characteristics.

Q3 What is an H-R diagram and what are the three main groups of stars that emerge when the diagram is plotted?

Exam Questions

Q1 The spectral classes of stars can be identified by examining the lines in their absorption spectra.

(a) Explain how temperature affects the strength of the Balmer lines in stellar absorption spectra. [3 marks]

(b) State the two spectral classes of star in which strong Balmer lines are observed. [2 marks]

(c) Describe the visible spectral characteristics and temperature of a star in class F. [3 marks]

Q2 The spectra of K and M stars have absorption bands corresponding to energy levels of molecules. Explain why this only occurs in the lowest temperature stars. [2 marks]

Q3 Draw the basic features of an H-R diagram, indicating where you would find main sequence stars, red giants and white dwarfs. [5 marks]

'Ospital Bound — A Furious Girl Kicked Me...

Spectral classes are another example of astronomers sticking with tradition. The classes used to be ordered alphabetically by the strength of the Balmer lines. When astronomers realised this didn't quite work, they just fiddled around with the old classes rather than coming up with a sensible new system. Just to make life difficult for people like you and me.

Stellar Evolution

Stars go through several different stages in their lives and move around the H-R diagram as they go (see p. 153).

Stars Begin as Clouds of **Dust** and **Gas**

1) Stars are born in a **cloud** of **dust** and **gas**, most of which was left when previous stars blew themselves apart in **supernovae**. The denser clumps of the cloud **contract** (very slowly) under the force of **gravity**.
2) When these clumps get dense enough, the cloud fragments into regions called **protostars**, that continue to contract and **heat up**.
3) Eventually the **temperature** at the centre of the protostar reaches a **few million degrees**, and **hydrogen nuclei** start to **fuse** together to form helium (see page 133).
4) This releases an **enormous** amount of **energy** and creates enough **pressure** (radiation pressure) to stop the **gravitational collapse**.
5) The star has now reached the **MAIN SEQUENCE** and will stay there, relatively **unchanged**, while it fuses hydrogen into helium.

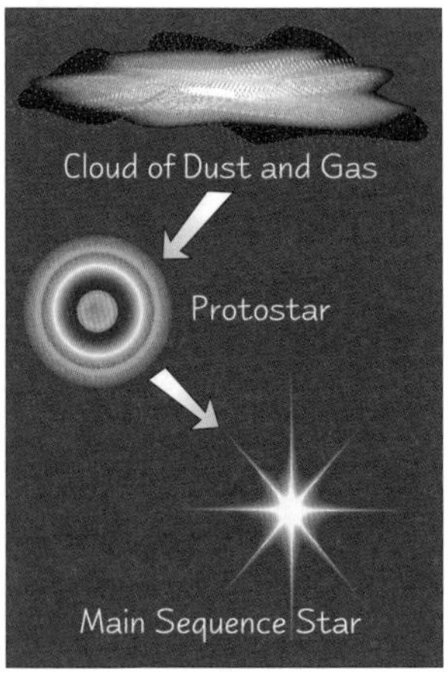

Main Sequence Stars become **Red Giants** when they **Run Out** of **Fuel**

1) Stars spend most of their lives as **main sequence** stars. The **pressure** produced from **hydrogen fusion** in their **core balances** the **gravitational force** trying to compress them. This stage is called **core hydrogen burning**.
2) When the **hydrogen** in the **core** runs out nuclear fusion **stops**, and with it the **outward pressure stops**. The core **contracts** and **heats up** under the **weight** of the star. The outer layers **expand** and **cool**, and the star becomes a **RED GIANT**.
3) The material **surrounding** the core still has **plenty of hydrogen**. The **heat** from the contracting **core** raises the **temperature** of this material enough for the hydrogen to **fuse**. This is called **shell hydrogen burning**. (Very low-mass stars stop at this point. They use up their fuel and slowly fade away...)
4) The core continues to contract until, eventually, it gets **hot** enough and **dense** enough for **helium** to **fuse** into **carbon** and **oxygen**. This is called **core helium burning**. This releases a **huge** amount of energy, which **pushes** the **outer layers** of the star further outwards.
5) When the **helium** runs out, the carbon-oxygen core **contracts again** and heats a **shell** around it so that helium can fuse in this region — **shell helium burning**.

Low Mass Stars (like the Sun) **Eject** their Shells, leaving behind a **White Dwarf**

1) In low-mass stars, the **carbon-oxygen core isn't hot enough** for any further **fusion** and so it continues to **contract** under its own **weight**. Once the core has shrunk to about **Earth-size**, **electrons** exert enough pressure (**electron degeneracy pressure**) to stop it collapsing any more (fret not — you don't have to know how).
2) The **helium shell** becomes more and more **unstable** as the core contracts. The star **pulsates** and **ejects** its outer layers into space as a **planetary nebula**, leaving behind the dense core.
3) The star is now a very **hot**, **dense solid** called a **WHITE DWARF**, which will simply **cool down** and **fade away**.

High Mass Stars have a **Shorter Life** and a more **Exciting Death**

1) Even though stars with a **large mass** have a **lot of fuel**, they use it up **more quickly** and don't spend so long as main sequence stars.
2) When they are **red giants** the **'core burning to shell burning'** process can continue beyond the fusion of helium, building up layers in an **onion-like structure**. For **really massive** stars this can go all the way up to **iron**.
3) Nuclear fusion **beyond iron** isn't **energetically favourable**, though, so once an iron core is formed then very quickly it's goodbye star.
4) The star explodes cataclysmically in a **SUPERNOVA**, leaving behind a **NEUTRON STAR** or (if the star was massive enough) a **BLACK HOLE**.

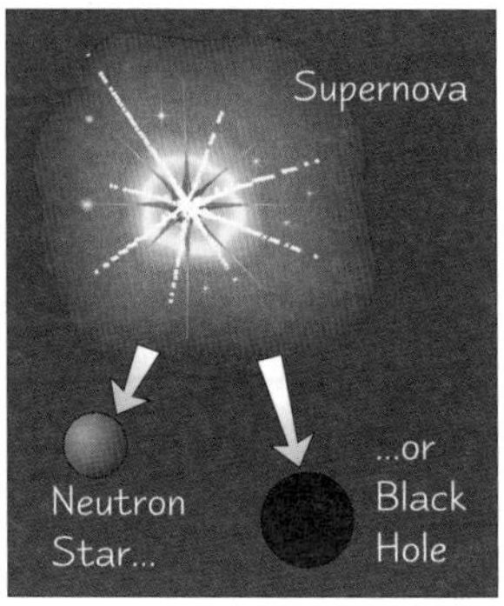

Stellar Evolution

*Massive Stars go out with a **Bit of a Bang***

1) When the core of a star runs out of fuel, it starts to **contract** — forming a white dwarf core.
2) If the star is **massive enough**, though, **electron degeneracy** can't stop the core contracting. This happens when the mass of the core is more than **1.4 times** the mass of the **Sun**.
3) The electrons get **squashed** onto the atomic **nuclei**, combining with protons to form **neutrons** and **neutrinos**.
4) The core suddenly collapses to become a **NEUTRON STAR**, which the outer layers then **fall** onto.
5) When the outer layers **hit** the surface of the **neutron star** they **rebound**, setting up huge **shockwaves**, ripping the star apart and causing a **supernova**. The light from a supernova can briefly outshine an **entire galaxy**.

> **Neutron stars** are incredibly **dense** (about 4×10^{17} kgm^{-3}).
> They're **very small**, typically about 20 km across, and they can **rotate very fast** (up to 600 times a second).
> They emit **radio waves** in two beams as they rotate. These beams sometimes sweep past the Earth and can be observed as **radio pulses** rather like the flashes of a lighthouse. These rotating neutron stars are called **PULSARS**.

*Neutron Stars are **Weird** but **Black Holes** are a **Lot Weirder***

1) If the **core** of the star is more than **3 times** the **Sun's mass**, the **neutrons** can't withstand the gravitational forces.
2) There are **no known mechanisms** left to stop the core collapsing to an **infinitely dense** point called a **singularity**. At that point, the **laws of physics** break down completely.
3) Up to a certain distance away (called the **Schwarzschild radius**) the gravitational pull is **so strong** that nothing, not even **light**, can escape its grasp. The **boundary** of this region is called the **event horizon**.

> **The Schwarzschild radius is the distance at which the escape velocity is the speed of light**
> An object moving at the **escape velocity** has **just enough kinetic energy** to overcome the black hole's gravitational field.
>
> From Newton's law of gravitation we get $\frac{1}{2}mv^2 = \frac{GMm}{r}$ where m = mass of object, M = mass of black hole, v = velocity of object, r = distance from centre of black hole, $G = 6.67 \times 10^{-11}$ Nm^2kg^{-2}
>
> Dividing through by m and making r the subject gives: $r = \frac{2GM}{v^2}$
>
> By replacing v with the speed of light, c, you get the Schwarzschild radius, R_S: $R_S = \frac{2GM}{c^2}$.

This derivation is a bit of a fudge (although it gives the right answer, as it happens) — Newton's law of gravity doesn't quite work in intense gravitational fields, so Einstein's general relativity should be used instead (the maths is way too hard to go into here though).

Practice Questions

Q1 Outline how the Sun was formed, how it will evolve and how it will die.

Q2 Describe a white dwarf and a neutron star. What are the main differences between them?

Exam Questions

Q1 Outline the main differences between the evolution of high mass and low mass stars, starting from when they first become main sequence stars. [6 marks]

Q2 (a) What is meant by the Schwarzschild radius of a black hole? [2 marks]

(b) Calculate the Schwarzschild radius for a black hole that has a mass of 6×10^{30} kg. [2 marks]

Live fast — die young...

The more massive a star, the more spectacular its life cycle. The most massive stars burn up the hydrogen in their core so quickly that they only live for a fraction of the Sun's lifetime — but when they go, they do it in style.

The Doppler Effect and Redshift

Everyone's heard of the Big Bang theory — well here's some evidence for it.

The **Doppler Effect** — the **Motion** of a Wave's **Source** Affects its **Wavelength**

1) You'll have experienced the Doppler effect **loads of times** with **sound waves**.
2) Imagine an ambulance driving past you. As it moves **towards you** its siren sounds **higher-pitched**, but as it **moves away**, its **pitch** is **lower**. This change in **frequency** and **wavelength** is called the **Doppler shift**.
3) The frequency and the wavelength **change** because the waves **bunch together** in **front** of the source and **stretch out behind** it. The **amount** of stretching or bunching together depends on the **velocity** of the **source**.
4) When a **light source** moves **away** from us, the wavelengths of its light become **longer** and the frequencies become lower. This shifts the light towards the **red** end of the spectrum and is called **redshift**.
5) When a light source moves **towards** us, the **opposite** happens and the light undergoes **blueshift**.
6) The amount of redshift or blueshift, z, is determined by the following formula:

$$z = \frac{\Delta f}{f} = -\frac{\Delta \lambda}{\lambda} = \frac{v}{c} \quad \text{if } v \ll c$$

λ is the emitted wavelength, f is the emitted frequency, $\Delta\lambda$ and Δf are the differences between the observed and emitted wavelengths/frequencies, v is the velocity of the source in the observer's direction and c is the speed of light. ($v \ll c$ means "v is much less than c".)

7) The way cosmologists tend to look at this stuff, the galaxies aren't actually moving **through space** away from us. Instead, **space itself** is expanding and the light waves are being **stretched** along with it. This is called **cosmological redshift** to distinguish it from **redshift** produced by sources that **are** moving through space.
8) The same formula works for both types of redshift as long as v is much less than c. If v is close to the speed of light, you need to use a nasty, relativistic formula instead (you don't need to know that one).

The **Red Shift** of Galaxies is **Strong Evidence** for the HBB

1) The **spectra** from **galaxies** (apart from a few very close ones) all show **redshift** — the **characteristic spectral lines** of the elements are all at a **longer wavelength** than you would expect. This shows they're all **moving apart**.
2) Hubble realised that the **speed** that **galaxies moved away** from us depended on **how far** they were away. This led to the idea that the Universe started out **very hot** and **very dense** and is currently **expanding**.

THE HOT BIG BANG THEORY (HBB): the Universe started off **very hot** and **very dense** (perhaps as an **infinitely hot, infinitely dense** singularity) and has been **expanding** ever since.

Redshift is Used to Study Spectroscopic **Binary Stars**

1) About half of the stars we observe are actually **two stars** that orbit each other. Many of them are too far away from us to be **resolved** with **telescopes** but the **lines** in their **spectra** show a binary star system. These are called **spectroscopic binary stars**.
2) By observing how the **absorption lines** in the spectrum change with **time** the **orbital period** can be calculated:

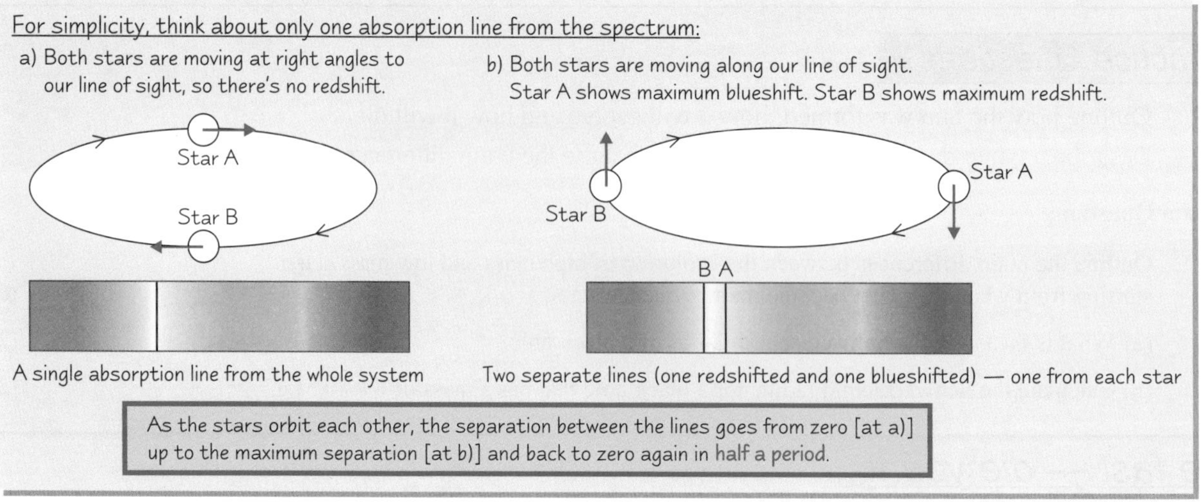

3) Astronomers have used a similar method to find **extrasolar planets**.

The Doppler Effect and Redshift

The Doppler effect explains quasars too. Read on.

Quasars — Quasi-Stellar Objects

1) **Quasars** were discovered in the late 1950s and were first thought to be **stars in our galaxy**.
2) The puzzling thing was that their spectra were **nothing like** normal stars. They sometimes shot out **jets** of material, and many of them were very active **radio sources**.
3) The 'stars' produced a **continuous spectrum** that was nothing like a black body radiation curve and instead of absorption lines, there were **emission lines** of elements that astronomers **had not seen before**.
4) However, these lines looked strangely familiar and in 1963 Maarten Schmidt realised that they were simply the **Balmer series** of hydrogen (see p.152) but **redshifted** enormously.

Quasars are a Very Long Way Away so they must be Very Bright

1) This **huge redshift** suggests they're a **huge distance away** (see next page) — in fact, the **most distant** objects seen.
2) The measured redshifts give us distances of **billions of light years**.
Using the **inverse square law** for intensity (see p. 150) gives an idea of just how **bright** quasars are:

Example A quasar has the same intensity as a star 20 000 ly away with the same luminosity as the Sun (4×10^{26} W). Its redshift gives a distance of 1×10^{10} ly. Calculate its luminosity.

$I_{quasar} = I_{star}$ so they cancel out of the equation.

$$L \propto Id^2 \Rightarrow \frac{L_{quasar}}{L_{star}} = \frac{d_{quasar}^2}{d_{star}^2} \Rightarrow L_{quasar} = L_{star} \cdot \frac{d_{quasar}^2}{d_{star}^2} = 4\times10^{26} \cdot \frac{1\times10^{20}}{4\times10^{8}} = 1\times10^{38}\ \text{W}$$

That's bright — about **10 times** the **luminosity** of the **entire Milky Way galaxy**!

3) And there's very good evidence to suggest that quasars are only about the size of the **Solar System**.
4) Let me run that past you again. **That's the power of a trillion Suns from something the size of the Solar System.**
5) These numbers caused a lot of controversy in the astrophysics community — they seemed crazy. Many astrophysicists thought there must be a more reasonable explanation. But then evidence for the distance of quasars came when **sensitive CCD** equipment detected the fuzzy cloud of **a galaxy around a quasar**.

The current consensus is that a quasar is a very powerful **galactic nucleus** — a huge **black hole** about 100 million times the mass of the Sun at the centre of a distant galaxy. (**All** galaxies are thought to have these 'supermassive' black holes at their centres).
This black hole is surrounded by a doughnut shaped mass of **whirling gas** falling into it, which produces the light. In the same way as a pulsar (see p.155), magnetic fields produce jets of radiation streaming out from the poles. The black hole must consume the mass of about **10 Suns per year** to produce the energy observed.

Practice Questions

Q1 What is the Doppler effect? Write down the formula for the redshift and blueshift of light.

Q2 Explain how the spectra of binary stars can be used to calculate their period of orbit.

Exam Questions

Q1 The spectra of three objects have been taken. What can you deduce from each of the following?

(a) The absorption lines from object A have been shifted towards the blue end of the spectrum. [1 mark]

(b) The absorption lines from object B oscillate either side of their normal position in the spectrum with a period of two weeks. [2 marks]

(c) The wavelength of the hydrogen alpha line in object C's spectrum is 667.83 nm. In the laboratory, the wavelength of the same line is measured as 656.28 nm. [3 marks]

Q2 (a) State one piece of evidence that suggests quasars are a very long distance away. [1 mark]

(b) Use the concept of the inverse square law to suggest why quasars must be very bright. [2 marks]

(c) Describe the main features of a quasar according to the current theory. [2 marks]

Long ago, in a galaxy far, far away — there was a radio-loud, supermassive black hole with a highly luminous arc...

Quasars are weird. There's still some disagreement in the astrophysics community about what they are. There's even some evidence (not generally accepted) that quasars are much nearer than the redshift suggests and are just moving very quickly.

The Big Bang Model of the Universe

Right, we're moving on to the BIG picture now — we all like a bit of cosmology...

The **Universe** is **The Same** in **Every Direction**

When you read that all the **galaxies** in the Universe are **moving away** from the **Earth** (see p.156 and below), it's easy to imagine that the Earth is at the **centre of the Universe**, or that there's something really **special** about it. **Earth** is special to us because we **live here** — but on a **universal scale**, it's just like any other lump of rock.

1) The **demotion** of **Earth** from anything special is taken to its logical conclusion with the **cosmological principle**...

> **COSMOLOGICAL PRINCIPLE**: on a **large scale** the Universe is **homogeneous** (every part is the same as every other part) and **isotropic** (everything looks the same in every direction) — so it doesn't have a **centre**.

2) Until the **1930s**, cosmologists believed that the Universe was **infinite** in both **space** and **time** (that is, it had always existed) and **static**. This seemed the **only way** that it could be **stable** using **Newton's law** of gravitation. Even **Einstein modified** his theory of **general relativity** to make it consistent with the **Steady-State Universe**.

Hubble Realised that the **Universe** is **Expanding**

1) The **spectra** from **galaxies** (apart from a few very close ones) all show **redshift**. The amount of **redshift** gives the **recessional velocity** — how fast the galaxy is moving away (see page 156).
2) A plot of **recessional velocity** against **distance** (found using standard candles — see p.149) showed that they were **proportional**, which suggests that the Universe is **expanding**. This gives rise to **Hubble's law**:

$$v = H_0 d$$

v = recessional velocity in **kms^{-1}**,
d = distance in **Mpc** and
H_0 = **Hubble's constant** in **$kms^{-1}Mpc^{-1}$**.

3) Since distance is very difficult to measure, astronomers disagree on the value of H_0. It's generally accepted that H_0 lies somewhere between 50 $kms^{-1}Mpc^{-1}$ and 100 $kms^{-1}Mpc^{-1}$.
4) The **SI unit** for H_0 is s^{-1}. To get H_0 in SI units, you need v in ms^{-1} and d in m (1 Mpc = 3.09×10^{22} m).

The **Expanding Universe** gives rise to the **Hot Big Bang Model**

1) The Universe is **expanding** and **cooling down** (because it's a closed system). So further back in time it must have been **smaller** and **hotter**. If you trace time back **far enough**, you get a **Hot Big Bang** (see page 156).
2) Since the Universe is **expanding uniformly** away from **us** it seems as though we're at the **centre** of the Universe, but this is an **illusion**. You would observe the **same thing** at **any point** in the Universe.

Hot big bang — the ultimate firework display?

The **Age** and **Observable Size** of the **Universe** Depend on H_0

1) If the Universe has been **expanding** at the **same rate** for its whole life, the **age** of the Universe is $t = 1/H_0$ (time = distance/speed). This is only an estimate though — see below.
2) Unfortunately, since no one knows the **exact value** of H_0 we can only guess the Universe's age. If $H_0 = 75$ $kms^{-1}Mpc^{-1}$, then the age of the Universe $\approx 1/(2.4 \times 10^{-18}\ s^{-1}) = 4.1 \times 10^{17}$ s = **13 billion years**.
3) The **absolute size** of the Universe is **unknown** but there is a limit on the size of the **observable Universe**. This is simply a **sphere** (with the Earth at its centre) with a **radius** equal to the **maximum distance** that **light** can travel during its **age**. So if $H_0 = 75$ $kms^{-1}Mpc^{-1}$ then this sphere will have a radius of **13 billion light years**.

> **THE RATE OF EXPANSION HASN'T BEEN CONSTANT**
>
> All the **mass** in the Universe is attracted together by **gravity**. This attraction tends to **slow down** the rate of expansion of the Universe. It's thought that the expansion **was** decelerating until about 5 billion years ago.
>
> But in the late 90s, astronomers found evidence that the expansion is now **accelerating**. Cosmologists are trying to explain this acceleration using **dark energy** — a type of energy that fills the whole of space. There are various theories of what this dark energy **is**, but it's really hard to test them.

The Big Bang Model of the Universe

Cosmic Microwave Background Radiation — More Evidence for the HBB

1) The Hot Big Bang model predicts that loads of **electromagnetic radiation** was produced in the **very early Universe**. This radiation should **still** be observed today (it hasn't had anywhere else to go).
2) Because the Universe has **expanded**, the wavelengths of this cosmic background radiation have been **stretched** and are now in the **microwave** region.
3) This was picked up **accidentally** by Penzias and Wilson in the 1960s.

Properties of the Cosmic Microwave Background Radiation (CMBR)

1) In the late 1980s a satellite called the **Cosmic Background Explorer** (**COBE**) was sent up to have a **detailed look** at the radiation.
2) It found a **perfect blackbody spectrum** corresponding to a **temperature** of **2.73 K** (see page 150).
3) The radiation is largely **isotropic** and **homogeneous**, which confirms the cosmological principle (see page 158).
4) There are **very tiny fluctuations** in temperature, which were at the limit of COBE's detection. These are due to tiny energy-density variations in the early Universe, and are needed for the initial '**seeding**' of galaxy formation.
5) The background radiation also shows a **Doppler shift**, indicating the Earth's motion through space. It turns out that the **Milky Way** is rushing towards an unknown mass (the **Great Attractor**) at over a **million miles an hour**.

Another Bit of Evidence is the Amount of Helium in the Universe

1) The HBB model also explained the **large abundance of helium** in the Universe (which had puzzled astrophysicists for a while).
2) The early Universe had been very hot, so at some point it must have been hot enough for **hydrogen fusion** to happen. This means that, together with the theory of the synthesis of the **heavier elements** in stars, the **relative abundances** of all of the elements can be accounted for.

Practice Questions

Q1 What is Hubble's law? How can it be used to find the age of the Universe?

Q2 What is the cosmic background radiation?

Q3 How do the relative amounts of hydrogen and helium in the Universe provide evidence for the HBB model?

Exam Questions

Q1 (a) State Hubble's law, explaining the meanings of all the symbols. [2 marks]

(b) What does Hubble's law suggest about the nature of the Universe? [2 marks]

(c) Assume $H_0 = +50\ \text{kms}^{-1}\text{Mpc}^{-1}$ (1 Mpc = 3.09×10^{22} m).

i) Calculate H_0 in SI units. [2 marks]

ii) Calculate an estimate of the age of the Universe, and hence the size of the observable Universe. [3 marks]

Q2 (a) A certain object has a redshift of 0.37. Estimate the speed at which it is moving away from us. [2 marks]

(b) Use Hubble's law to estimate the distance (in light years) that the object is from us.
(Take $H_0 = 2.4 \times 10^{-18}\ \text{s}^{-1}$, 1 ly = 9.5×10^{15} m.) [2 marks]

(c) With reference to the speed of the object, explain why your answers to a) and b) are estimates. [1 mark]

Q3 Describe the main features of the cosmic background radiation and explain why its discovery was considered strong evidence for the Hot Big Bang model of the Universe.
The quality of your written answer will be assessed in this question. [7 marks]

My Physics teacher was a Great Attractor — everyone fell for him...

The simple Big Bang model doesn't actually work — not quite, anyway. There are loads of little things that don't quite add up. Modern cosmologists are trying to improve the model using a period of very rapid expansion called inflation.

Physics of the Eye

*The eyes contain **converging lenses**, which focus light rays to form images. Page 142 tells you all about lenses — make sure you're confident with that before you start. Here are the two big equations you need to know:*

power of a lens: $P = \frac{1}{f}$

lens equation: $\frac{1}{f} = \frac{1}{u} + \frac{1}{v}$

You Need to Know the Basic **Structure** of the **Eye**

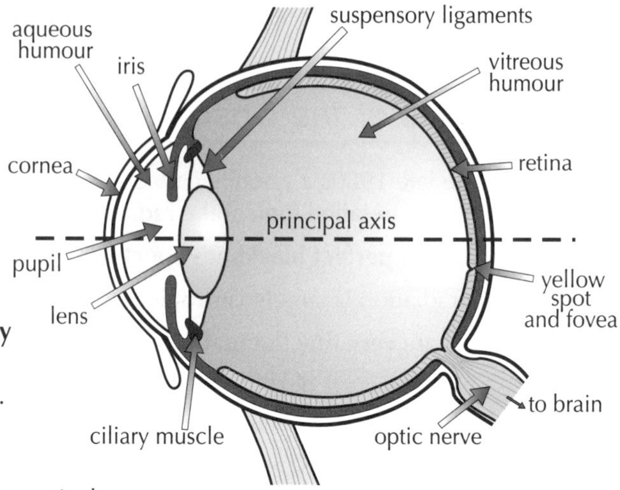

1) The **cornea** is a **transparent** 'window' with a **convex** shape, and a **high refractive index**. The cornea does most of the eye's focusing.
2) The **aqueous humour** is a **watery** substance that lets light pass through the pupil to the lens.
3) The **iris** is the coloured part of the eye. It consists of **radial** and **circular muscles** that control the size of the **pupil** — the hole in the middle of the iris. This regulates the intensity of light entering the eye.
4) The **lens** acts as a **fine focus** and is controlled by the **ciliary muscles**. When the ciliary muscles **contract**, tension is released and the lens takes on a **fat**, more **spherical** shape. When they **relax**, the **suspensory ligaments** pull the lens into a **thin**, **flatter** shape.
5) The **vitreous humour** is a **jelly-like** substance that keeps the eye's shape.
6) Images are formed on the **retina**, which contains **light-sensitive cells** called **rods** and **cones** (see below).
7) The **yellow spot** is a particularly sensitive region of the retina. In the centre of the yellow spot is the **fovea**. This is the part of the retina with the highest concentration of **cones**.
8) The **optic nerve** carries signals from the rods and cones to the **brain**.

The image is upside down but it's interpreted by the brain to seem the right way up.

The **Eye** is an **Optical Refracting System**

1) The **far point** is the **furthest distance** that the eye can focus comfortably. For normally sighted people that's **infinity**. When your eyes are focusing at the far point, they're **'unaccommodated'**. The **near point** is the **closest distance** that the eye can focus on. For young people it's about 9 cm.
2) The **cornea** and **aqueous humour** act as a **fixed converging lens** with a **power** of about **41 D**.
3) The power of the eye's **lens** itself is about **18 D** when **unaccommodated**. By changing shape, it can increase to about **29 D** in young people to view objects at the **near point**.
4) You can **add together** the **powers** of the cornea, aqueous humour and lens. That means you can think of the eye as a **single converging lens** of **power 59 D** at the far point. This gives a **focal length** of **1.7 cm**.
5) When looking at nearer objects, the eye's power **increases**, as the lens changes shape and the **focal length decreases** — but the distance between the lens and the image, ***v***, stays the same, at 1.7 cm.

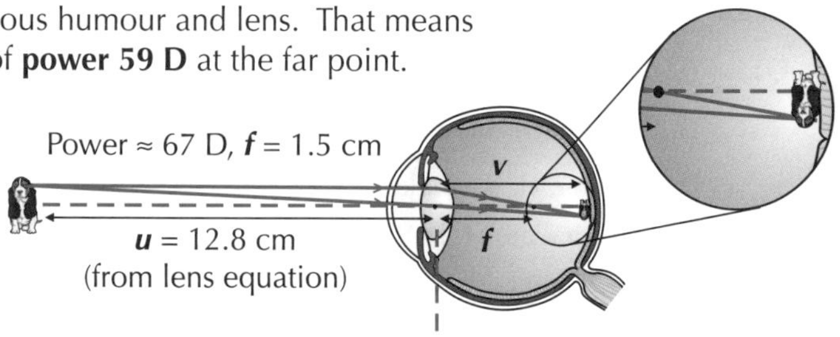

The **Retina** has **Rods** and **Cones**

1) **Rods** and **cones** are cells at the back of the **retina** that respond to **light**. Light travels **through the retina** to the rods and cones at the back.
2) Rods and cones all contain chemical **pigments** that **bleach** when **light** falls on them. This bleaching stimulates the cell to send signals to the **brain** via the **optic nerve**.
3) The cells are **reset** (i.e. unbleached) by enzymes using **vitamin A** from the blood.
4) There's only **one** type of **rod** but there are **three** types of **cone**, which are sensitive to **red**, **green** and **blue** light.

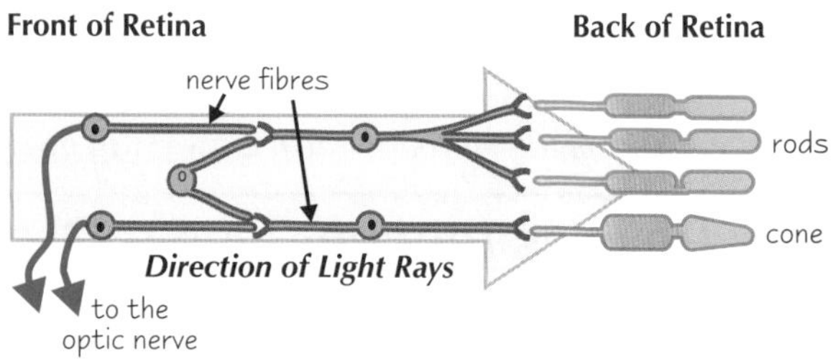

Physics of the Eye

The **Cones** let you See in **Full Colour**

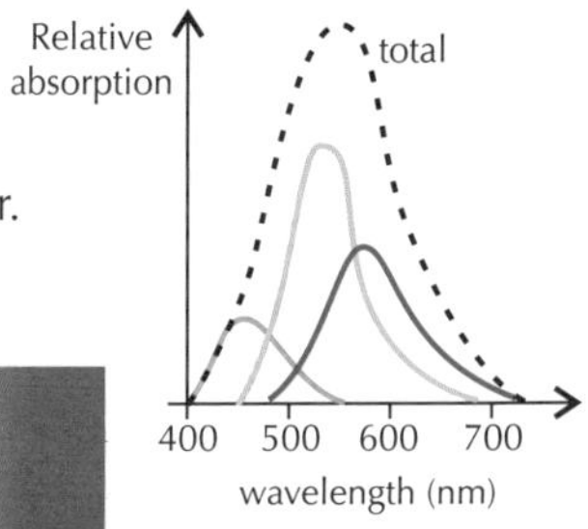

1) The red, green and blue **cones** each absorb a **range of wavelengths**.
2) The eye is **less responsive** to blue light than to red or green, so blues often look dimmer.
3) The brain receives signals from the three types of cone and interprets their **weighted relative strengths** as **colour**...

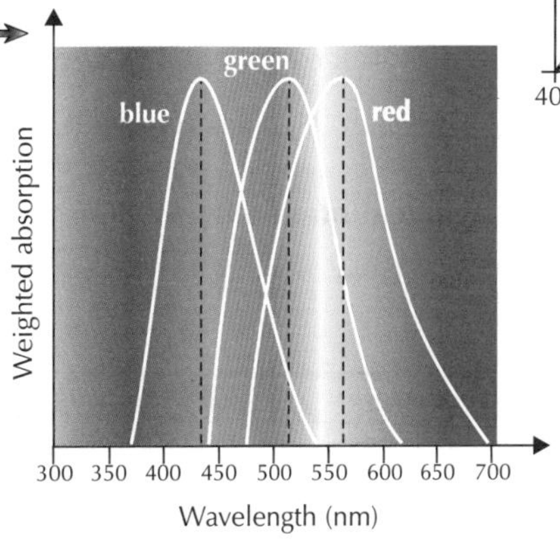

> **Example**
> Yellow light produces almost equal responses from the red and green cones.
> Yellow light can therefore be 'faked' by combining red and green light of almost equal intensity — the electrical signal from the retina will be the same and the brain interprets it as 'yellow'.

4) **Any** colour can be produced by **combining** different intensities of **red**, **green** and **blue** light. Colour televisions work like this.

You Need Good **Spatial Resolution** to See **Details**

1) Two objects can only be distinguished from each other if there's **at least one rod** or **cone between** the light from each of them. Otherwise the brain can't **resolve** the two objects and it 'sees' them as one.

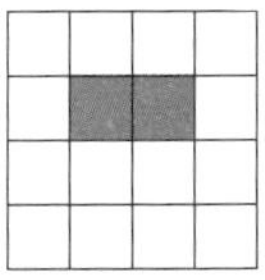

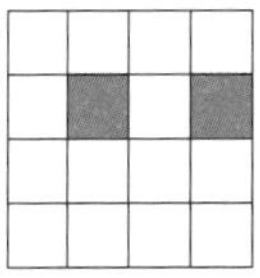

2) **Spatial resolution** is **best** at the **yellow spot** — the **cones** are very **densely packed** here and each cone always has its **own nerve fibre**. There are **no rods in the yellow spot**, though. This means that in **dim light**, when **cones don't work**, resolution is best slightly off the direct line of sight, where the **rods** are more **densely packed**.
3) Away from the yellow spot, resolution is much worse. The light-sensitive cells are **not** as **densely packed** and the rods **share nerve fibres** — there are up to 600 rods per fibre at the edges of the retina.

Persistence of Vision means you **Don't See Rapid Flickering**

1) **Nerve impulses** from the eye take about a fifth of a second to **decay**. So a very dim light flashing faster than **five times per second** (5 Hz) seems to be **on continuously**. This is called **flicker fusion**.
2) At **higher light intensity**, more nerve cells are 'firing' so a **higher frequency** is needed for flicker fusion to occur.
3) Cinema and TV rely on **persistence of vision** to give the illusion of **smooth** rather than **jerky movements**.

Practice Questions

Q1 Draw a ray diagram to show a young person's eye focusing at the near point.
Assume that the eye's total power is 70 D. Mark on your diagram the distances ***u***, ***v*** and ***f***.

Q2 Describe the differences and similarities between rods and cones.

Q3 Sketch a graph showing how the cone cells in the retina respond to different wavelengths of light.

Q4 What is meant by 'persistence of vision'? Give a situation where it is useful.

Exam Questions

Q1 The power of an unaccommodated eye is 60 D.
(a) Determine the image distance, ***v***, when the eye focuses at infinity. [2 marks]
(b) Calculate the extra power that the lens must produce for the eye to focus on an object that is 30 cm away. [3 marks]

Q2 The eye is designed to receive incoming light and focus it on the retina.
(a) Outline the path of light through the eye, explaining the purpose of the structures it passes through. [4 marks]
(b) Describe the relationship between the structure of the retina and spatial resolution. [2 marks]

The eyes are the window on the soul...

Or so they said in the 16th century. Sadly, that won't get you far with a question about the spectral response of the retina.

Defects of Vision

*Plenty of people don't have perfect vision, and need **auxiliary lenses** to correct their sight.*

Real is Positive, Virtual is Negative

(see p. 142 for more on real and virtual images)

Lenses can produce **real** or **virtual** images, and you need to follow the "**real** is **positive**, **virtual** is **negative**" rule.

1) A **converging lens** produces a **real image**, so its **focal length**, ***f***, is **positive**.
2) A **diverging lens** produces a **virtual image**, so it has a **negative focal length**.
3) As you know, the focal length is related to object and image distances by the **lens equation**... $\frac{1}{f} = \frac{1}{u} + \frac{1}{v}$
4) The **linear magnification** of a lens is $m = \frac{\text{size of image}}{\text{size of object}}$ and $m = \frac{v}{u}$

Myopia is Corrected with Diverging Lenses

1) **Short-sighted** (myopic) people are unable to focus on distant objects — this happens if their **far point** is **closer** than infinity (see p. 160).
2) Myopia occurs when the **cornea** and **lens** are too **powerful** or the **eyeball** is too **long**.
3) The focusing system is **too powerful** and images of distant objects are brought into focus in **front** of the retina.
4) A lens of **negative power** is needed to correct this defect — so a **diverging** lens is placed in front of the eye.

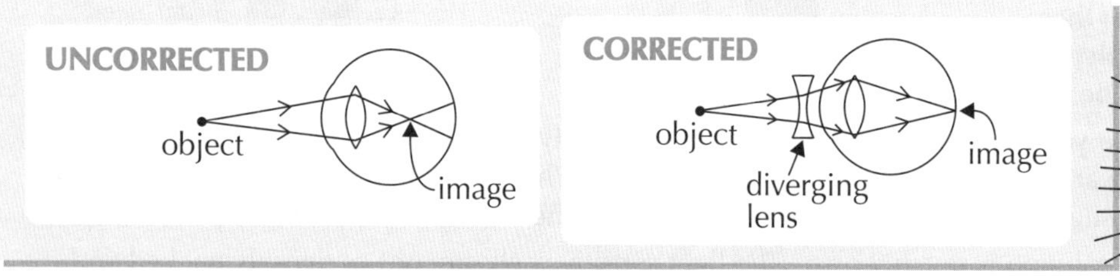

As well as correcting the far point, the diverging lens also makes the near point a little further away than it was. This isn't usually a problem — short-sighted people usually have a near point that is closer than normal anyway.

Hypermetropia is corrected with Converging Lenses

1) **Long-sighted** (hypermetropic) people are unable to focus clearly on near objects. This happens if their **near point** is **further** away than normal (25 cm or more).
2) Long sight occurs because the **cornea** and **lens** are too **weak** or the **eyeball** is too **short**.
3) The focusing system is **too weak** and images of near objects are brought into focus **behind** the retina.
4) A lens of **positive power** is needed to correct the defect — so a **converging** lens is placed in front of the eye.

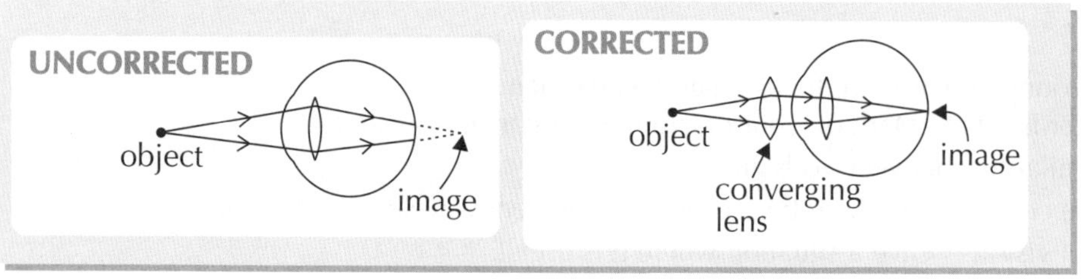

Long-sightedness is common among young children whose lenses have grown quicker than their eyeballs.

Astigmatism is Corrected with Cylindrical Lenses

1) **Astigmatism** is caused by an irregularly shaped **cornea** or **lens** which has **different focal lengths** for different **planes**. For instance, when **vertical lines** are in focus, **horizontal** lines might not be.
2) The condition is corrected with **cylindrical lenses**.

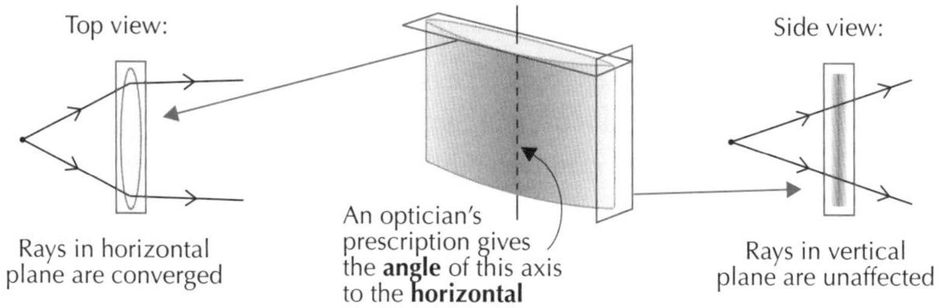

Defects of Vision

Choosing a **Lens** to Correct for **Short Sight** Depends on the **Far Point**

1) To correct for **short sight**, a **diverging** lens is chosen which has its **principal focus** at the eye's **faulty far point**.
2) The **principal focus** is the point that rays from a distant object **appear** to have come from (see p.142).
3) The lens must have a **negative focal length** which is the same as the **distance to the eye's far point**. This means that objects at **infinity**, which were out of focus, now seem to be in focus at the far point.

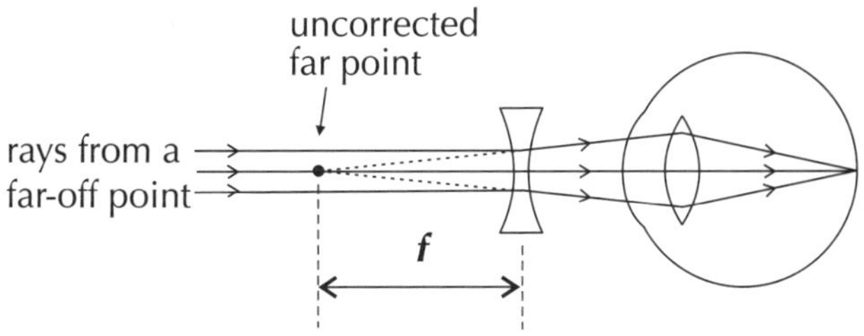

Example

Ben is short-sighted. His far point is 5 m.
Calculate the power of lens he needs to correct his vision.

Focal length, f = far point = −5 m

Power needed = $\frac{1}{f}$ = **−0.2 D**

The power's always **negative** to correct for **short** sight.

Calculations Involving **Long Sight** Use the **Lens Equation**

1) People with these conditions have a near point which is too far away. An 'acceptable' near point is 25 cm.
2) A **converging lens** is used to produce a **virtual image** of objects 0.25 m away **at the eye's near point.** This means that close objects, which were out of focus, now seem to be in focus at the near point.
3) You can work out the **focal length**, and hence the **power** of lens needed, using the **lens equation** $\frac{1}{f} = \frac{1}{u} + \frac{1}{v}$

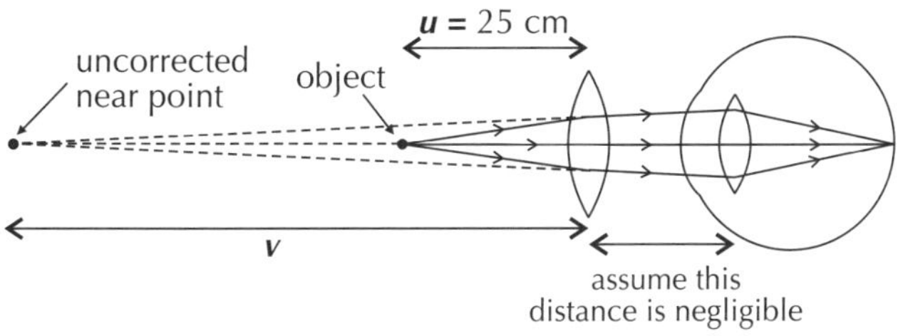

Example

Mavis can't read her book — her near point is 5 m.
What power of lens does she need?

$u = 0.25$ m, $v = -5$ m (real is positive, virtual is negative)

$$\frac{1}{f} = \frac{1}{0.25} - \frac{1}{5}$$

$\frac{1}{f} = 3.8$ so power = **+3.8 D**

The power's always **positive** to correct for **long** sight.

Practice Questions

Q1 Define the terms myopia, hypermetropia and astigmatism.

Q2 What type of auxiliary lenses are used to correct these conditions?

Q3 Define the terms near point and far point.

Exam Questions

Q1 A man with short sight has a far point of 4 m.
Calculate the power of auxiliary lens needed to correct his far point. [3 marks]

Q2 A girl has a near point of 2 m.
Calculate the power of lens required to correct her near point to 25 cm. [3 marks]

Q3 Claire suffers from astigmatism.

(a) State the type of lenses that are used to correct astigmatism. [1 mark]

(b) Draw a diagram of a lens that would converge rays in the vertical plane but not in the horizontal plane. [2 marks]

You can't fly fighter planes if you wear glasses...

There's a hidden bonus to having dodgy eyes — in the exam, you can take your specs off (discreetly) and have a look at the lenses to remind yourself what type is needed to correct short sight, long sight, or whatever it is you have. Cunning.

Physics of the Ear

Ears are pretty amazing — they convert sound into electrical energy, using some tiny bones and lots of even tinier hairs.

The **Intensity** of **Sound** is **Power** per **Unit Area**

The **intensity** of a sound wave is defined as the amount of sound **energy** that passes **per second per unit area** (perpendicular to the direction of the wave). That's **power per unit area.**

1) If the sound energy arriving at the ear per second is ***P***, then the intensity of the sound is:

$$I = \frac{P}{A}$$

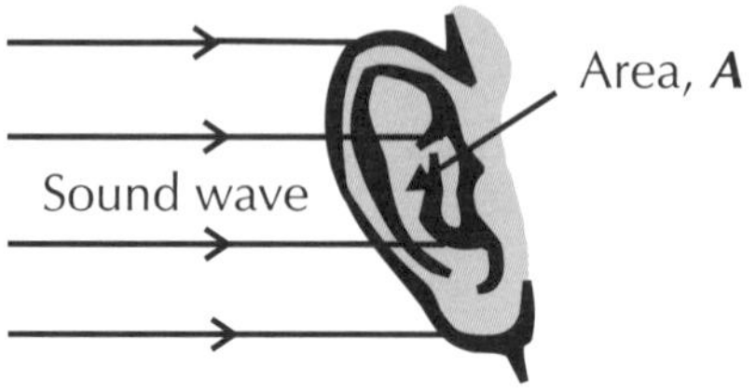

Felicity thought her waves were sound.

2) The SI unit of intensity is Wm^{-2}, but you'll often see decibels used instead (see p. 166).
3) For any wave, **intensity** $\propto$ **amplitude²** — so doubling the amplitude will result in four times the intensity.
4) Intensity is related to the **loudness** of sound (see p. 166).

The **Ear** has **Three Main Sections**

The ear consists of three sections: the **outer** ear (**pinna** and **auditory canal**), the **middle ear** (**ossicles** and **Eustachian tube**) and the **inner ear** (**semicircular canals**, **cochlea** and **auditory nerve**).

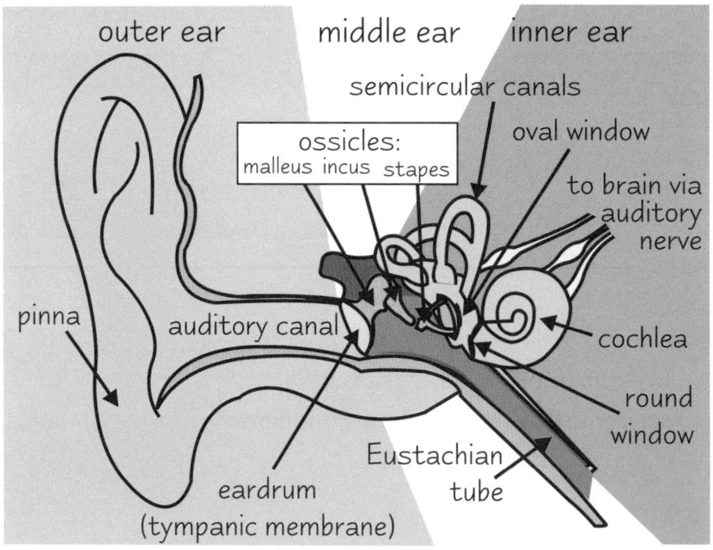

1) The **tympanic membrane** (eardrum) separates the **outer** and **middle** ears.
2) Although separated, the **outer** and **middle** ears both contain **air** at **atmospheric pressure**, apart from slight pressure variations due to sound waves. This pressure is maintained by **yawning** and **swallowing** — the middle ear is opened up to the outside via the **Eustachian tube** (which is connected to the throat).
3) The **oval** and **round windows** separate the **middle** and **inner** ears.
4) The **inner ear** is filled with fluid called **perilymph** (or **endolymph** in the **cochlear duct**). This fluid allows **vibrations** to pass to the basilar membrane in the **cochlea**.
5) The **semicircular canals** are involved with **maintaining balance**.

Physics of the Ear

The Ear acts as a Transducer, converting Sound Energy...

1) The **pinna** (external ear) acts like a funnel, channelling sound waves into the auditory canal — this **concentrates** the energy onto a **smaller area**, which increases the **intensity**.
2) The sound waves consist of **variations** in **air pressure**, which make the **tympanic membrane** (eardrum) **vibrate**.
3) The tympanic membrane is connected to the **malleus** — one of the **three tiny bones** (**ossicles**) in the middle ear. The malleus then passes the **vibrations** of the eardrum on to the **incus** and the **stapes** (which is connected to the **oval window**).
4) As well as **transmitting vibrations**, the **ossicles** have **two** other functions — **amplifying** the sound signal and **reducing** the **energy reflected back** from the inner ear.
5) The **oval window** has a much **smaller area** than the **tympanic membrane**. Together with the **increased force** produced by the ossicles, this results in **greater pressure variations** at the oval window.
6) The **oval window** transmits vibrations to the **fluid** in the **inner ear**.

...into Electrical Energy

1) Pressure waves in the fluid of the **cochlea** make the **basilar membrane** vibrate. Different regions of this membrane have different **natural frequencies**, from 20 000 Hz near the middle ear to 20 Hz at the other end.
2) When a sound wave of a particular **frequency** enters the inner ear, one part of the basilar membrane **resonates** and so vibrates with a **large amplitude**.
3) **Hair cells** attached to the basilar membrane trigger **nerve impulses** at this point of greatest vibration.
4) These **electrical impulses** are sent, via the **auditory nerve**, to the **brain**, where they are interpreted as **sounds**.

THE INNER EAR (UNROLLED):

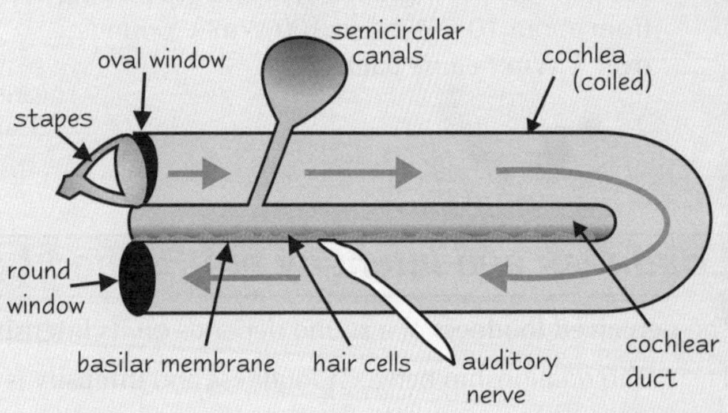

Practice Questions

Q1 What is meant by the 'intensity' of sound? What is the formula for intensity?

Q2 Sketch a diagram of the ear, labelling the structures within it.

Q3 Describe the function of the ossicles.

Q4 Explain why the relative size of the oval window and tympanic membrane is important.

Exam Question

Q1 The ear is designed to transduce sound energy into electrical energy.

(a) State the function of the pinna. [1 mark]

(b) Describe how sound energy is transmitted through the middle ear. [3 marks]

(c) The surface area of the tympanic membrane is around 14 times the area of the oval window. Show that this increases the amplitude of vibrations in the ear by a factor of approximately 3.74. [3 marks]

(d) Describe how pressure waves in the cochlea are converted to electrical impulses. [2 marks]

(e) Explain how the ear is able to encode the frequency of a sound in the information sent to the brain. [2 marks]

Ears are like essays — they have a beginning, middle and end...

Or outer, middle and inner, if we're being technical. Learn what vibrates where, and you'll be fine.

Intensity and Loudness

*The ear's sensitivity depends on the **frequency** and **intensity** of sounds, and deteriorates as you get older.*

Humans** can Hear a Limited Range of **Frequencies

1) Young people can hear frequencies ranging from about **20 Hz** (low pitch) up to **20 000 Hz** (high pitch). As you get older, the upper limit decreases.
2) Our ability to **discriminate between frequencies** depends on how **high** that frequency is. For example, between 60 and 1000 Hz, you can hear frequencies 3 Hz apart as **different pitches**. At **higher** frequencies, a **greater difference** is needed for frequencies to be distinguished. Above 10 000 Hz, pitch can hardly be discriminated at all.
3) The **loudness** of sound you hear depends on the **intensity** and **frequency** of the sound waves.
4) The **weakest intensity** you can hear — the **threshold of hearing**, I_0 — depends on the **frequency** of the sound wave.
5) The ear is **most sensitive** at around **3000 Hz**. For any given intensity, sounds of this frequency will be **loudest**.
6) Humans can hear sounds at intensities ranging from about 10^{-12} Wm^{-2} to 100 Wm^{-2}. Sounds **over 1 Wm^{-2}** cause **pain**.

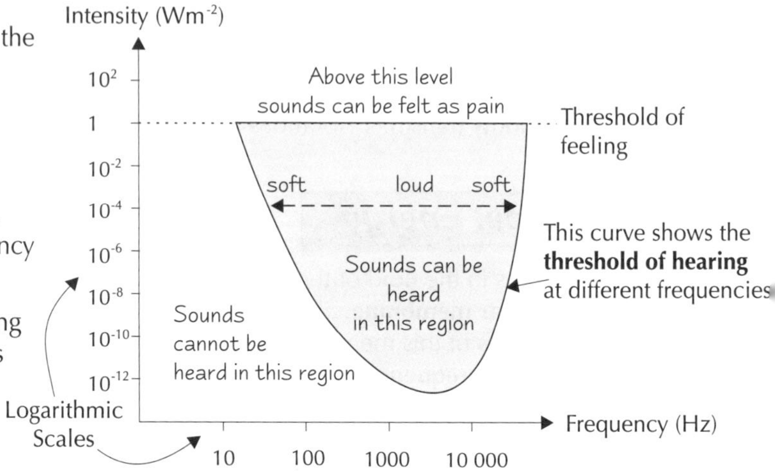

Loudness** and **Intensity** are Related **Logarithmically

The **perceived loudness** of a sound depends on its **intensity** (and its frequency — see above).

1) The relationship between **loudness** and **intensity** is **logarithmic**.
2) This means that **loudness**, L, goes up in **equal intervals** if **intensity**, I, increases by a **constant factor** (provided the frequency of the sound doesn't change).
3) E.g. if you **double** the intensity, **double it again** and so on, the **loudness** keeps going up in **equal steps**.

$$\Delta L \propto \log\left(\frac{I_2}{I_1}\right)$$

I_1 is the original intensity
I_2 is the new intensity
ΔL is increase in loudness

*The **Decibel Scale** is used for Measuring **Intensity Level***

1) You can often measure loudness using a **decibel meter**. The **decibel scale** is a **logarithmic scale** which actually measures **intensity level**.
2) The **intensity level**, ***IL***, of a sound of intensity I is defined as

$$IL = 10\log\left(\frac{I}{I_0}\right)$$

I = intensity
I_0 = threshold of hearing

3) I_0 is the **threshold of hearing** (the **lowest intensity** of sound that can be heard) at a frequency of **1000 Hz**.
4) The value of I_0 is **1×10^{-12} Wm^{-2}**.
5) The units of ***IL*** are **decibels** (dB). Intensity level can be given in **bels** — one decibel is a tenth of a bel — but decibels are usually a more convenient size.

*The **dBA Scale** is an **Adjusted Decibel Scale***

1) The **perceived loudness** of a sound depends on its **frequency** as well as its intensity. Two different frequencies with the **same loudness** will have **different intensity levels** on the dB scale.
2) The **dBA** scale is an **adjusted decibel scale** which is designed to take into account the **ear's response** to **different frequencies**.
3) On the **dBA scale**, sounds of the **same intensity level** have the **same loudness** for the average human ear.

Intensity and Loudness

You can Generate *Curves of Equal Loudness*

1) Start by generating a **control frequency** of **1000 Hz** at a particular **intensity level**.
2) Generate another sound at a different frequency. Vary the volume of this sound until it appears to have the **same loudness** as the 1000 Hz frequency. Measure the **intensity level** at this volume.
3) Repeat this for several different frequencies, and plot the resulting curve on a graph.
4) Change the **intensity level** of the **control frequency** and repeat steps two and three.
5) If you measure **intensity level** in **decibels**, then the **loudness** of the sound is given in **phons**.

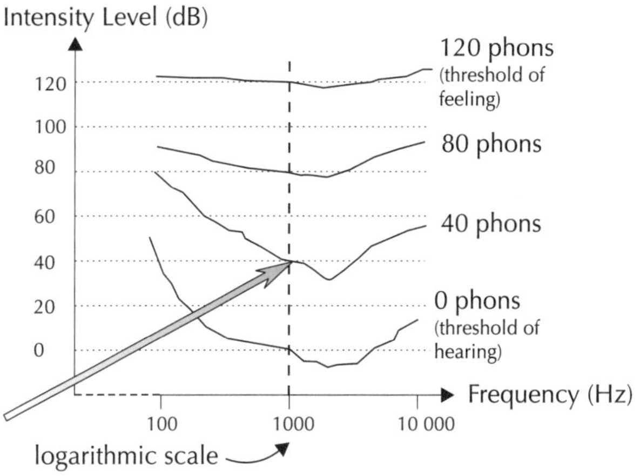

At 1000 Hz, the loudness in phons is the same value as the intensity level in decibels.

Hearing Deteriorates with Age and Exposure to Excessive Noise

1) As you get **older**, your hearing deteriorates **generally**, but **higher frequencies** are affected **most**.
2) Your ears can be damaged by **excessive noise**. This results in general hearing loss, but frequencies around **4000 Hz** are usually worst affected.
3) People who've worked with very **noisy machinery** have most hearing loss at the **particular frequencies** of the noise causing the damage.
4) **Equal loudness curves** can show hearing loss.
5) For a person with hearing loss, **higher intensity levels** are needed for the **same loudness**, when compared to a normal ear. A **peak** in the curve shows damage at a **particular** range of **frequencies**.

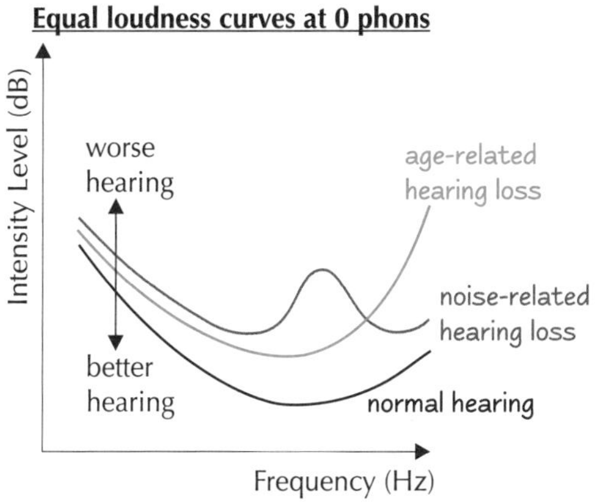

Practice Questions

Q1 Define the threshold of hearing and sketch a graph that shows how it depends on frequency.

Q2 How are curves of equal loudness generated?

Q3 What is the dB scale? How is the dBA scale different?

Exam Questions

Q1 A small siren, which can be regarded as a point source, emits sound waves at a frequency of 3000 Hz. The intensity of the sound is 0.94 Wm^{-2} at a distance of 10 m.

(a) State the accepted value of the threshold of hearing at 1000 Hz, I_0, in Wm^{-2}. [1 mark]

(b) Calculate the intensity level of the sound of the siren. [2 marks]

(c) Explain why the siren uses a frequency of 3000 Hz. [1 mark]

Q2 The diagram shows an equal loudness curve for a person suffering hearing loss and a person with normal hearing. The patient believes his hearing may have been damaged by working with noisy machinery. Does his equal loudness curve support this? Explain your answer. [3 marks]

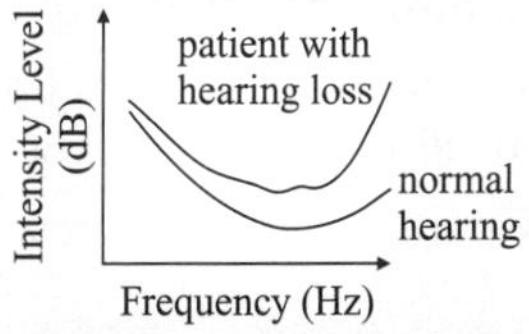

Saved by the decibel....

It's medical fact that prolonged loud noise damages your hearing, so you should really demand ear protection before you agree to do the housework — some vacuum cleaners are louder than 85 dBA — the 'safe' limit for regular exposure.

Physics of the Heart

There's a bit of biochemistry on this page — my, my, aren't you lucky people...

The **Heart** is a **Double Pump**

1) The heart is a **large muscle**. It acts as a **double pump**, with the **left**-hand side pumping blood from the **lungs** to the **rest of the body** and the **right**-hand side pumping blood from the **body** back to the **lungs**.
2) Traditionally, a diagram of the heart is drawn as though you're looking at it **from the front**, so the **right**-hand side of the heart is drawn on the **left**-hand side of the **diagram** and vice versa (just to confuse you).
3) Each side of the heart has **two chambers** — an **atrium** and a **ventricle** — separated by a **valve**.
4) **Blood** enters the **atria** from the veins, then the atria **contract**, squeezing blood into the **ventricles**. The **ventricles** then **contract**, squeezing the blood **out** of the heart into the **arteries**. The **valves** are there so that the blood doesn't go back into the atria when the ventricles contract.

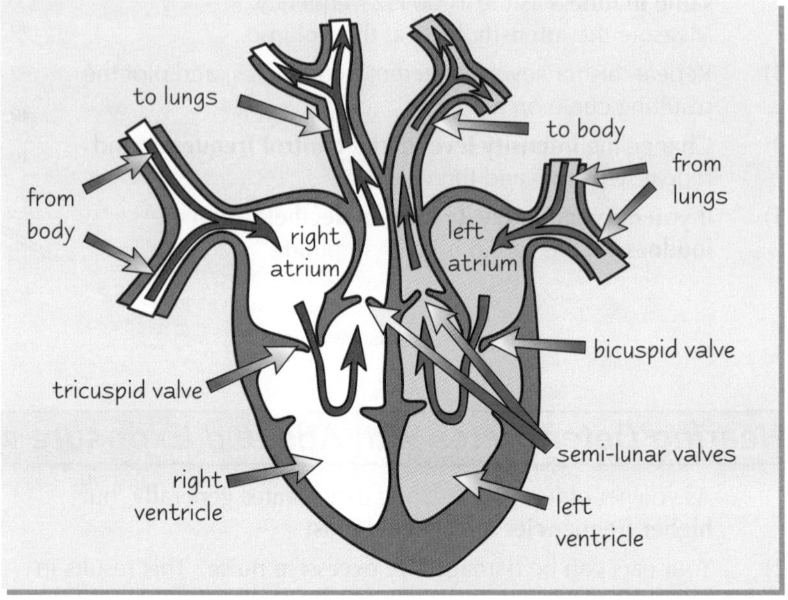

The Movement of **Sodium** and **Potassium Ions** Generates **Electrical Signals**

1) The movement of **sodium** and **potassium ions** across **cell membranes** in the heart is controlled by **diffusion** and a **sodium-potassium pump**.
2) The **pump** uses a series of **chemical reactions** to move **sodium ions** (Na^+) through the membrane and out of the cell, and to move **potassium ions** (K^+) into the cell.
3) By **diffusion**, particles tend to move from **high** to **low concentrations** (along a **concentration gradient**). So, as the pump builds up the ion concentrations, the ions try to **diffuse back** in the **opposite** direction.
4) So far, so good — but here's where it gets a bit more complicated.
5) For most of the time, the membrane **only lets K^+ ions** diffuse through, and **not** the Na^+ ions. That means there's a **net flow** of **positive charge out of** the cell and a **voltage** builds up until equilibrium is reached.
6) By convention, the potential on the **outside** of the membrane (where there's lots of sodium) is taken to be **0 V**, so the membrane potential on the inside is **negative** (–80 mV for heart muscle). You've now got a **polarised** membrane.

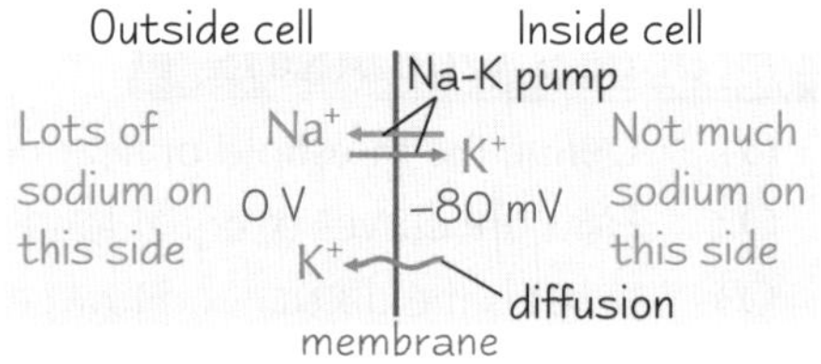

7) If the membrane is **stimulated** by an **electrical signal** it suddenly becomes **permeable** to **Na^+** as well. The sodium ions **rush** across the membrane towards the **negative side** (inside the cell), pushed by both the **Na^+ concentration gradient** and **electrostatic forces**.
8) This first **depolarises** the membrane, then **charges it up** the **other way** (called reverse-polarisation), reaching a **positive potential** of about 40 mV. This makes the heart muscle **CONTRACT**.
9) The membrane then becomes **impermeable** to **Na^+** ions again, but **very permeable** to **K^+** ions. The K^+ ions **diffuse** very quickly, repolarising the membrane, and the heart muscle **RELAXES**.
10) The Na-K pump then slowly restores the potential back to its **equilibrium polarised state**.

The sudden flip in potential is called the **action potential**. When this happens at one part of a membrane, it triggers the part next to it to do the same and so an electrical signal passes down the membrane.

Physics of the Heart

The Heart's **Pacemaker** is the **Sinoatrial Node**

1) A group of specialised cells at the **sinoatrial (SA) node** (in the wall of the right atrium) produce **electrical signals** that pulse about **70 times a minute**.
2) These signals spread through the **atria** and make them **contract** via the **action potential** (see previous page).
3) The signals then pass to the **atrioventricular (AV) node**, which **delays** the pulse for about **0.1 seconds** before passing it on to the **ventricles**.
4) The ventricles **contract** and the process repeats.

The Heart can be **Monitored** by an **Electrocardiograph (ECG)**

1) The **potential difference** between the **polarised** and **depolarised** heart cells produces a **weak electrical signal** at the surface of the body. This is plotted against time to give an **electrocardiogram (ECG)**, which can provide useful information about the **condition** of the heart.
2) A **normal** ECG, covering a **single heartbeat**, has **three** separate parts: a **P** wave, a **QRS** wave and a **T** wave.
3) The **P wave** corresponds to the **depolarisation** and **contraction** of the **atria**.
4) The **QRS wave** (about 0.2 seconds later) corresponds to the **depolarisation** and **contraction** of the **ventricles**. This completely swamps the trace produced by the repolarisation and relaxation of the atria.
5) Finally, the **T wave** (another 0.2 seconds later) corresponds to the **repolarisation** and **relaxation** of the **ventricles**.

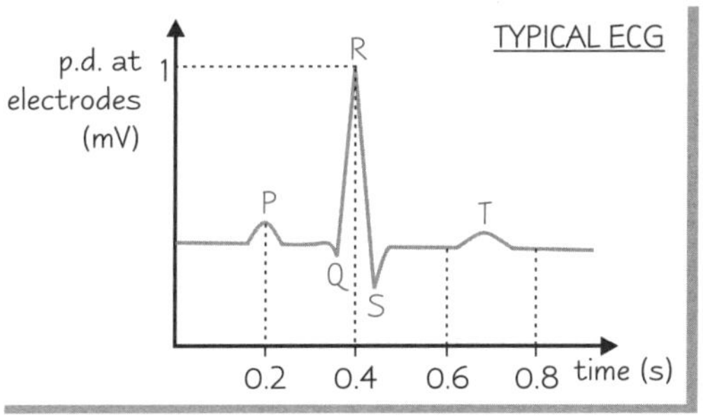

6) There are **12** standard ways of placing electrodes on the body to obtain an **ECG**, each producing a slightly different waveform. In all cases the signal is **heavily attenuated** (absorbed and weakened) by the body and needs to be amplified by a high impedance **amplifier**.
7) **Electrodes** are placed on the **chest** and the **limbs** where the arteries are close to the surface. The **right leg** is **never** used since it is **too far away** from the heart.
8) In order to get a **good electrical contact**, **hairs** and **dead skin** cells are removed and a **conductive gel** is used.

Practice Questions

Q1 Describe the basic structure of the heart, and the passage of blood through the heart, lungs and the body.

Q2 Describe how a membrane in heart muscle is initially polarised and what happens when it is stimulated.

Q3 Sketch a typical ECG trace and indicate the main features.

Exam Questions

Q1 A patient with a suspected heart condition is given an ECG.

(a) Describe how the patient's skin is prepared to ensure a good electrical contact with the electrodes. [2 marks]

(b) The patient's ECG shows a pause between the P and QRS waves.
Describe how the action of the heart corresponds to these waves. [3 marks]

Q2 Explain the processes of depolarisation and repolarisation in the formation of an action potential.
The quality of your written answer will be assessed in this question. [6 marks]

Q3 Outline how electrical signals are involved in the function of the heart. [4 marks]

Be still my beating sinoatrial node...

If you rely on the cast of ER to get your heart beating faster, console yourself that it's all very educational. Listen out for the machine that goes 'bip, bip, bip', and look for the P waves, QRS waves and T waves on the screen. If there aren't enough waves, the brave docs have to start shouting 'clear' and waving defibrillators around.

X-Ray Production

X-ray imaging is one kind of non-invasive diagnostic technique — these techniques let doctors see what's going on (or going wrong) inside your body, without having to open you up and have a look.

X-rays are Produced by Bombarding Tungsten with High Energy Electrons

1) In an X-ray tube, **electrons** are emitted from a **heated filament** and **accelerated** through a high **potential difference** (the **tube voltage**) towards a **tungsten anode**.

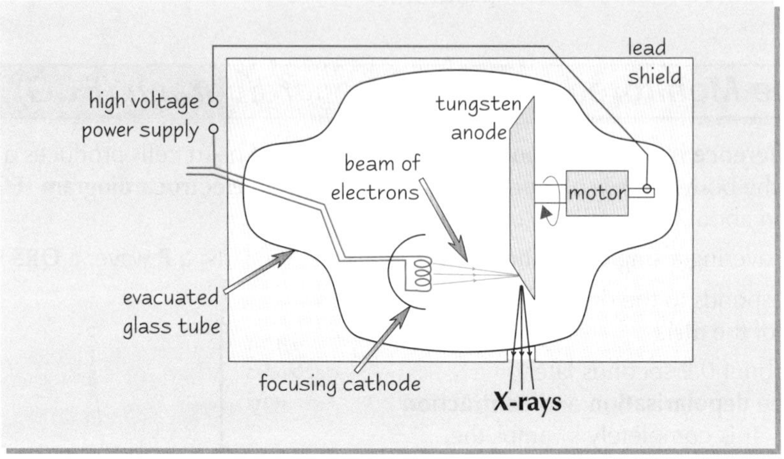

2) When the **electrons** smash into the **tungsten anode**, they **decelerate** and some of their **kinetic energy** is converted into **electromagnetic energy**, as **X-ray photons**. The tungsten anode emits a **continuous spectrum** of **X-ray radiation** — this is called **bremsstrahlung** ('braking radiation').

3) 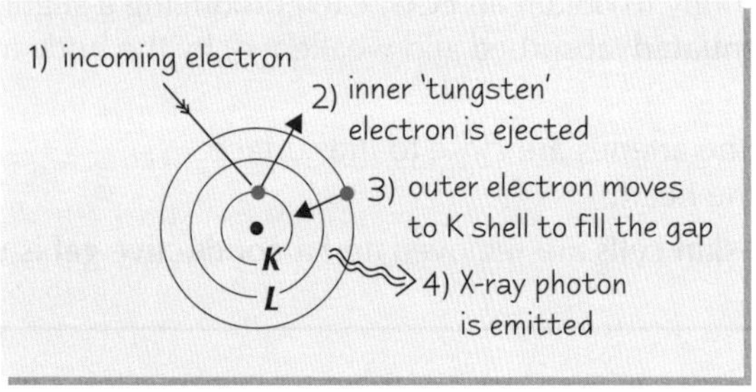

X-rays are also produced when beam electrons **knock out** other electrons from the **inner shells** of the **tungsten atoms**.

Electrons in the atoms' **outer shells** move into the **vacancies** in the **lower energy levels**, and **release energy** in the form of **X-ray photons**.

4) This process results in **line spectra** superimposed on a **continuous spectrum**.

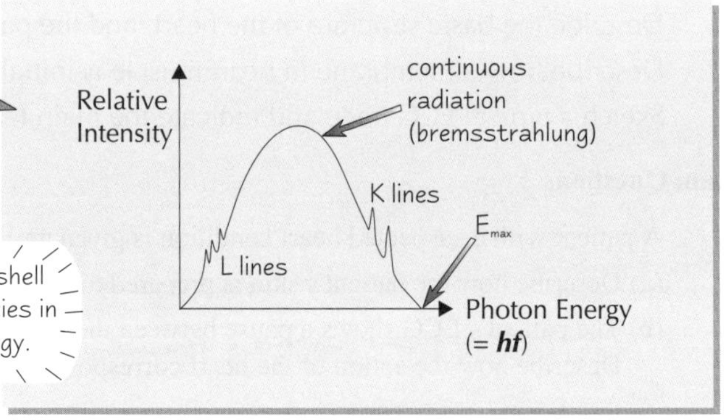

The K and L lines show which shell the electrons are filling vacancies in to cause the release of energy.

Brian's glasses gave a whole new meaning to 'undressing with your eyes'.

5) Only about **1%** of the electrons' **kinetic energy** is converted into **X-rays**. The rest is converted into **heat**, so, to avoid overheating, the tungsten anode is **rotated** at about 3000 rpm. It's also **mounted** on **copper** — this **conducts** the heat away effectively.

X-Ray Production

Beam Intensity and **Photon Energy** can be Varied

The **intensity** of the X-ray beam is the **energy per second per unit area** passing through a surface (at right angles). There are two ways to increase the **intensity** of the X-ray beam:

1) Increase the **tube voltage**. This gives the electrons **more kinetic energy**. Higher energy electrons can **knock out** electrons from shells **deeper** within the tungsten atoms — giving more 'spikes' on the graphs. Individual **X-ray photons** also have **higher maximum energies**.

   Intensity is approximately **proportional** to **voltage squared**.

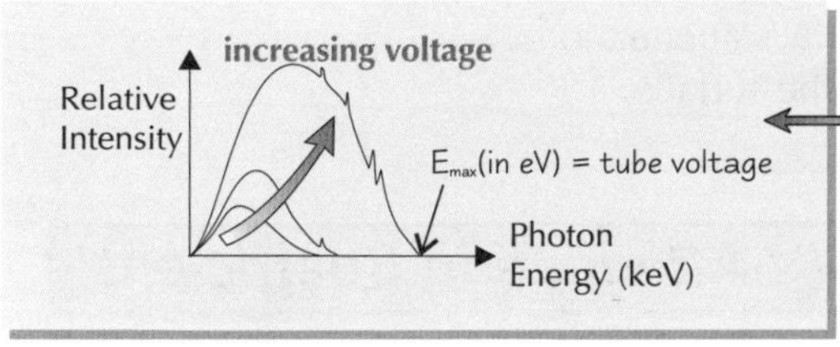

The intensity of the X-ray beam is related to the area under the graph.

2) Increase the **current** supplied to the filament. This liberates **more electrons per second**, which then produce **more X-ray photons per second**. Individual **photons** have the **same energy** as before.

   Intensity is approximately **proportional** to **current**.

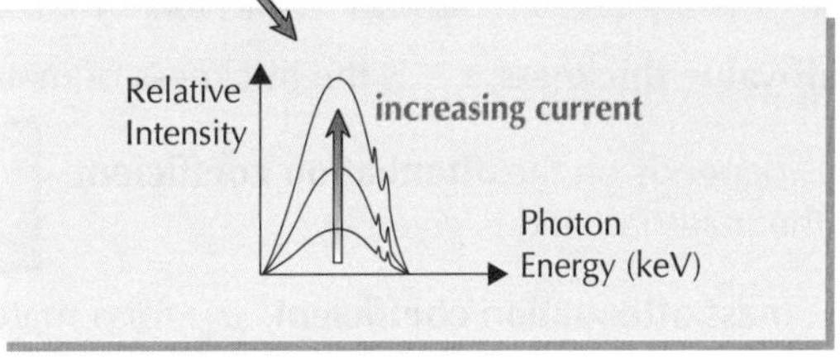

Radiographers try to Produce a **Sharp Image** and **Minimise** the **Radiation Dose**

Medical X-rays are a compromise between producing really sharp, clear images, whilst keeping the amount of radiation the patient is exposed to as low as possible. To do this, radiographers:

1) Put the **detection plate close** to the patient and the **X-ray tube far** away from the patient.
2) Make sure the patient **keeps still** — if they move around, the image will be blurred.
3) Put a **lead grid** between the patient and film to **stop** scattered radiation '**fogging**' the film and **reducing contrast**.
4) Use an **intensifying screen** next to the film surface. This consists of crystals that **fluoresce** — they **absorb X-rays** and re-emit the energy as **visible light**, which helps to develop the photograph quickly. A shorter exposure time is needed, keeping the patient's radiation dose lower.

Practice Questions

Q1 Draw a diagram of an X-ray tube and explain how a typical X-ray spectrum is produced.

Q2 What is meant by the 'intensity'of an X-ray beam?

Q3 Give two ways in which the intensity of an X-ray beam can be increased.

Q4 What measures can be taken to produce a high quality X-ray image while reducing the patient's radiation dose?

Exam Question

(Charge on an electron (e) = -1.6×10^{-19} J. Mass of an electron (m_e) = 9.11×10^{-31} kg.)

Q1 An X-ray tube is connected to a potential difference of 30 kV.

(a) Sketch a graph of relative intensity against photon energy (in eV) for the resulting X-ray spectrum, and indicate its main features. [3 marks]

(b) Show how the graph in (a) would change if the tube voltage were increased. [2 marks]

(c) Calculate the velocity of the electrons arriving at the anode. [4 marks]

Situation vacant — electron needed for low energy position...

I have a question — why, when something could have a nice, simple name like 'braking radiation', do scientists insist on giving it a much fancier one? 'Bremsstrahlung' just sounds baffling — well, unless you speak German of course.

X-Ray and MRI Imaging

So, you know how X-rays are produced and what the radiographer does — but why, I hear you cry, do some bits of you (i.e. your bones) show up nicely in an X-ray image, while others fade into the background? Attenuation, that's why...

X-Rays are Attenuated when they Pass Through Matter

When X-rays pass through matter (e.g. a patient's body), they are **absorbed** and **scattered**. The intensity of the X-ray beam **decreases** (attenuates) **exponentially** with the **distance from** the **surface**, according to the material's **attentuation coefficient**.

$$I = I_0 e^{-\mu x}$$

where I is the intensity of the X-ray beam, I_0 is the initial intensity, μ is the material's attenuation coefficient and x is the distance from the surface.

Half-value Thickness Depends on a Material's Attenuation Coefficient

1) **Half-value thickness**, $x_{\frac{1}{2}}$, is the thickness of material required to **reduce** the **intensity** to **half** its **original value**.

   This depends on the **attenuation coefficient** of the material, and is given by:

   $$x_{\frac{1}{2}} = \frac{\ln 2}{\mu}$$

2) The **mass attenuation coefficient**, μ_m, for a material of density ρ is given by:

   $$\mu_m = \frac{\mu}{\rho}$$

X-rays are Absorbed More by Bone than Soft Tissue

1) X-rays are **attenuated** by **absorption** and **scattering**. How much **energy is absorbed** by a **material** depends on its **atomic number**.
2) So tissues containing atoms with **different atomic numbers** (e.g. **soft tissue** and **bone**) will **contrast** in the X-ray image.
3) If the tissues in the region of interest have similar attenuation coefficients then artificial **contrast media** can be used — e.g. **barium meal**.
4) **Barium** has a **high atomic number**, so it shows up clearly in X-ray images and can be followed as it moves along the patient's digestive tract.

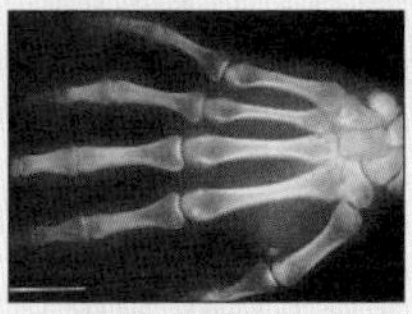

Bones show up brightly in X-ray images because they absorb more X-rays than the surrounding soft tissue.

CT Scans use X-rays to Produce High-Quality Images

1) **Computed tomography** (CT) scans produce an image of a **two-dimensional slice** through the body.
2) A narrow **X-ray beam** consisting of a single wavelength (**monochromatic**) **rotates** around the body and is picked up by thousands of **detectors**. The detectors feed the signal to a **computer**.
3) The computer works out how much attenuation has been caused by each part of the body and produces a very **high quality** image.
4) However, the machines are **expensive** and the scans involve a **high radiation dose** for the patient.

Fluoroscopy is used to Create Moving Images

1) **Moving images** can be created using a **fluorescent screen** and an **image intensifier**. This is useful for **imaging organs** as they **work**.
2) **X-rays** pass through the patient and hit the **fluorescent screen**, which **emits light**.
3) The **light** causes **electrons** to be emitted from the **photocathode**.
4) The **electrons** travel through the **glass tube** towards the **fluorescent viewing screen**, where they form an image. Electrodes in the glass tube **focus** the **image** onto the screen.

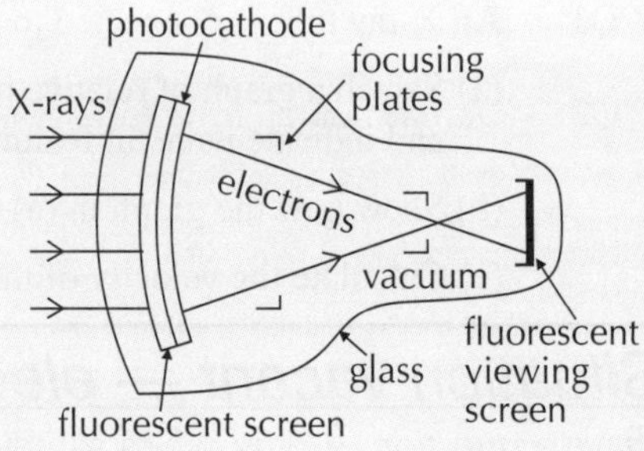

X-Ray and MRI Imaging

Magnetic Resonance Imaging, or MRI to you and me, is another form of non-invasive diagnostic imaging — enjoy.

Magnetic Resonance can be used to Create Images

1) The patient lies in the centre of a huge **superconducting magnet** that produces a **uniform magnetic field.** The magnet needs to be **cooled** by **liquid helium** — this is partly why the scanner is so **expensive**.
2) Radio frequency **coils** are used to transmit **radio waves**, which **excite hydrogen nuclei** in the patient's body.
3) When the radio waves are switched off, the hydrogen nuclei relax and emit electromagnetic energy — this is the **MRI signal** (more details below). The radio frequency coils **receive the signal** and send it to a **computer**.
4) The computer **measures** various quantities of the MRI signal — amplitude, frequency, phase — and **analyses** them to generate an **image** of a **cross-section** through the body.

Contrast can be Controlled by Varying the Pulses of Radio Waves

1) Radio waves are applied in **pulses**. Each short pulse **excites** the hydrogen nuclei and then allows them to **relax** and emit a signal. The response of **different tissue types** can be enhanced by varying the **time between pulses**.
2) Tissues consisting of **large molecules** such as **fat** are best imaged using **rapidly repeated pulses**. This technique is used to image the internal **structure** of the body.
3) Allowing **more time** between pulses enhances the response of **watery** substances. This is used for **diseased** areas.

MRI has Advantages and Disadvantages

ADVANTAGES:
1) There are **no** known **side effects**.
2) An image can be made for any slice in any **orientation** of the body.
3) High quality images can be obtained for **soft tissue** such as the **brain**.
4) **Contrast** can be **weighted** in order to investigate different situations.

DISADVANTAGES:
1) The imaging of **bones** is very **poor**.
2) Some people suffer from **claustrophobia** in the scanner.
3) Scans can be **noisy** and take a **long time**.
4) MRI can't be used on people with **pacemakers** or some **metal implants** — the strong magnetic fields would be very harmful.
5) Scanners **cost millions** of pounds.

Practice Questions

Q1 Outline why bone shows up in X-ray images.

Q2 What are 'artificial contrast media' used for? Give an example.

Q3 Give two advantages of MRI compared to X-ray imaging.

Exam Questions

Q1 The half-value thickness for aluminium is 3 mm for 30 keV X-ray photons.

(a) Explain what is meant by the term 'half-value thickness'. [1 mark]

(b) Determine the thickness of aluminium needed to reduce the intensity of a homogeneous beam of X-rays at 30 keV to 1% of its initial value. [4 marks]

Q2 Outline how an MRI scanner is used to produce an image of a section of a patient's body.
The quality of your written answer will be assessed in this question. [6 marks]

Q3 Discuss the advantages and disadvantages of MRI scanning as a medical imaging technique. [6 marks]

I've got attenuation coefficient disorder — I get bored really easily...

OK, so it hasn't been the easiest of pages. But at least now you know why people sit in vats of baked beans to raise money for their local hospital to buy an MRI scanner. Though perhaps you need A2 Psychology to understand the beans part.

Ultrasound Imaging

Ultrasound is a 'sound' with higher frequencies than we can hear.

Ultrasound has a Higher Frequency than Humans can Hear

1) Ultrasound waves are **longitudinal** waves with **higher frequencies** than humans can hear (>20 000 Hz).
2) For **medical** purposes, frequencies are usually from **1** to **15 MHz**.
3) When an ultrasound wave meets a **boundary** between two **different materials**, some of it is **reflected** and some of it passes through (undergoing **refraction** if the **angle of incidence** is **not 90°**).
4) The **reflected waves** are detected by the **ultrasound scanner** and are used to **generate an image**.

The Amount of Reflection depends on the Change in Acoustic Impedance

1) The **acoustic impedance**, Z, of a medium is defined as: $Z = \rho c$
Z has units of $kgm^{-2}s^{-1}$.
ρ = density of the material, in kgm^{-3}
c = speed of sound in the medium, in ms^{-1}
2) Say an ultrasound wave travels through a material with an impedance Z_1. It hits the boundary between this material and another with an impedance Z_2. The incident wave has an intensity of I_i.
3) If the two materials have a **large difference** in **impedance**, then **most** of the energy is **reflected** (the intensity of the reflected wave I_r will be high). If the impedance of the two materials is the **same** then there is **no reflection**.
4) The **fraction** of wave **intensity** that is reflected is called the **intensity reflection coefficient**, α.

$$\alpha = \frac{I_r}{I_i} = \left(\frac{Z_2 - Z_1}{Z_2 + Z_1}\right)^2$$

You don't need to learn this equation. Just practise using it.

There are Advantages and Disadvantages to Ultrasound Imaging

ADVANTAGES:
1) There are **no** known **hazards** — in particular, **no** exposure to **ionising radiation**.
2) It's good for imaging **soft tissues**, since you can obtain **real-time** images — X-ray fluoroscopy can achieve this, but involves a huge dose of radiation.
3) Ultrasound devices are relatively **cheap** and **portable**.

DISADVANTAGES:
1) Ultrasound **doesn't penetrate bone** — so it **can't** be used to **detect fractures** or examine the **brain**.
2) Ultrasound **cannot** pass through **air spaces** in the body (due to the **mismatch** in **impedance**) — so it can't produce images from behind the lungs.
3) The **resolution** is **poor** (about 10 times worse than X-rays), so you **can't see** fine **detail**.

Ultrasound Images are Produced Using the Piezoelectric Effect

1) **Piezoelectric crystals** produce a **potential difference** when they are **deformed** (squashed or stretched) — the rearrangement in structure displaces the **centres of symmetry** of their electric **charges**.

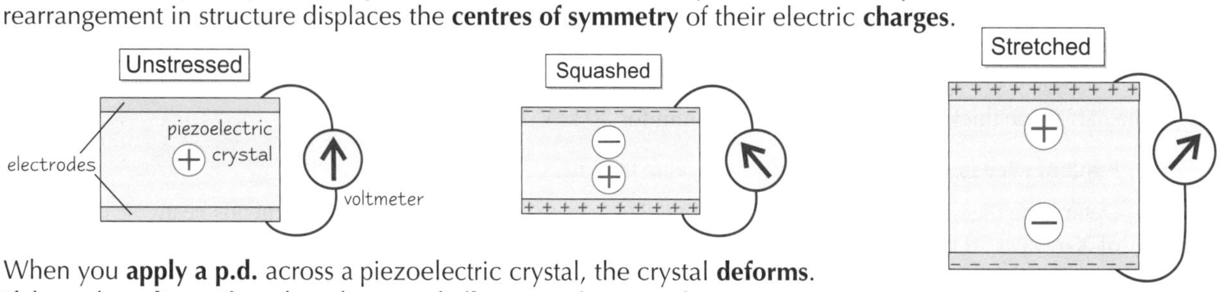

2) When you **apply a p.d.** across a piezoelectric crystal, the crystal **deforms**.
If the p.d. is **alternating**, then the crystal **vibrates** at the **same frequency**.
3) A piezoelectric crystal can act as a **receiver** of **ultrasound**, converting **sound waves** into **alternating voltages**, and also as a **transmitter**, converting **alternating voltages** into **sound waves**.
4) Ultrasound devices use **lead zirconate titanate** (**PZT**) crystals. The **thickness** of the crystal is **half the wavelength** of the ultrasound that it produces. Ultrasound of this frequency will make the crystal **resonate** (like air in an open pipe — see p.102) and produce a large signal.
5) The PZT crystal is **heavily damped**, to produce **short pulses** and **increase** the **resolution** of the device.

Ultrasound Imaging

You need a **Coupling Medium** between the **Transducer** and the **Body**

1) **Soft tissue** has a very different **acoustic impedance** from **air**, so almost all the ultrasound **energy** is **reflected** from the surface of the body if there is air between the **transducer** and the **body**.
2) To avoid this, you need a **coupling medium** between the transducer and the body — this **displaces** the **air** and has an impedance much closer to that of body tissue. The use of **coupling media** is an example of **impedance matching**.
3) The coupling medium is usually an **oil** or **gel** that is smeared onto the skin.

The **A-Scan** is a **Range Measuring** System

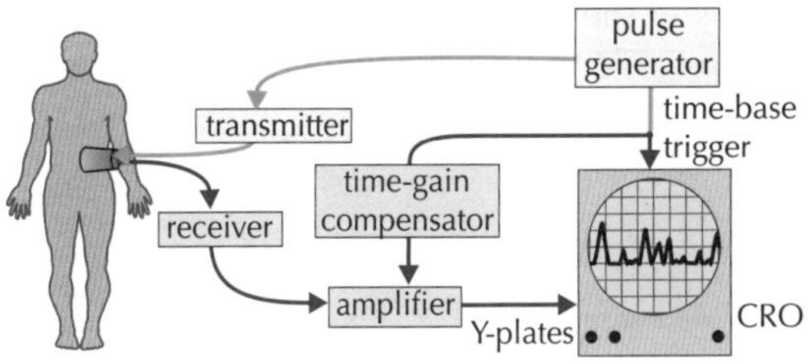

1) The **amplitude scan** (**A-Scan**) sends a short **pulse** of ultrasound into the body simultaneously with an **electron beam** sweeping across a cathode ray oscilloscope (**CRO**) screen.
2) The scanner receives **reflected** ultrasound pulses that appear as **vertical deflections** on the CRO screen.
3) **Weaker** pulses (that have travelled further in the body and **arrive later**) are **amplified** more to avoid the loss of valuable data — this process is called **time-gain compensation** (**TGC**).
4) The **horizontal positions** of the reflected pulses indicate the **time** the 'echo' took to return, and are used to work out **distances** between structures in the body (e.g. the **diameter** of a **baby's head** in the uterus).
5) A **stream** of pulses can produce a **steady image** on the screen (due to **persistence of vision** — see p.161), although modern CROs can store a digital image after just one exposure.

In a **B-Scan**, the **Brightness** Varies

1) In a **brightness scan** (**B-Scan**), the electron beam sweeps **down** the screen rather than across.
2) The amplitude of the reflected pulses is displayed as the **brightness** of the spot.
3) You can use a **linear array** of transducers to produce a **two-dimensional** image.

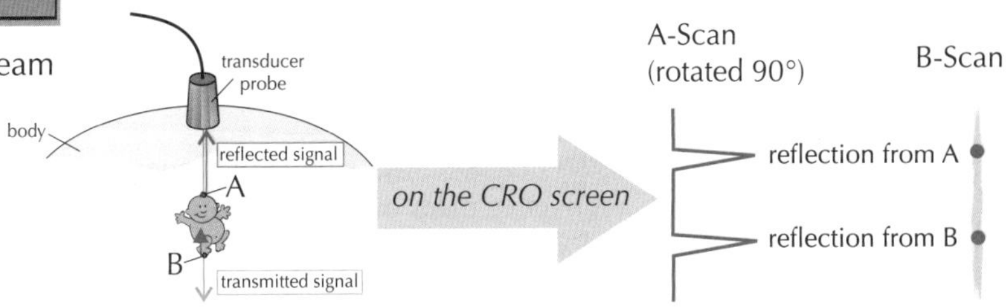

Practice Questions

Q1 What are the main advantages and disadvantages of imaging using ultrasound?

Q2 How are ultrasound waves produced and received in an ultrasound transducer?

Q3 Define acoustic impedance.

Exam Questions

Q1 (a) What fraction of intensity is reflected when ultrasound waves pass from air to soft tissue? Use $Z_{air} = 0.430 \times 10^3$ kgm^{-2}s^{-1}, $Z_{tissue} = 1630 \times 10^3$ kgm^{-2}s^{-1}. [2 marks]

(b) Calculate the ratio between the intensity of the ultrasound that **enters** the body when a coupling gel is used ($Z_{gel} = 1500 \times 10^3$ kgm^{-2}s^{-1}) and when none is used. Give your answer to the nearest power of ten. [4 marks]

Q2 (a) The acoustic impedance of a certain soft tissue is 1.63×10^6 kgm^{-2}s^{-1} and its density is 1.09×10^3 kgm^{-3}. Show that ultrasound travels with a velocity of 1.50 kms^{-1} in this medium. [2 marks]

(b) The time base on a CRO was set to be 50 μscm^{-1}. Reflected pulses from either side of a fetal head are 2.4 cm apart on the screen. Calculate the diameter of the fetal head if the ultrasound travels at 1.5 kms^{-1}. [4 marks]

Ultrasound — Mancunian for 'très bien'

You can use ultrasound to make images in cases where X-rays would do too much damage — like to check up on the development of a baby in the womb. You have to know what you're looking for though, or it just looks like a blob.

Endoscopy

Phew, that ultrasound stuff wasn't exactly a walk in the park — luckily, endoscopes are easier to understand...

Optical Fibres Use Total Internal Reflection to Transmit Light

1) **Optical fibres** are a bit like electric wires — but instead of carrying current they **transmit light**.
2) A typical optical fibre consists of a **glass core** (about 5 μm to 50 μm in diameter) **surrounded** by a **cladding**, which has a slightly **lower refractive index**.
3) The **difference** in refractive index means that light travelling along the fibre will be **reflected** at the **cladding-core interface**.

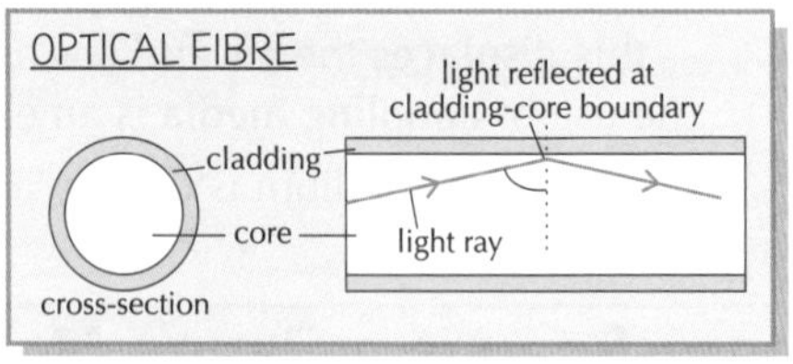

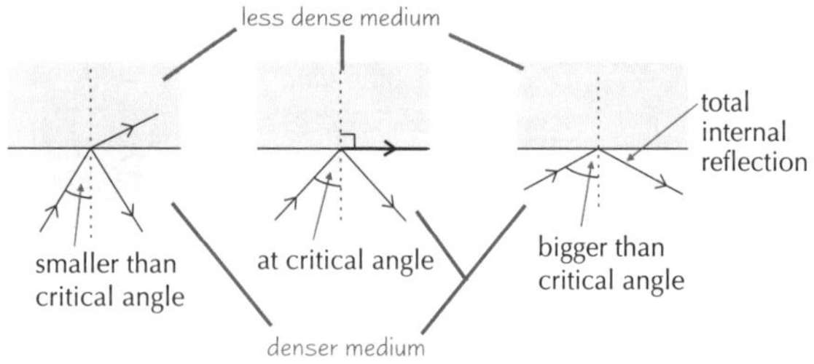

4) If the light ray's **angle of incidence** is **less than or equal** to a **critical angle**, some light will be **lost** out of the fibre.
5) But if the **angle of incidence** is **larger** than the **critical angle**, the light ray will be **completely reflected** inside the fibre.
6) This phenomenon is called **total internal reflection** and means that the ray **zigzags** its way along the fibre — so long as the fibre isn't too curved.

The Critical Angle for an Optical Fibre can be Worked Out

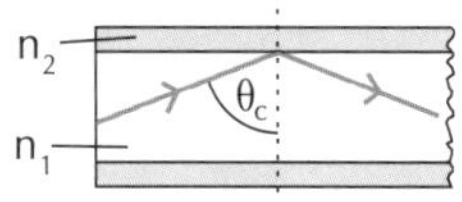

1) The **critical angle**, θ_c, depends on the **refractive index** of the **core**, n_1, and **cladding**, n_2, in an optical fibre.
2) You can work out this value using the formula:

$$\sin\theta_c = \frac{n_2}{n_1}$$

Example

An optical fibre consists of a core with a refractive index of 1.5 and cladding with a refractive index of 1.4.

a) What is the critical angle at the core-cladding boundary?

$$\theta_c = \sin^{-1}\left(\frac{n_2}{n_1}\right) = \sin^{-1}\left(\frac{1.4}{1.5}\right) = 69°$$

b) Would total internal reflection occur if the incident angle of light is 70°?

70° > θ_c, so total internal reflection would occur.

If some of this is sounding familiar, it's because you did it in AS. The equation for critical angle is derived from Snell's law of refraction.

Lots of Optical Fibres can be Bundled Together

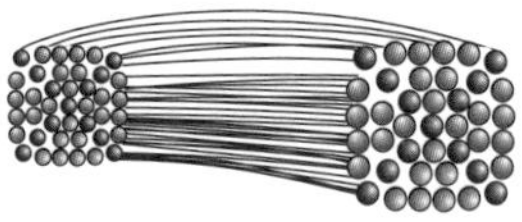

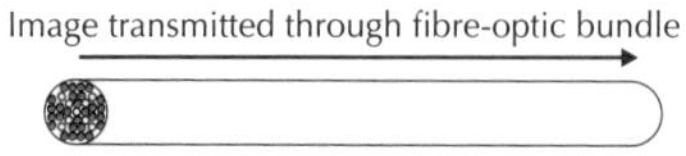

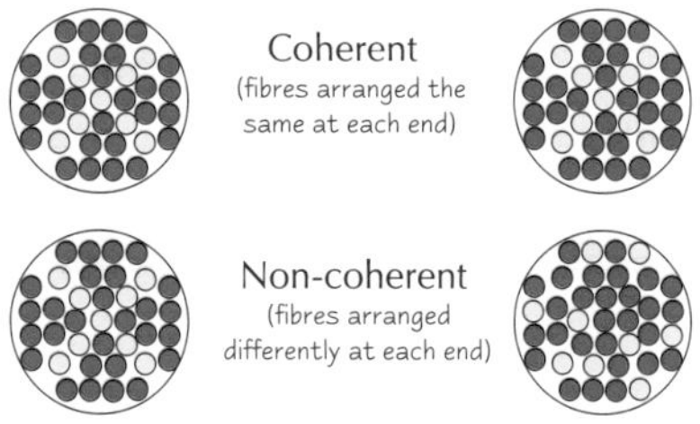

1) An **image** can be transmitted along a **bundle** of optical fibres.
2) This can only happen if the **relative positions** of fibres in a bundle are the **same** at each end (otherwise the image would be jumbled up) — a fibre-optic bundle in this arrangement is said to be **coherent**.
3) The **resolution** (i.e. how much detail can be seen) depends on the **thickness** of the fibres. The thinner the fibres, the **more detail** that can be resolved — but thin fibres are more **expensive** to make.
4) Images can be **magnified** by making the diameters of the fibres get **gradually larger** along the length of the bundle.
5) If the relative **position** of the fibres **does not** remain the same between each end the bundle of fibres is said to be **non-coherent**.
6) **Non-coherent bundles** are much easier and **cheaper** to make. They **can't** transmit an **image** but they can be used to get **light** to hard-to-reach places — kind of like a flexible **torch**.

Endoscopy

Endoscopes Use Optical Fibres to Create an **Image**

1) An **endoscope** consists of a **long tube** containing **two bundles** of fibres — a **non-coherent** bundle to carry **light** to the area of interest and a **coherent** bundle to carry an **image** back to the eyepiece.
2) Endoscopes are widely used by surgeons to examine inside the body
3) An **objective lens** is placed at the **distal** end (**furthest from the eye**) of the **coherent** bundle to form an image, which is then transmitted by the fibres to the **proximal** end (**closest to the eye**) where it can be **viewed** through an **eyepiece**.
4) The **endoscope tube** can also contain a **water channel**, for cleaning the objective lens, a **tool aperture** to perform **keyhole surgery** and a **CO_2 channel** which allows CO_2 to be pumped into the area in front of the endoscope, making more room in the body.

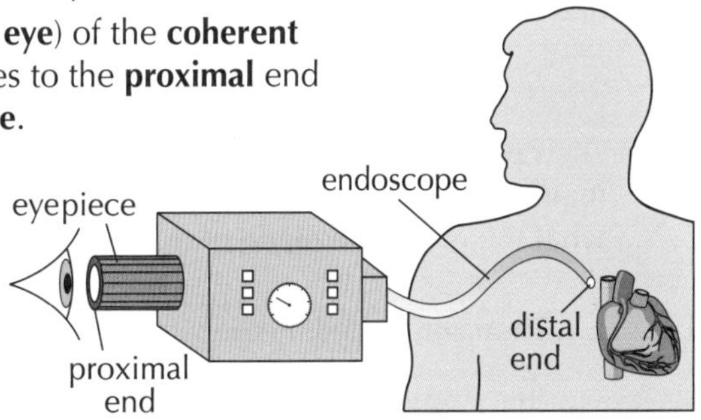

Endoscopes are Used in **Keyhole Surgery**

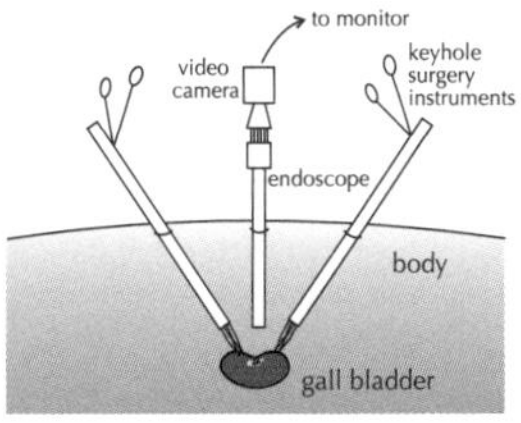

1) **Traditional** surgery needs a **large cut** to be made in the body so that there's **room** for the surgeons to get in and perform an **operation**.
2) This means that there's a **large risk of infection** to the exposed tissues and that permanent **damage** could be done to the patient's **body**.
3) New techniques in **minimally invasive surgery** (MIS or **keyhole surgery**) mean that only a **few small holes** need to be cut into the body.
4) An **endoscope** can be used in keyhole surgery to show the surgeon an **image** of the area of interest. **Surgical instruments** are passed through the endoscope tube, or through additional **small holes** in the body, so that the **operation** can be carried out.
5) **Common procedures** include the removal of the **gall bladder**, investigation of the **middle ear**, and removal of abnormal polyps in the **colon** so that they can be investigated for **cancer**.
6) **Recovery times** tend to be **quicker** for keyhole surgery, so the **patient** can usually **return home** on the **same day** — which makes it much **cheaper** for the hospital and **nicer** for the patient.

Practice Questions

Q1 What condition must be satisfied for total internal reflection to occur?

Q2 Explain the difference between a coherent and a non-coherent bundle of fibres.

Q3 What are the main features of an endoscope?

Q4 How have endoscopes revolutionised some surgical techniques?

Exam Questions

Q1 A beam of light is transmitted through an optical fibre.
The refractive index of the fibre's core is 1.35 and the refractive index of its cladding 1.30.
(a) Determine the critical angle for the core-cladding boundary. [1 mark]
(b) Explain why the angle of incidence of the beam of light should be kept at or above the critical angle. [2 marks]

Q2 Coherent fibre-optic bundles can be used to transmit images.
Describe the main features of the structure of a coherent fibre-optic bundle, and explain why each feature is important for the bundle's function. [4 marks]

If you ask me, physics is a whole bundle of non-coherentness...

If this is all getting too much, and your brain is as fried as a pork chipolata, just remember the wise words of revision wisdom from the great Spike Milligan — Ying tong, ying tong, ying tong, ying tong, ying tong, iddly-I-po, iddly-I-po...

Charge/Mass Ratio of the Electron

e/m was known for quite a long time before anyone came up with a way to measure e or m separately.

Cathode Ray is an Old-Fashioned name for a Beam of Electrons

1) The phrase '**cathode ray**' was first used in 1876, to describe the **glow** that appears on the wall of a discharge tube like the one in the diagram, when a **potential difference** is applied across the terminals.
2) The **rays** seemed to come from the **cathode** (hence their name) and there was a lot of argument about **what** the rays were made of.

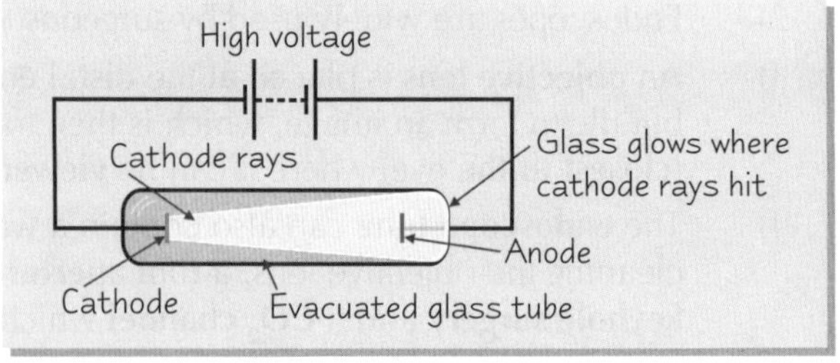

3) **J. J. Thomson** ended the debate in 1897, when he demonstrated (see opposite) that cathode rays:

a) have **energy**, **momentum** and **mass**,
b) have a **negative charge**,
c) have the **same properties**, no matter **what gas** is in the tube and what the **cathode** is made of,
d) have a **charge to mass ratio** much **bigger** than that of **hydrogen** ions. So they either have a **tiny mass**, or a much higher charge — Thomson assumed they had the same size charge as hydrogen ions.

Thomson concluded that **all atoms** contain these 'cathode ray particles', that were later called **electrons**.

Electron Beams are Produced by Thermionic Emission

1) When you **heat** a **metal**, its **free electrons** gain a load of **thermal energy**.
2) Give them **enough energy** and they'll **break free** from the surface of the metal — this is called **thermionic emission**. (Try breaking the word down — think of it as '**therm**' [to do with heat] + '**ionic**' [to do with charge] + '**emission**' [giving off] — so it's 'giving off charged particles when you heat something'.)
3) Once they've been emitted, the electrons can be **accelerated** by an **electric field** in an **electron gun**:

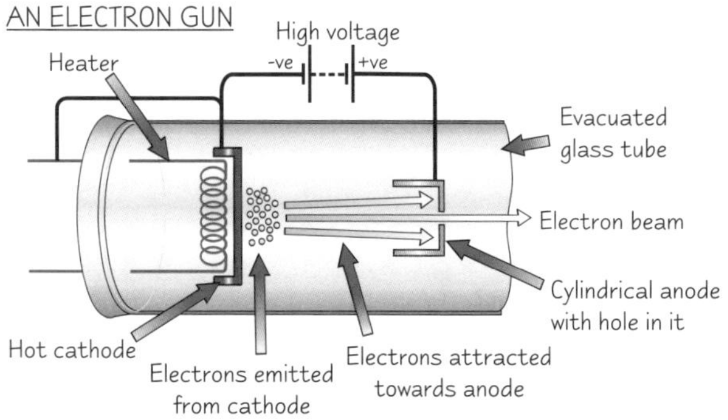

A **heating coil** heats the metal cathode. The electrons that are emitted are **accelerated** towards the **cylindrical anode** by the electric field set up by the high voltage.

Some electrons pass through a **little hole** in the **anode**, making a narrow electron beam. The electrons in the beam move at a **constant velocity** because there's **no field** beyond the anode — i.e., there's **no force**.

The Electronvolt is Defined Using Accelerated Charges

1) The **kinetic energy** that a particle with charge Q **gains** when it's **accelerated** through a p.d. of V volts is QV joules. That just comes from the definition of the **volt** (JC^{-1}).
2) If you replace Q in the equation with the charge of a **single electron**, **e**, you get: $\frac{1}{2}mv^2 = eV$
3) From this you can define a new **unit of energy** called the **electronvolt** (**eV**):

1 electronvolt is the **kinetic energy carried** by an **electron** after it has been **accelerated** through a **potential difference** of **1 volt**.

4) So, the **energy in eV** of an electron accelerated by a potential difference is:

energy gained by electron (eV) = accelerating voltage (V)

The unit MeV is the mega-electronvolt (equal to 1.6×10^{-13} J) and GeV is the giga-electronvolt (1.6×10^{-10} J).

Conversion factor: $1 \text{ eV} = 1.6 \times 10^{-19}$ J

Charge/Mass Ratio of the Electron

Thomson *Measured the* ***Specific Charge*** *of the* ***Electron***

1) The **specific charge** or **charge/mass ratio** of a charged particle is just its **charge** per unit **mass**.
2) There are a **few different ways** of measuring it, and you need to know about **one** of them. This isn't the method that Thomson used, but that's not important.

Check out Unit 4: Section 4 — Magnetic Fields, if you're having trouble with the experiment.

Measuring the Charge/Mass Ratio of an Electron

1) Electrons are charged particles, so they can be deflected by an **electric** or a **magnetic field**. This method uses a magnetic field in a piece of apparatus called a **fine beam tube**.
2) When the beam of electrons from the **electron gun** (see previous page) passes through the low-pressure gas, the hydrogen atoms along its path **absorb energy**. As the electrons in these **excited hydrogen atoms** fall back to the ground state, they **emit light**. The electron beam is seen as a **glowing trace** through the gas.

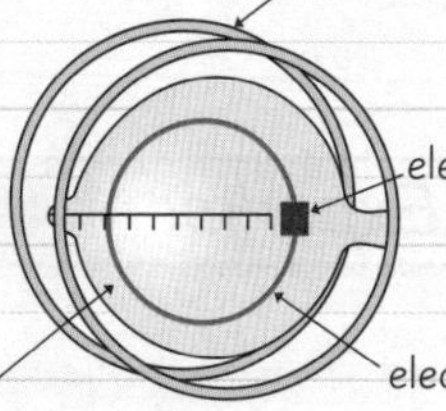

3) Two circular **magnetic field coils** either side generate a **uniform magnetic field** inside the tube.
4) The electron beam is initially fired at **right angles** to the **magnetic field**, so the beam curves round in a **circle**.
5) This means that the **magnetic force** on the electron (see p. 117) is acting as a **centripetal force** (see p. 97). So the radius of the circle is given by:

$$\frac{mv^2}{r} = Bev$$

where m is the mass of an electron, e is the charge on an electron, B is the magnetic field strength, v is the velocity of the electron and r is the radius of the circle.

6) From the previous page, you've got an equation that you can rearrange to give v in terms of the **accelerating potential** of the electron gun. If you substitute that expression for v into the equation above (and tidy it all up a bit) you get:

$$\frac{e}{m} = \frac{2V}{B^2r^2}$$

where m is the mass of an electron, e is the charge on an electron, B is the magnetic field strength, V is the accelerating potential and r is the radius of the circle.

You can **measure** all the quantities on the **right-hand side** of the equation using the **fine beam tube**, leaving you with the **specific charge, *e/m***. It turns out that ***e/m*** (1.76×10^{11} Ckg^{-1}) is about **1800 times greater** than the **specific charge of a hydrogen ion** or **proton** (9.58×10^7 Ckg^{-1}). And the **mass** of a **proton** is about **1800 times greater** than the **mass** of an **electron** — **Thomson was right**, electrons and protons do have the **same size charge**.

Practice Questions

Q1 What is meant by thermionic emission?

Q2 Sketch a labelled diagram of an electron gun that could be used to accelerate electrons.

Q3 What was Thomson's main conclusion following his measurement of e/m for electrons?

Exam Questions

Q1 An electron of mass 9.1×10^{-31} kg and charge -1.6×10^{-19} C is accelerated through a potential difference of 1 kV.

(a) State its energy in eV. [1 mark]

(b) Calculate its energy in joules. [1 mark]

(c) Calculate its speed in ms^{-1} and express this as a percentage of the speed of light (3.0×10^8 ms^{-1}). [3 marks]

Q2 Explain the main features of an experiment to determine the specific charge of the electron.
The quality of your written answer will be assessed in this question. [5 marks]

New Olympic event — the electronvault...

Electronvolts are really handy units — they crop up all over the rest of this book, particularly in nuclear and particle physics. It stops you having to mess around with a load of nasty powers of ten. Cathode ray tubes (CRTs) are pretty handy too — there might be one in your telly... unless it's one of those new-fangled flat-screen plasma thingummy-do-dahs...

Millikan's Oil-Drop Experiment

Thomson had already found the charge/mass ratio of the electron in 1897 — now it was down to Robert Millikan, experimenter extraordinaire, to find the absolute charge...

Millikan's Experiment used ***Stoke's Law***

1) Before you start thinking about Millikan's experiment, you need a bit of **extra theory**.
2) When you drop an object into a fluid, like air, it experiences a **viscous drag** force. This force acts in the **opposite direction** to the velocity of the object, and is due to the **viscosity** of the fluid.
3) You can calculate this viscous force on a spherical object using **Stoke's law**:

$$F = 6\pi\eta r v$$

where η is the viscosity of the fluid, r is the radius of the object and v is the velocity of the object.

Millikan's Experiment — the ***Basic Set-Up***

Millikan's Oil-Drop Experiment — Apparatus

Millikan's Set-Up

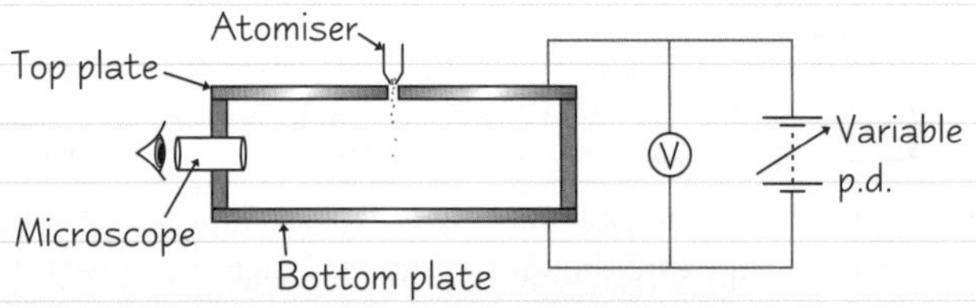

1) The **atomiser** created a **fine mist** of oil drops that were **charged** by **friction** as they left the atomiser (positively if they lost electrons, negatively if they gained electrons).
2) Some of the drops fell through a **hole** in the top plate and could be viewed through the **microscope**. (The eyepiece carried a **scale** to measure distances — and so **velocities** — accurately.)
3) When he was ready, Millikan could apply a **potential difference** between the two plates, producing a **field** that exerted a **force** on the charged drops. By **adjusting** the p.d., he could vary the strength of the field.

To give you a feel for the **size** of the apparatus, Millikan's plates were circular, with a diameter of about the width of this page. They were separated by about 1.5 cm.

Before the ***Field*** is Switched on, there's only ***Gravity*** and the ***Viscous Force***

1) With the electric field turned off, the forces acting on each oil drop are:

   a) the **weight** of the drop — acting downwards
   b) the **viscous force** from the air — acting upwards

Millikan had to take account of things like upthrust as well, but you don't have to worry about that — keep it simple.

2) The drop will reach **terminal velocity** (i.e. it will stop accelerating) when these two forces are equal. So, from Stoke's law (see above):

$$mg = 6\pi\eta r v$$

3) Since the **mass** of the drop is the **volume** of the drop multiplied by the **density**, ρ, of the oil, this can be rewritten as:

$$\frac{4}{3}\pi r^3 \rho g = 6\pi\eta r v \quad \Rightarrow \quad r^2 = \frac{9\eta v}{2\rho g}$$

Millikan measured η and ρ in separate experiments, so he could now calculate r — ready to be used when he switched on the electric field...

Millikan's Oil-Drop Experiment

*Then he **Turned On** the **Electric Field**...*

1) The field introduced a **third major factor** — an **electric force** on the drop.
2) Millikan adjusted the applied p.d. until the drop was **stationary**. Since the **viscous force** is proportional to the **velocity** of the object, once the drop stopped moving, the viscous force **disappeared**.
3) Now the only two forces acting on the oil drop were:

   a) the **weight** of the drop — acting downwards
   b) the force due to the **uniform electric field** — acting upwards

4) The **electric force** is given by: $F = \frac{QV}{d}$ where Q is the charge on the oil drop, V is the p.d. between the plates and d is the distance between the plates. **See p. 108-110**
5) Since the drop is **stationary**, this electric force must be equal to the weight, so:

$$\frac{QV}{d} = \frac{4}{3}\pi r^3 \rho g$$

   The first part of the experiment gave a value for r, so the **only unknown** in this equation is Q.
6) So Millikan could find the **charge on the drop**, and repeated the experiment for hundreds of drops. The charge on any drop was always a **whole number multiple** of -1.6×10^{-19} C.

*These Results Suggested that **Charge was Quantised***

1) This result was **really significant**. Millikan concluded that charge can **never exist** in **smaller** quantities than 1.6×10^{-19} C. He assumed that this was the **charge** carried by an **electron**.
2) Later experiments confirmed that **both** these things are true.

Charge is "**quantised**". It exists in "packets" of size **1.6×10^{-19} C** — the **fundamental unit of charge**. This is the size of the charge carried by **one electron**.

Practice Questions

Q1 Write down the equation for Stoke's law, defining any variables.

Q2 List the forces that act on the oil drop in Millikan's experiment:
(a) with the drop drifting downwards at terminal velocity but with no applied electrical field,
(b) when the drop is stationary, with an electrical field applied.

Q3 Briefly explain the significance of Millikan's oil-drop experiment in the context of quantum physics.

Exam Question

Q1 An oil drop of mass 1.63×10^{-14} kg is held stationary in the space between two charged plates 3.00 cm apart. The potential difference between the plates is 5000 V. The density of the oil used is 880 kgm^{-3}.

(a) Describe the relative magnitude and direction of the forces acting on the oil drop. [2 marks]

(b) Calculate the charge on the oil drop using $g = 9.81$ Nkg^{-1}.
Give your answer in terms of e, the charge on an electron. [3 marks]

The electric field is switched off and the oil drop falls towards the bottom plate.

(c) Explain why the oil drop reaches terminal velocity as it falls. [3 marks]

(d) Calculate the terminal velocity of the oil drop using $\eta = 1.84 \times 10^{-5}$ kgm^{-1}s^{-1}. [3 marks]

So next time you've got a yen for 1.59×10^{-19} coulombs — tough...

This was a huge leap. Along with the photoelectric effect (see p. 184) this experiment marked the beginning of quantum physics. The world wasn't ruled by smooth curves any more — charge now jumped from one allowed step to the next...

Light — Newton vs Huygens

Newton was quite a bright chap really, but even he could make mistakes — and this was his biggest one. The trouble with being Isaac Newton is that everyone just assumes you're right...

Newton had his **Corpuscular Theory**

1) In 1672, Newton published his **theory of colour**. In it he suggested that **light** was made up of **tiny particles** that he called '**corpuscles**'.
2) One of his major arguments was that light was known to travel in **straight lines**, yet waves were known to **bend** in the shadow of an **obstacle** (diffraction). Experiments weren't **accurate enough** then to detect the diffraction of light. Light was known to **reflect** and **refract**, but that was it.
3) His theory was based on the principles of his **laws of motion** — that all particles, including his 'corpuscles', will 'naturally' travel in **straight lines**.
4) Newton believed that **reflection** was due to a force that **pushed** the particles away from the surface — just like a ball bouncing back off a wall.
5) **Refraction** worked if the corpuscles travelled **faster** in a **denser** medium.

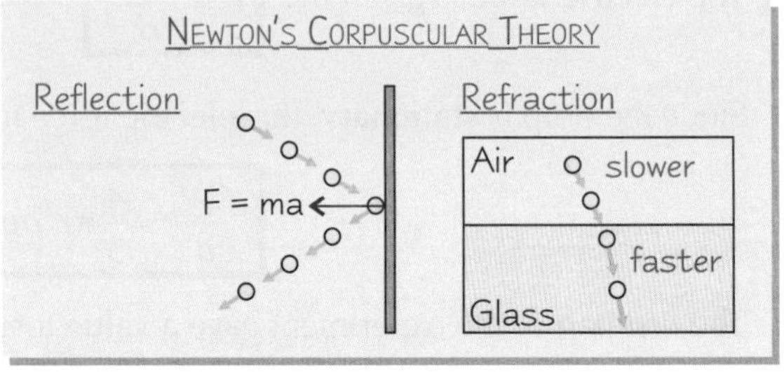

Huygens thought Light was a **Wave**

1) The idea that light might be a **wave** had existed for some time before it was formalised by Huygens in 1678.
2) At the time, nobody took much notice of him because his theory was **different** from Newton's.
3) Huygens developed a **general model** of the propagation of **waves** in what is now known as **Huygens' principle**:

> **HUYGENS' PRINCIPLE:** Every point on a wavefront may be considered to be a **point source** of **secondary wavelets** that spread out in the forward direction at the speed of the wave. The **new wave front** is the surface that is **tangential** to all of these **secondary wavelets**.

The diagram below shows how this works:

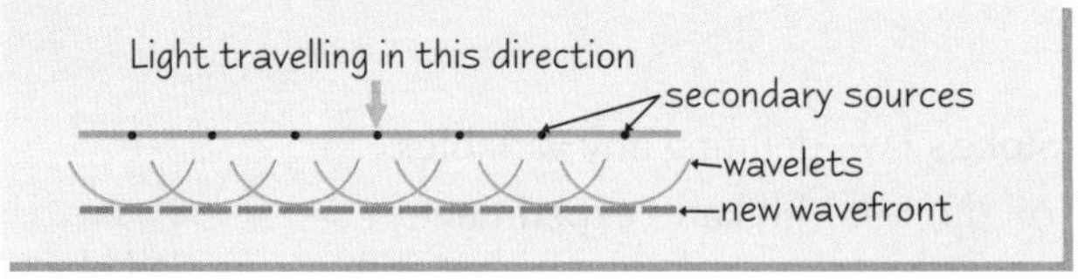

4) By applying his theory to **light**, he found that he could explain **reflection** and **refraction** easily.
Huygens predicted that light should **slow down** when it entered a **denser medium**, rather than speed up.

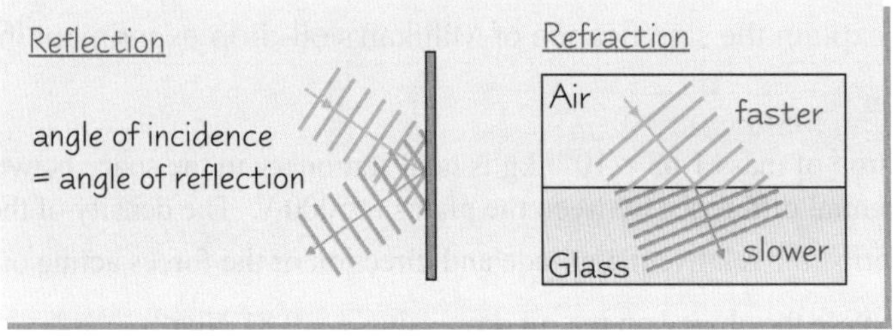

5) Huygens also predicted that light should **diffract** around tiny objects and that two coherent light sources should **interfere** with each other.

> Up until the end of the 18th century, most scientists sided with **Newton**. He'd been right about so many things before, so it was generally assumed that he **must be right** about light being corpuscular. The debate raged for **over 100 years** until **Thomas Young** carried out experiments on the **interference** of light in Cambridge around **1800**...

Light — Newton vs Huygens

Young Proved *Huygens Right* with his *Double-Slit Experiment*

1) **Diffraction** and **interference** are both uniquely **wave** properties. If it could be shown that **light** showed **interference** patterns, that would help decide once and for all between corpuscular theory and wave theory.
2) The problem with this was getting two **coherent** light sources, as **light** is emitted in **random bursts**.
3) Young solved this problem by using only **one point source of light** (a narrow slit with a light source behind it). In front of this was a **screen** with **two narrow slits** in it. Light spreading out by **diffraction** from the slits was equivalent to **two coherent point sources**.
4) In the area on the screen where light from the two slits **overlapped**, bright and dark '**fringes**' were formed. This was **proof** that light could both **diffract** (through the narrow slits) and **interfere** (to form the interference pattern on the screen) — **Huygens** was right all along.

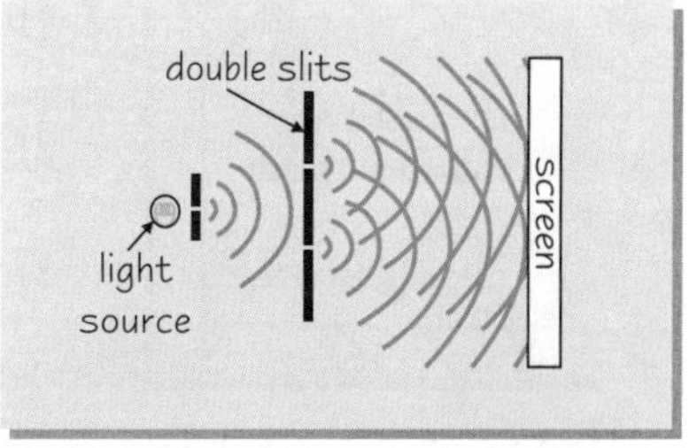

Observations and *Theories* Developed *Rapidly* during the *19th Century*

1) In 1808, Etienne-Louis Malus discovered that light was **polarised** by reflection. Physicists at the time thought that light spread like sound, as a longitudinal wave, so they struggled to explain polarisation.
2) In 1817, Young suggested that light was a **transverse wave** consisting of **vibrating electric** and **magnetic fields** at **right angles** to each other and the **direction of travel**. This explained why light could be **polarised**.
3) In the second half of the 19th century, James Clerk Maxwell showed theoretically that **all electromagnetic waves** should travel at the same speed in a vacuum, **c**.

> **James Clerk Maxwell** calculated the **speed of light** in a vacuum using:
>
> $$c = \frac{1}{\sqrt{\mu_0 \varepsilon_0}}$$
>
> where **c** is the speed of the wave in ms^{-1}, μ_0 ("mu-nought") is the permeability of free space (a constant — $4\pi \times 10^{-7}\ Hm^{-1}$) and ε_0 ("epsilon-nought") is the permittivity of free space (another constant — $8.85 \times 10^{-12}\ Fm^{-1}$).
>
> μ_0 relates to the **magnetic flux density** due to a current-carrying wire in free space, while ε_0 relates to the **electric field strength** due to a charged object in free space.

4) By that time, the **velocity of light** could be measured quite accurately, and it was found to be very close to Maxwell's value of **c**. This suggested that **light** is an **electromagnetic wave**. We now know that all electromagnetic waves, including light, travel in a **vacuum** at a **speed** of $\mathbf{2.998 \times 10^8\ ms^{-1}}$.
5) In 1887, **Heinrich Hertz** produced and detected **radio waves** using electric sparks. He showed by **experiment** that they could be reflected, refracted, diffracted and polarised, and show interference. This helped confirm that radio waves, like light, are electromagnetic waves.
6) This was the accepted theory up until the very end of the 19th century, when the **photoelectric effect** was discovered. Then the particle theory had to be resurrected, and it was all up in the air again...

Practice Questions

Q1 What was the main argument that Newton used to support his corpuscular theory of light?

Q2 What part does diffraction play in a Young's double-slit experiment?

Q3 Sketch a diagram showing an experiment to demonstrate Young's fringes for white light in a laboratory.

Exam Questions

Q1 Describe Newton's corpuscular theory of light. [2 marks]

Q2 Give a brief history of the understanding of the nature of light in the 18th and 19th centuries. You should mention the various theories that have been proposed and the evidence that has been used to support them.
The quality of your written answer will be assessed in this question. [6 marks]

In the blue corner — the reigning champion... IsaaaAAAAC NEWton...

So, light's a wave, right? We've got that sorted — the double-slit experiment laid the whole argument to rest... or did it?

The Photoelectric Effect

You did the photoelectric effect at AS, so it should be familiar. You need it again though, so here it is...

Shining Light on a Metal can Release Electrons

If you shine **light** of a **high enough frequency** onto the **surface of a metal**, it will **emit electrons**. For **most** metals, this **frequency** falls in the **UV** range.

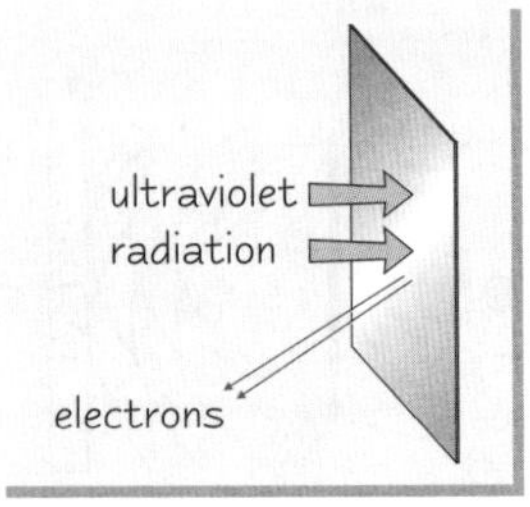

1) **Free electrons** on the **surface** of the metal **absorb energy** from the light, making them **vibrate** or move faster.
2) If an electron **absorbs enough** energy, the **bonds** holding it to the metal can be **broken** and the electron **released**.
3) This is called the **photoelectric effect** and the electrons emitted are called **photoelectrons**.

You don't need to know the details of any experiments on this — you just need to learn the three main conclusions:

Conclusion 1	For a given metal, **no photoelectrons are emitted** if the radiation has a frequency **below** a certain value — called the **threshold frequency**.
Conclusion 2	The photoelectrons are emitted with a variety of kinetic energies ranging from zero to some maximum value. This value of **maximum kinetic energy** increases with the **frequency** of the radiation, and is **unaffected** by the **intensity** of the radiation.
Conclusion 3	The **number** of photoelectrons emitted per second is **directly proportional** to the **intensity** of the radiation.

These are the two that had people puzzled. They can't be explained using wave theory.

The Photoelectric Effect Couldn't be Explained by Wave Theory

According to wave theory:

1) For a particular frequency of light, the **energy** carried is **proportional** to the **intensity** of the beam.
2) The energy carried by the light would be **spread evenly** over the wavefront.
3) **Each** free electron on the surface of the metal would gain a **bit of energy** from each incoming wave.
4) Gradually, each electron would gain **enough energy** to be able to leave the metal.

SO... If the light had a **lower frequency** (i.e. was carrying less energy) it would take **longer** for the electrons to gain enough energy — but it would happen eventually. There is **no explanation** for the **threshold frequency**.

The **higher the intensity** of the wave, the **more energy** it should transfer to each electron — so the kinetic energy should increase with **intensity**. There's **no explanation** for the **kinetic energy** depending only on the **frequency**.

Einstein came up with the Photon Model of Light

1) When Max Planck was investigating **black body radiation** (don't worry, you don't need to know about that just yet), he suggested that **EM waves** can **only** be **released** in **discrete packets**, or **quanta**.
2) The **energy carried** by one of these **wave-packets** had to be:

$$E = hf = \frac{hc}{\lambda}$$

where h = Planck's constant = 6.63×10^{-34} Js and c = speed of light in a vacuum = 3.00×10^{8} ms^{-1}

3) **Einstein** went **further** by suggesting that **EM waves** (and the energy they carry) can only **exist** in discrete packets. He called these wave-packets **photons**.
4) He saw these photons of light as having a **one-on-one**, **particle-like** interaction with **an electron** in a **metal surface**. It would **transfer all** its **energy** to that **one**, **specific electron**.

The Photoelectric Effect

The *Photon Model* Explained the *Photoelectric Effect* Nicely

According to the photon model, when light hits its surface, the metal is **bombarded** by photons. If one of these photons **collides** with a free electron, the electron will gain energy equal to ***hf***.

Before an electron can **leave** the surface of the metal, it needs enough energy to **break the bonds holding it there**. This energy is called the **work function** (symbol ϕ) and its **value** depends on the **metal**.

It Explains the *Threshold Frequency*...

1) If the energy **gained** from the photon is **greater** than the **work function**, the electron can be **emitted**.
2) If it **isn't**, the electron will just **shake about a bit** or move faster, then release the energy as another photon or in collisions. The metal will heat up, but **no electrons** will be emitted.
Since for **electrons** to be released, $hf \geq \phi$, the **threshold frequency** must be: ⟹

$$f = \frac{\phi}{h}$$

... and the *Maximum Kinetic Energy*

1) The **energy transferred** to an electron is ***hf***.
2) The **kinetic energy** it will be carrying when it **leaves** the metal is ***hf*** **minus** any energy it's **lost** on the way out (there are loads of ways it can do that, which explains the **range** of energies).
3) The **minimum** amount of energy it can lose is the **work function**, so the **maximum kinetic energy** is given by the equation: ⟹
The **kinetic energy** of the electrons is **independent of intensity**, as they can **only absorb one photon** at a time.

$$\frac{1}{2}mv^2_{max} = hf - \phi$$

4) The maximum kinetic energy can be **measured** using the idea of **stopping potential**. The emitted electrons are made to lose their **energy** by **doing work against** an applied **potential difference**. The work done by the p.d. in **stopping** the **fastest electrons** is equal to the energy they were carrying:

$$\frac{1}{2}mv^2_{max} = eV_s$$

where **e** = charge on the electron = 1.6×10^{-19} C,
V_s = stopping potential in V, and the maximum kinetic energy is measured in J.
The kinetic energy in eV is just equal to the stopping potential — see p. 88.

5) The stopping potential for a frequency of light can be measured using a circuit containing a **vacuum photocell** (two metal plates in a vacuum). The light is directed on to one plate, causing **photoelectron**s to be released.
6) If the photoelectrons reach the other plate, a **current** flows in the circuit. The p.d. across the plates is **increased** from 0 using a **potential divider.** When the current **stops**, the **stopping potential** has been reached.

Practice Questions

Q1 What three main conclusions were drawn from detailed experimentation on the photoelectric effect?

Q2 What is meant by the work function energy of a metal?

Q3 How is the energy of a photon related to its frequency?

Exam Questions *Use Planck's constant = 6.63×10^{-34} Js, speed of light = 3.00×10^{8} ms^{-1}, electron charge = 1.6×10^{-19} C.*

Q1 An isolated zinc plate with neutral charge is exposed to high-frequency ultraviolet light.
State and explain the effect of the ultraviolet light on the charge of the plate. [2 marks]

Q2 Light of wavelength 0.50 μm hits a metal surface and causes electrons to be released.
Their maximum kinetic energy is 2.0×10^{-19} J.
(a) Calculate the work function of the metal. [3 marks]
(b) Explain why a beam of light with a wavelength of 1.5 μm would not cause electrons to be emitted. [3 marks]

Q3 Potassium has a work function of 2.2 eV. A potassium anode is illuminated with light of wavelength 350 nm. What potential must be applied to stop the electrons leaving the anode? [4 marks]

And that's all there is to it — *sob*...

Learn why the light theory can't explain the photoelectric effect, and how photon theory does — then tell someone else.

Wave-Particle Duality

Is it a wave? Is it a particle?

Interference and ***Diffraction*** show ***Light*** as a ***Wave***

1) Light produces **interference** and **diffraction** patterns — **alternating bands** of **dark** and **light**.
2) These can **only** be explained using **waves interfering constructively** (when two waves overlap in phase) or **interfering destructively** (when two waves are out of phase).

The ***Photoelectric Effect*** Shows ***Light*** Behaving as a ***Particle***

1) **Einstein** explained the results of **photoelectricity experiments** (see p. 184) by thinking of the **beam of light** as a series of **particle-like "photons"**.
2) If a **photon** of light is a **discrete** bundle of energy, then it can **interact** with an **electron** in a **one-to-one way**.
3) **All** the **energy** in the **photon** is **given** to one **electron**.

Neither the **wave theory** nor the **particle theory** describe what light actually **is**.
They're just two different **models** that help to explain the way light behaves.

I'm not impressed — this is just speculation. What do you think Dad?

De Broglie came up with the ***Wave-Particle Duality Theory***

1) Louis de Broglie made a **bold suggestion** in his **PhD thesis**:

> If "**wave-like**" **light** showed **particle properties** (photons), "**particles**" like **electrons** should be expected to show **wave-like properties**.

2) The **de Broglie equation** relates a **wave property** (**wavelength**, λ) to a **moving particle property** (**momentum**, mv). h = Planck's constant = 6.63×10^{-34} Js.

$$\lambda = \frac{h}{mv}$$

3) The **de Broglie wave** of a particle can be interpreted as a "**probability wave**". The **probability** of finding a particle at a point is **directly proportional** to the **square of the wave's amplitude**.
4) Later experiments **confirmed** the wave nature of electrons.

Most physicists at the time weren't very impressed — his ideas were just speculation.

Electron Diffraction shows the ***Wave Nature*** of ***Electrons***

1) De Broglie's suggestions prompted a lot of experiments to try to show that **electrons** can have **wave-like** properties. In **1927**, Davisson and Germer succeeded in **diffracting electrons**.
2) They saw **diffraction patterns** when **accelerated electrons** in a vacuum tube **interacted** with the **spaces** in a graphite **crystal**.
3) According to wave theory, the **spread** of the **lines** in the diffraction pattern **increases** if the **wavelength** of the wave **increases**.
4) In electron diffraction experiments, a **small accelerating voltage**, i.e. **slow** electrons, gives **widely spaced** rings.
5) **Increase** the **electron speed** and the diffraction pattern circles **squash together** towards the **middle**. This fits in with the **de Broglie** equation above — if the **velocity** is **higher**, the **wavelength** is **shorter** and the **spread** of the lines is **smaller**.

Electron diffraction patterns look like this.

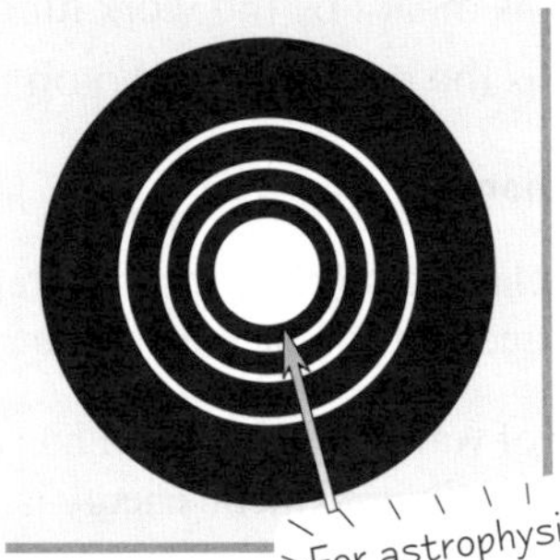

For astrophysics students, this circle is called the Airy disc (see p. 143).

> In general, λ for **electrons** accelerated in a **vacuum tube** is about the **same size** as λ for **electromagnetic waves** in the **X-ray** part of the spectrum.

6) The de Broglie wavelength of an electron (λ) is related to the **accelerating voltage** (V) by:

$$\lambda = \frac{h}{\sqrt{2meV}}$$

where e is the charge on the electron and m is its mass

Wave-Particle Duality

Particles Don't Show Wave-Like Properties All the Time

You **only** get **diffraction** if a particle interacts with an object of about the **same size** as its **de Broglie wavelength**.
A **tennis ball**, for example, with **mass 0.058 kg** and **speed 100 ms^{-1}** has a **de Broglie wavelength** of **10^{-34} m**.
That's **10^{19} times smaller** than the **nucleus** of an **atom**! There's nothing that small for it to interact with.

Example An electron of mass 9×10^{-31} kg is fired from an electron gun at $7 \times 10^6 ms^{-1}$. What size object will the electron need to interact with in order to diffract?

Momentum of electron = $mv = 6.3 \times 10^{-24}$ kg ms^{-1}
$\lambda = h/mv = 6.63 \times 10^{-34} / 6.3 \times 10^{-24} = 1 \times 10^{-10}$ m

Only crystals with atom layer spacing around this size are likely to cause the diffraction of this electron.

You can also calculate the anode voltage needed to produce this wavelength using the formula at the bottom of the previous page: $\lambda = \frac{h}{\sqrt{2meV}} \Rightarrow V = \frac{h^2}{2me\lambda^2} = \mathbf{153\ V}$

A **shorter wavelength** gives **less diffraction**. This is important in **microscopes** where diffraction **blurs out details**.
The **tiny** wavelength of electrons means an **electron microscope** can resolve **finer detail** than a **light** microscope.

Electron Microscopes use Electrons Instead of Light

In electron microscopes:

1) A **stream of electrons** is accelerated towards the sample using a **positive electric potential** — an **electron gun**.
2) To **resolve detail** around the size of an **atom**, the **electron wavelength** needs to be similar to the **diameter** of an **atom** (**0.1 nm**) — which (using the equation in the example above) means an **anode voltage** of **at least 150 V**.
3) The **stream of electrons** from the electron gun is confined into a thin **beam** using a **magnetic field**.
4) The beam is **focused** onto the sample and any interactions are transformed into an **image**.
The sort of image you get depends on the **type of microscope** you're using:

A **transmission electron microscope** (TEM) works a bit like a **slide** projector, but uses electrons instead of light. A **very thin** specimen is used and the parts of the beam that pass through the specimen are projected onto a **screen** to form an image.

A **scanning tunnelling microscope** (STM) is a different kind of microscope that uses principles of **quantum mechanics**. A very fine **probe** is moved over the surface of the sample and a **voltage** is applied between the probe and the surface. Electrons "**tunnel**" from the probe to the surface, resulting in a weak **electrical current**. The smaller the **distance** between the probe and the surface, the **greater the current**. By scanning the probe over the surface and measuring the current, you produce an **image** of the **surface** of the sample.

Practice Questions

Q1 What name is normally given to "particles" of light?
Q2 What observation showed that electrons could behave as waves?
Q3 What is the advantage of an electron microscope over a light microscope?

Exam Questions *Use $h = 6.63 \times 10^{-34}$ Js, $e = 1.6 \times 10^{-19}$ C, $m_e = 9.1 \times 10^{-31}$ kg.*

Q1 An electron is accelerated through a p.d. of 500 V.
(a) Calculate:
i) the velocity of the electron, ii) its de Broglie wavelength. [4 marks]
(b) In which region of the electromagnetic spectrum does this fall? [1 mark]

Q2 (a) Describe how a transmission electron microscope (TEM) uses a beam of electrons to produce an image. [3 marks]
(b) Show that an anode voltage of at least 150 V is needed for a TEM to resolve detail around the size of an atom (0.1 nm). [3 marks]

Wave-Particle duelity — pistols at dawn...

*You're getting into the weird bits of quantum physics now — it says that light isn't a wave, and it isn't a particle, it's **both**... at the **same time**. And if you think that's confusing, just wait till you get onto relativity — not that I want to put you off.*

The Speed of Light and Relativity

First — a bit of a history lesson.

Michelson and Morley tried to find the Absolute Speed of the Earth

1) During the 19th century, most physicists believed in the idea of **absolute motion**. They thought everything, including light, moved relative to a **fixed background** — something called the **ether**.
2) **Michelson** and **Morley** tried to measure the **absolute speed** of the **Earth** through the ether using a piece of apparatus called an **interferometer**.
3) They expected the motion of the Earth to affect the **speed of light** they measured in **certain directions**. According to Newton, the speed of light measured in a **lab** moving parallel to the light would be ($c + v$) or ($c - v$), where v is the speed of the lab. By measuring the speed of light **parallel** and **perpendicular** to the motion of the Earth, Michelson and Morley hoped to find v, the absolute speed of the Earth.

They used an Interferometer to Measure the Speed of the Earth

The interferometer was basically **two mirrors** and a **partial reflector** (a beam-splitter). When you shine light at a partial reflector, some of the light is **transmitted** and the rest is **reflected**, making **two separate beams**.

The mirrors were at **right angles** to each other, and an **equal distance**, L, from the beam-splitter.

The Michelson-Morley Interferometer

1) **Monochromatic light** is sent towards the **partial reflector**.
2) The light is split into **two beams** travelling at **right angles** to each other.
3) The beams are reflected at **mirrors M1** and **M2**.
4) When the reflected beams meet back at the partial reflector, they form an **interference pattern**.
5) This interference pattern is **recorded** by the observer.
6) Then the whole interferometer is **rotated** through **90°** and the experiment **repeated**.

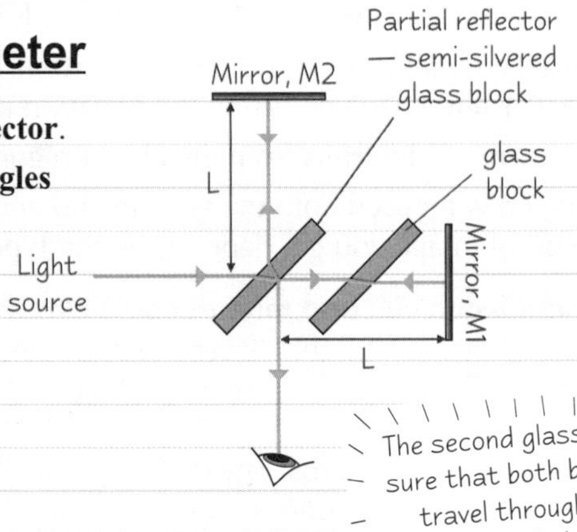

The second glass block makes sure that both beams of light travel through the same amount of air and glass.

EXPECTED OUTCOME

According to Newton's laws, light moving **parallel** to the motion of the Earth should take **longer** to travel to the mirror and back than light travelling at **right angles** to the Earth's motion. So **rotating** the apparatus should have changed the **travel time** for the two beams.

This would cause a **tiny shift** in the **interference pattern**.

They Didn't get the Result they were Expecting

They **repeated** the experiment **over** and **over** again — at different **times of day** and at different points in the **year**. Taking into account any **experimental errors**, there was **absolutely no shift** in the interference pattern.

The time taken by each beam to travel to each mirror was **unaffected** by rotating the apparatus.

So, Newton's laws **didn't work** in this situation.

Most scientists were really puzzled by this "null result". Eventually, the following **conclusions** were drawn:

a) It's **impossible** to detect **absolute motion** — the ether doesn't exist.
b) The **speed of light** has the **same value** for all observers.

The Speed of Light and Relativity

The **invariance** of the speed of light is one of the cornerstones of special relativity. The other is based on the concept of an **inertial frame of reference**.

Anything Moving with a **Constant Velocity** *is in an* **Inertial Frame**

A reference frame is just a **space** that we decide to use to describe the **position of an object** — you can think of a reference frame as a **set of coordinates**.

An **inertial reference frame** is one in which **Newton's 1st law** is obeyed. (Newton's 1st law says that objects won't accelerate unless they're acted on by an external force.)

1) Imagine sitting in a carriage of a train **waiting at a station**. You put a **marble** on the table. The marble **doesn't move**, since there aren't any horizontal **forces** acting on it. **Newton's 1st law** applies, so it's an **inertial frame**.
2) You'll get the **same result** if the carriage moves at a **steady speed** (as long as the track is **smooth, straight and level**) — another inertial frame.
3) As the train **accelerates** out of the station, the marble **moves** without any force being applied. Newton's 1st law **doesn't apply**. The accelerating carriage **isn't an inertial frame**.
4) **Rotating** or **accelerating** reference frames **aren't** inertial. In most cases, though, you can think of the **Earth** as an inertial frame — it's near enough.

A stationary train carriage is an inertial frame

An accelerating train carriage is NOT an inertial frame

Einstein's **Postulates** *of* **Special Relativity**

Einstein's theory of **special relativity** only works in **inertial frames** and is based on **two postulates** (assumptions):

1) **Physical laws have the same form in all inertial frames.**
2) **The speed of light in free space is invariant.**

1) The first postulate says that if we do **any physics experiment** in any inertial frame we'll always get the **same result**. That means it's **impossible** to use the result of **any experiment** to work out if you're in a **stationary reference frame** or one moving at a **constant velocity**.
2) The second postulate says that the **speed of light** (in a vacuum) always has the **same value**. It isn't affected by the **movement** of the **person measuring it** or by the movement of the **light source**.

Practice Questions

Q1 Draw a labelled diagram showing the apparatus used by Michelson and Morley to determine the absolute speed of the Earth. Include the light source, mirrors, partial reflector, glass block and the position of the observer.

Q2 State the postulates of Einstein's theory of special relativity.

Q3 Explain why a carriage on a rotating Ferris wheel is not an inertial frame.

Exam Questions

Q1 In the Michelson-Morley interferometer experiment, interference fringes were observed. When the apparatus was rotated through 90 degrees the expected result was not observed.

(a) State the result that was expected. [1 mark]

(b) Describe the conclusions that were eventually drawn from these observations. [2 marks]

Q2 (a) Using a suitable example, explain what is meant by an inertial reference frame. [2 marks]

(b) Explain what is meant by the invariance of the speed of light. [2 marks]

The speed of light is always the same — whatever your reference frame...

Michelson and Morley showed that Newton's laws didn't always work. This was a huge deal. Newton's laws of motion had been treated like gospel by the physics community since the 17th century. Then along came Herr Einstein...

Special Relativity

Special relativity ONLY WORKS IN INERTIAL FRAMES — it doesn't work in an accelerating frame.

A Moving Clock Runs Slow

1) Time runs at **different speeds** for two observers **moving relative** to each other.
2) A **stationary** observer measures the interval between two events as t_0, the **proper time**. (Since there's no such thing as absolute motion, a "stationary" observer means someone that's stationary relative to the reference frame the events are happening in.) An observer moving at a **constant velocity**, v, will measure a **longer** interval, t, between the two events. t is given by the equation:

$$t = \frac{t_0}{\sqrt{1-\frac{v^2}{c^2}}}$$

where $\sqrt{1-\frac{v^2}{c^2}}$ is called the relativity factor and c is the speed of light.

3) This is called **time dilation**.

A THOUGHT EXPERIMENT TO ILLUSTRATE TIME DILATION

*Anne is on a high-speed train travelling at 0.9**c**. She switches on a torch for exactly 2 seconds.*

Claire is standing on the platform and sees the same event, but records a longer time. It appears to Claire that Anne's clock is running slow.

In this experiment, **Anne** is the **stationary observer**, so she measures the **proper time**, t_0. Claire is **moving** at **0.9c relative to the events**, and so measures a time t given by:

$$t = \frac{t_0}{\sqrt{1-\frac{v^2}{c^2}}} = \frac{2}{\sqrt{1-\frac{(0.9c)^2}{c^2}}} = \frac{2}{\sqrt{1-0.9^2}} = \mathbf{4.59\ s}$$

To the **external observer** (e.g. Claire) **moving clocks** run **slowly**.

It's really important that you get the "stationary observer" right.

There's Proof of Time Dilation from Muon Decay

1) **Muons** are **particles** created in the **upper atmosphere** that move towards the ground at speeds close to **c**.
2) In the laboratory (**at rest**) they have a **half-life** of less than **2 μs**. From this half-life, you would expect most muons to **decay** between the top of the atmosphere and the Earth's surface, but that **doesn't happen**.

Experiment to Measure Muon Decay

1) Measure the speed, v, of the muons (this is about **0.99*c***).
2) Place a detector (MR1) at high altitude and measure the muon count rate.
3) Use another detector (MR2) to measure the count rate at ground level.
4) Compare the two figures.

Here are some typical results:
MR1 = 500 per minute
MR2 = 325 per minute
Distance between detectors (d) = 2000 m
Time as measured by an observer = d/v = 6.73 μs
Half-life of muons at rest = 1.53 μs

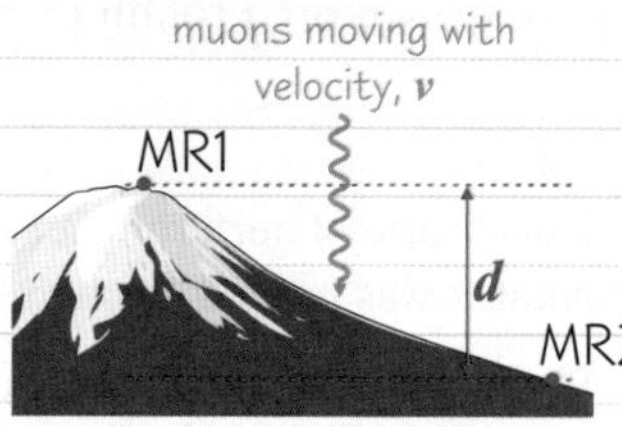

3) We can do some calculations using the data above. In the reference frame of the **observer** the muons seemed to have travelled for **4.4 half-lives** between the two detectors. You would expect the count rate at the **second detector** to be only about **25 counts per minute**.
4) However, in a **muon's reference frame**, travelling at 0.99**c**, the time taken for the journey is just t_0 **= 0.94 μs**. From the point of view of the muons, the time elapsed is **less** than their **half-life**. From the point of view of the **observer**, it appears that the half-life of the muons has been **extended**.

Special Relativity

A Moving Rod Looks Shorter

1) A **rod** moving in the **same direction** as its **length** looks **shorter** to an external observer.
2) A **stationary** observer measures the length of an object as l_0. An observer moving at a **constant velocity**, v, will measure a **shorter** length, l. l is given by the equation:

$$l = l_0\sqrt{1-\frac{v^2}{c^2}}$$

This is called **length contraction.**

A THOUGHT EXPERIMENT TO ILLUSTRATE LENGTH CONTRACTION

*Anne (still in the train moving at 0.9**c**) measures the length of her carriage as 3 m. Claire, on the platform, measures the length of the carriage as it moves past her.*

Claire measures a length: $l = l_0\sqrt{1-\frac{v^2}{c^2}} = 3\sqrt{1-\frac{(0.9c)^2}{c^2}} = 3\sqrt{1-0.9^2} = \mathbf{1.3\,m}$

The Mass of an Object Increases with Speed

1) The **faster** an object **moves**, the **more massive** it gets.
2) An object with rest mass m_0 moving at a **velocity** v has a **relativistic mass** m given by the equation:

$$m = \frac{m_0}{\sqrt{1-\frac{v^2}{c^2}}}$$

So increasing an object's kinetic energy increases its mass — but it's only noticeable near the speed of light.

3) As the relative speed of an object approaches **c**, the mass approaches **infinity**. So, in practice, no massive object can move at a speed **greater than** or **equal to** the speed of light.

Mass and Energy are Equivalent

1) Einstein extended his idea of **relativistic mass** to write down the most famous equation in physics: $E = mc^2$
2) This equation says that **mass** can be **converted** into **energy** and vice versa. Or, alternatively, **any energy** you supply to an object **increases** its **mass** — it's just that the increase is usually **too small** to measure.
3) The **total energy** of a relativistic object is given by the equation:

$$E = \frac{m_0 c^2}{\sqrt{1-\frac{v^2}{c^2}}}$$

This is just substituting the relativistic mass into $E = mc^2$.

Practice Questions

Q1 State the equations for time dilation and length contraction, carefully defining each symbol.

Q2 Using the results from the muon experiment (page 100), show that the time elapsed in the reference frame of the muon is 0.94 μs.

Q3 A particle accelerated to near the speed of light gains a very large quantity of energy. Describe how the following quantities change as the particle gains more and more energy: a) the mass; b) the speed.

Exam Questions

Q1 A subatomic particle has a half-life of 20 ns when at rest. If a beam of these particles is moving at 0.995**c** relative to an observer, calculate the half-life of these particles in the frame of reference of the observer. [3 marks]

Q2 Describe a thought experiment to illustrate time dilation. [4 marks]

Q3 For a proton ($m_0 = 1.67 \times 10^{-27}$ kg) travelling at 2.8×10^8 ms^{-1} calculate:

(a) the relativistic mass, [1 mark]

(b) the total energy. [1 mark]

Have you ever noticed how time dilates when you're revising physics...

*In a moving frame, time stretches out, lengths get shorter and masses get bigger. One of the trickiest bits is remembering which observer's which — t_0, m_0 and l_0 are the values you'd measure if the object was **at rest**.*

Exponentials and Natural Logs

Mwah ha ha ha... you've hacked your way through the rest of the book and think you've finally got to the end of A2 Physics, but no, there's this tasty titbit of exam fun to go. You can get asked to look at and work out values from log graphs all over the shop, from astrophysics to electric field strength. And it's easy when you know how...

*Many Relationships in Physics are **Exponential***

A fair few of the relationships you need to know about in A2 Physics are **exponential** — where the **rate of change** of a quantity is **proportional** to the **amount** of the quantity left. Here are just a few you should have met before (if they don't ring a bell, go have a quick read about them)...

Charge on a capacitor — the decay of charge on a capacitor is proportional to the amount of charge left on the capacitor: $Q = Q_0 e^{(-t/RC)}$ (see p 115)

Radioactive decay — the rate of decay is proportional to the **number of nuclei left** to decay in a sample: $N = N_0 e^{(-\lambda t)}$ (see p 129)

The **activity** of a radioactive sample behaves in the same way: $A = A_0 e^{(-\lambda t)}$ (see p 129)

*You can **Plot** Exponential Relations Using the **Natural Log, ln***

1) Say you've got two variables, ***x*** and ***y***, which are related to each other by the formula $y = ke^{-ax}$ (where ***k*** and ***a*** are constants).
2) The inverse of e is the natural logarithm, **ln**.
3) By definition, $\ln(e^x) = x$. So far so good... now you need some **log rules**:

$\ln(ab) = \ln a + \ln b$ $\quad \ln\left(\frac{a}{b}\right) = \ln a - \ln b$ $\quad \ln a^b = b \ln a$

When it came to logs, Geoff always took time to smell the flowers...

4) So, if you take the natural log of the exponential function you get:
   $\ln y = \ln(ke^{-ax}) = \ln k + \ln(e^{-ax})$ ⟹ $\ln y = \ln k - ax$
5) Then all you need to do is plot (ln ***y***) against ***x***, and Eric's your aunty:

You get a **straight-line** graph with (**ln *k***) as the ***y*-intercept**, and **–*a*** as the **gradient**.

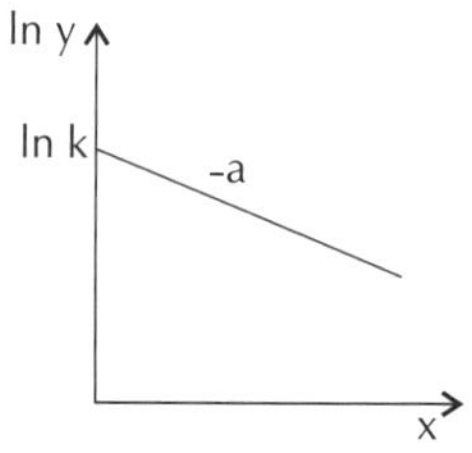

*You Might be Asked to find the **Gradient** of a Log Graph...*

This log business isn't too bad when you get your head around which bit of the log graph means what. On the plus side, they won't ask you to plot a graph like this (yipee) — they'll just want you to find the **gradient** or the **y-intercept**.

Example — finding the radioactive half-life of material X

The graph shows the radioactive decay of substance X.
(a) Find the initial number of atoms, N_0, in the sample.

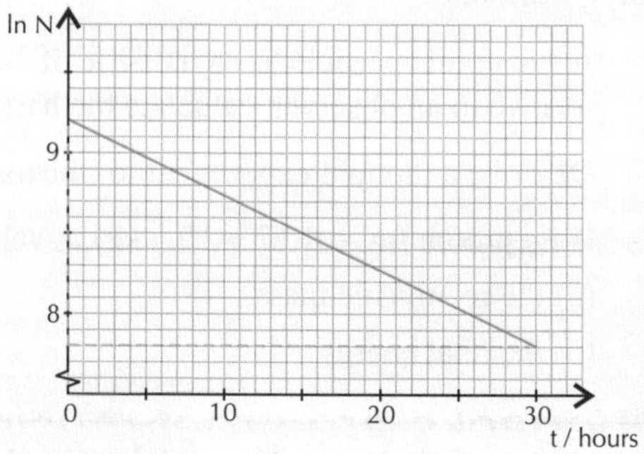

You know that the number of radioactive atoms in a sample, ***N***, is related to the initial number of atoms by the equation $N = N_0 e^{-\lambda t}$.
So, $(\ln N) = (\ln N_0) - \lambda t$ and $\ln N_0$ is the y-intercept of the graph = 9.2, $N_0 = e^{9.2} \approx$ **9900 atoms**.

(b) Find the decay constant **λ** of substance X.

$-\lambda$ is the gradient of the graph, so: $\lambda = \frac{\Delta \ln N}{\Delta t} = \frac{9.2 - 7.8}{30 \times 60 \times 60} = \frac{1.4}{108\,000} = 1.3 \times 10^{-5}\ s^{-1}$ (2s.f.)

Log Graphs and Long Answer Questions

You can Plot **Any Power Law** as a **Log-Log Graph**

You can use logs to plot a straight-line graph of **any power law** — it doesn't have to be an exponential.
Take the relationship between the energy stored in a spring, ***E***, and the spring's extension, ***x***:

$$E = kx^n$$

Take the log (base 10) of both sides to get:

$$\log E = \log k + n \log x$$

So **log *k*** will be the ***y***-intercept and ***n*** the gradient of the graph.

Example

The graph shows how the intensity of radiation from the Sun, ***I***, varies with its distance, ***d***.
I is related to ***d*** by the power law $I = kd^n$. Find ***n***.

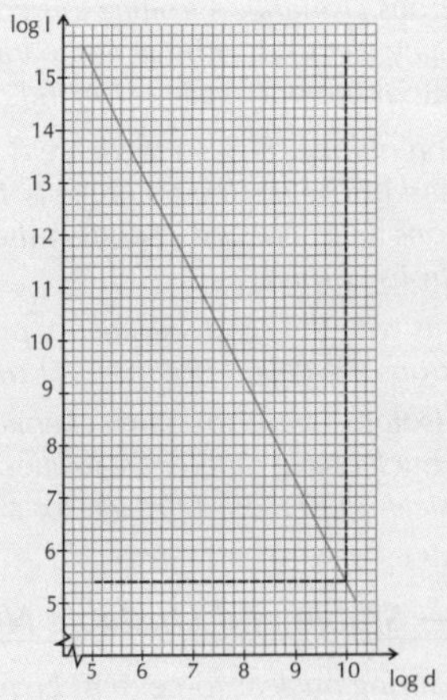

$\log I = \log (kd^n) = \log k + \log d^n$
$= \log k + n \log d$.
so ***n*** is the **gradient** of the graph.
Reading from the graph:

$$n = \frac{\Delta \log I}{\Delta \log d} = \frac{15.4 - 5.4}{5 - 10} = \frac{10}{-5} = -2$$

And that's the End of Logs... Now **Explain Yourself...**

In A2, they often give a couple of marks for 'the quality of written communication' when you're writing a slightly long answer (and not just pumping numbers into an equation). You can pick up a couple of easy marks just by making sure that you do the things in the fetching blue box.

1) **Explain** your ideas or argument **clearly** as this is usually what you'll get a mark for. And make sure you **answer the question** being asked — it's dead easy to go off on a tangent. Like my mate Phil always says... have I ever told you about Phil? Well he...
2) Write in **whole sentences**.
3) Use **correct spelling**, **grammar** and **punctuation**.
4) Also check how many marks the question is worth.
   If it's only a two-marker, they don't want you to spend half an hour writing an essay about it.

Example

Human hearing is most sensitive around the resonant frequency of the ear passages — around 3000 Hz. Explain fully what is meant by this statement and describe two circumstances in which you would expect a **higher** resonant frequency in a human.
The quality of your written answer will be assessed in this question. *[6 marks]*

Good Answer

Resonance happens when the driving frequency is the same as the natural frequency of the oscillator.

The amplitude of the oscillations will be greater at the resonant frequency for any given amplitude of driving wave.

The resonant frequency of the ear passages depends on their length. In a child, for example, the ear passages are shorter than average, so the resonant frequency is higher.

You would also expect the ear to have a higher resonant frequency if the auditory canal was filled with e.g. water rather than air. Sound travels faster through water than air, so the resonant frequency of the canal would be higher.

Bad Answer

driving frequency = nat. freq.
shorter ear passages = higher res. freq.
length varies between humans.
sound faster in water.

There's nothing wrong with the physics in the bad answer, but you'd miss out on some nice easy marks just for not bothering link you thoughts together properly or put your answer into proper sentences.

Lumberjacks are great musicians — they have a natural logarithm...

Well, that's it folks. Crack open the chocolate bar of victory and know you've earnt it. Only the tiny detail of the actual exam to go... ahem. Make sure you know which bit means what on a log graph and you'll pick up some nice easy marks. Other than that, stay calm, be as clear as you can and good luck — I've got my fingers, toes and eyes crossed for you.

AS Answers

Unit 1: Section 1 — Particles and Radiation

Page 5 — Atomic Structure

1) *Inside every atom there is a nucleus which contains protons and neutrons [1 mark]. Orbiting this core are the electrons [1 mark]. The tiny electrons orbit at comparatively vast distances, so most of the atom is empty space [1 mark].*

2) *Proton number is 8, so there are 8 protons and 8 electrons [1 mark]. The nucleon number is 16. This is the total number of protons and neutrons. Subtract the 8 protons and that leaves 8 neutrons [1 mark].*

3) *Atoms with the same number of protons but different numbers of neutrons are called isotopes. [1 mark for definition]*
E.g. Isotopes have the same chemical reactions and behaviour. / The nuclei have different stabilities. / Isotopes have different physical properties. [2 marks for any 2 sensible points]

Page 7 — Stable and Unstable Nuclei

1) a) *The strong nuclear force must be repulsive at very small nucleon separations to prevent the nucleus being crushed to a point. [1 mark]*
b) *Beyond a few fm the strong nuclear force is smaller than the electrostatic force [1 mark]. This means the protons in the nucleus would be forced apart. So a nucleus bigger than this would be unstable [1 mark].*

2) a) $^{226}_{88}\text{Ra} \longrightarrow {}^{222}_{86}\text{Rn} + {}^{4}_{2}\alpha$
[3 marks available — 1 mark for alpha particle, 1 mark each for proton number and nucleon number of radon]
b) $^{40}_{19}\text{K} \longrightarrow {}^{40}_{20}\text{Ca} + {}^{0}_{-1}\beta + \overline{\nu}_e$
[4 marks available — 1 mark for beta particle, 1 mark each for proton number and nucleon number of calcium, 1 mark for antineutrino]

Page 9 — Particles and Antiparticles

1) $e^+ + e^- \rightarrow \gamma + \gamma$ *[1 mark]*
This is called annihilation. [1 mark]
Remember that there are two photons and they go off in opposite directions, just like two bits from an explosion.

2) *The protons, neutrons and electrons which make up the iron atoms would need to annihilate with their antiparticles [1 mark].*
No antiparticles are available for this to happen in the iron block [1 mark]

3) *The creation of a particle of matter also requires the creation of its antiparticle. In this case no antineutron has been produced. [1 mark]*

Page 12 — Classification of Particles

1) *Proton, electron and electron-antineutrino. [1 mark]*
The electron and the electron-antineutrino are leptons. [1 mark]
Leptons are not affected by the strong interaction, so the decay can't be due to the strong interaction. [1 mark]
Remember that this is really just the same as beta decay. Some books might leave out the antineutrino, so don't be misled.

2) *Mesons are hadrons but the muon is a lepton. [1 mark]*
The muon is a fundamental particle but mesons are not. Mesons are built up from simpler particles. [1 mark]
Mesons interact via the strong interaction but the muon does not. [1 mark]
You need to classify the muon correctly first and then say why it's different from a meson because of what it's like and what it does.

Page 15 — Quarks

1) $\pi^- = d\bar{u}$ *[1 mark]*
Charge of down quark = −1/3 unit.
Charge of anti-up quark = −2/3 unit
Total charge = −1 unit [1 mark]

2) *The weak interaction converts a down quark into an up quark plus an electron and an electron-antineutrino. [1 mark]*
The neutron (udd) becomes a proton (uud). [1 mark]
The lepton number L_e is conserved in this reaction.

3) *The baryon number changes from 2 to 1, so baryon number is not conserved [1 mark]. The strangeness changes from 0 to 1, so strangeness is not conserved [1 mark].*

Page 17 — Exchange Particles and Feynman Diagrams

1) *The electrostatic force is due to the exchange of virtual photons that only exist for a very short time. [1 mark]*
The force is due to the momentum transferred to or gained from the photons as they are emitted or absorbed by a proton. [1 mark]

2)

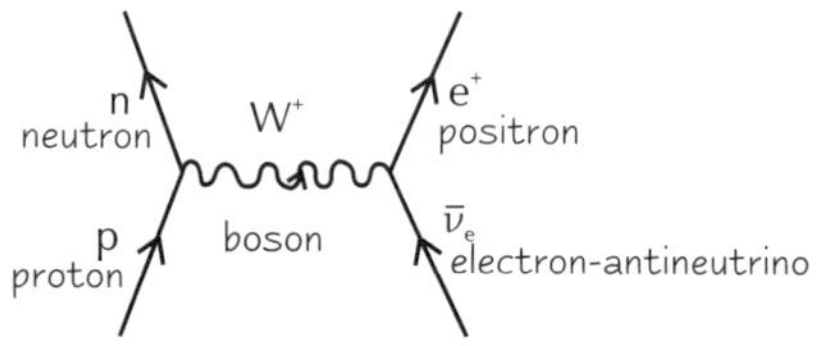

OR

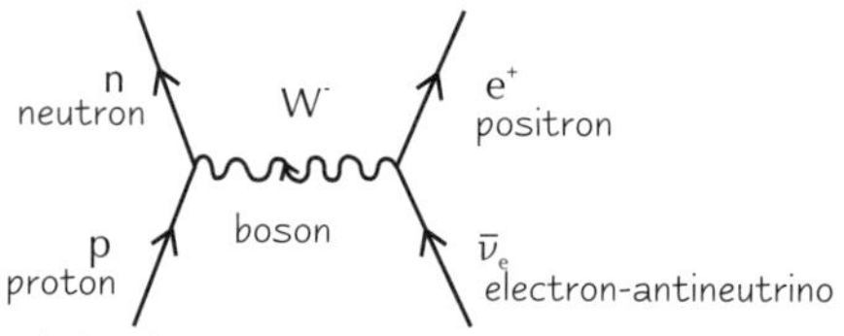

[1 mark for showing a W^+ boson (or a W^- boson) being exchanged in the correct direction, 1 mark for showing a neutron is produced]
This is a weak interaction. [1 mark]
Remember that only the weak interaction can change protons into neutrons or neutrons into protons.

Unit 1: Section 2 — Electromagnetic Radiation and Quantum Phenomena

Page 19 — The Photoelectric Effect

1) *The plate becomes positively charged [1 mark]*
Negative electrons in the metal absorb energy from the UV light and leave the surface [1 mark].

2) *An electron needs to gain a certain amount of energy (the work function energy) before it can leave the surface of the metal. [1 mark] If the energy carried by each photon is less than this work function energy, no electrons will be emitted. [1 mark]*

Page 21 — Energy Levels and Photon Emission

1) a) i) $\mathbf{E} = \mathbf{V} = 12.1$ eV *[1 mark]*
ii) $\mathbf{E} = \mathbf{V} \times 1.6 \times 10^{-19} = 12.1 \times 1.6 \times 10^{-19} = 1.9 \times 10^{-18}$ J *[1 mark]*
b) i) *The movement of an electron from a lower energy level to a higher energy level by absorbing energy. [1 mark]*
ii) $-13.6 + 12.1 = -1.5$ eV. *This corresponds to* $\mathbf{n} = 3$. *[1 mark]*
iii) $\mathbf{n} = 3 \rightarrow \mathbf{n} = 2$: $3.4 - 1.5 = 1.9$ eV *[1 mark]*
$\mathbf{n} = 2 \rightarrow \mathbf{n} = 1$: $13.6 - 3.4 = 10.2$ eV *[1 mark]*
$\mathbf{n} = 3 \rightarrow \mathbf{n} = 1$: $13.6 - 1.5 = 12.1$ eV *[1 mark]*

Page 23 — Wave-Particle Duality

AS Answers

Page 23 — Wave-Particle Duality

1) a) Electromagnetic radiation can show characteristics of both a particle and a wave. [1 mark]

b) i) $E_{photon} = \frac{hc}{\lambda} = \frac{6.63\times10^{-34}\times3.00\times10^{8}}{590\times10^{-9}}$ [1 mark]

$= 3.37 \times 10^{-19} J$ [1 mark]

$E \text{ (in eV)} = \frac{E \text{ (in J)}}{1.6\times10^{-19}} = \frac{3.37\times10^{-19}}{1.6\times10^{-19}} = 2.11 \text{ eV}$ [1 mark]

ii) $\lambda = \frac{h}{mv}$ [1 mark]

$\therefore v = \frac{h}{m\lambda} = \frac{6.63\times10^{-34}}{9.11\times10^{-31}\times590\times10^{-9}} = 1230 \text{ ms}^{-1}$ [1 mark]

2) a) $\lambda = \frac{h}{mv}$ [1 mark]

$= \frac{6.63\times10^{-34}}{9.11\times10^{-31}\times3.5\times10^{6}} = 2.08\times10^{-10} \text{ m}$ [1 mark]

b) Either $v = \frac{h}{m\lambda}$ with $m_{proton} = 1840 \times m_{electron}$

or momentum of protons = momentum of electrons

$1840\times \cancel{m_e} \times v_p = \cancel{m_e} \times 3.5\times10^{6}$

$v_p = 1900 \text{ ms}^{-1}$

[1 mark for either method with correct substitution, 1 mark for correct answer]

c) The two have the same kinetic energy if the voltages are the same. The proton has a larger mass, so it will have a smaller speed. [1 mark] Kinetic energy is proportional to the square of the speed, while momentum is proportional to the speed, so they will have different momenta. [1 mark] Wavelength depends on the momentum, so the wavelengths are different. [1 mark]

This is a really hard question. If you didn't get it right, make sure you understand the answer fully. Do the algebra if it helps.

3) a) $E_k = 6 \times 10^3 \text{ eV}$ [1 mark]

$= 6000 \times 1.6 \times 10^{-19} = 9.6 \times 10^{-16} J$ [1 mark]

b) $E_k = \frac{1}{2}mv^2$ [1 mark]

$9.6\times10^{-16} = \frac{1}{2}\times9.11\times10^{-31}\times v^2$

$v = \sqrt{\frac{2\times9.6\times10^{-16}}{9.11\times10^{-31}}} = 4.6\times10^{7} \text{ ms}^{-1}$ [1 mark]

c) $\lambda = \frac{h}{mv}$ [1 mark]

$= \frac{6.63\times10^{-34}}{9.11\times10^{-31}\times4.6\times10^{7}} = 1.56\times10^{-11} \text{ m}$ [1 mark]

Unit 1: Section 3 — Current Electricity

Page 25 — Current, Potential Difference and Resistance

1) Time in seconds = 10 × 60 = 600 s.
Use the formula $I = Q / t$ [1 mark]
which gives you $I = 4500 / 600 = 7.5$ A [1 mark]

Write down the formula first. Don't forget the unit in your answer.

2) Work done = 0.75 × electrical energy input
so the energy input will be 90 / 0.75 = 120 J. [1 mark]
Rearrange the formula $V = W / Q$ to give $Q = W / V$ [1 mark]
so you get $Q = 120 / 12 = 10$ C. [1 mark]

The electrical energy input to a motor has to be greater than the work it does because motors are less than 100% efficient.

3) a) $R = V/I = 10/60 = 0.166...$
$= 0.17 \ \Omega$ (to 2 d.p.) [1 mark]

b) $I = V/R = 35/0.166...$
$= 210$ A [1 mark]

c)

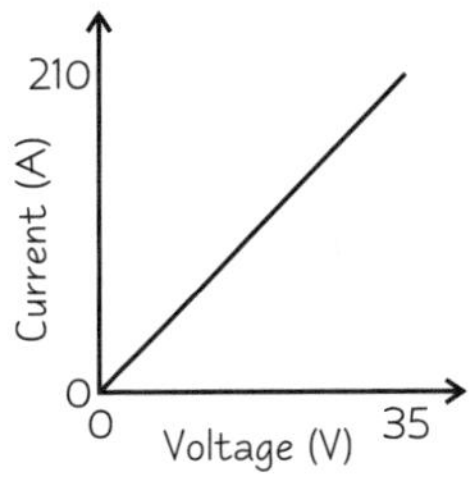

[1 mark]

Page 27 — I/V Characteristics

1) a)

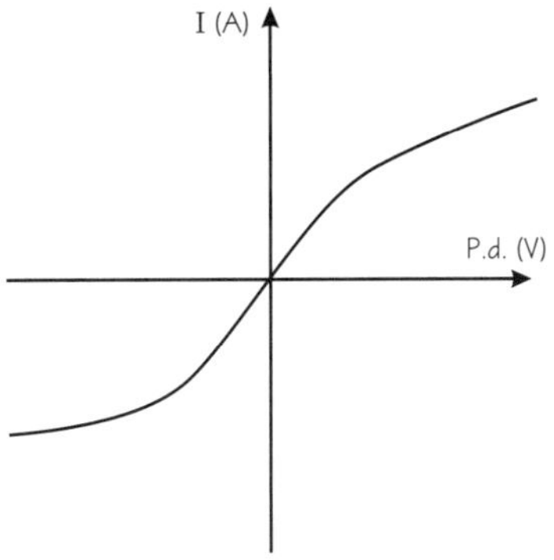

[1 mark]

b) Resistance increases as the temperature increases [1 mark]

Page 29 — Resistivity and Superconductivity

1) a) Area = $\pi(d/2)^2$ and $d = 1.0 \times 10^{-3}$ m
so Area = $\pi \times (0.5 \times 10^{-3})^2$
$= 7.85 \times 10^{-7} m^2$ [1 mark for correct evaluation of area]

$R = \frac{\rho l}{A} = \frac{2.8\times10^{-8}\times4}{7.85\times10^{-7}} = 0.14 \ \Omega$

[1 mark for use of correct equation or correct working, 1 mark for answer and unit.]

b) Resistance will now be zero [1 mark] because aluminium is a superconductor below its transition temperature of 1.2K [1 mark]

2) a) $R = V / I = \frac{2}{2.67\times10^{-3}} = 749 \ \Omega$

[1 mark for either step of working, 1 mark for correct answer and units.]

b) Two further resistance calculations give 750 Ω for each answer [1 mark]. No significant change in resistance for different potential differences [1 mark]. Component is an ohmic conductor because its resistance is constant for different potential differences. [1 mark]

Page 31 — Electrical Energy and Power

1) a) Current = power/p.d.
= 920/230 [1 mark for working] = 4 A [1 mark]

b) Current = p.d./resistance
= 230/190 [1 mark for working] = 1.21 A [1 mark]

c) Motor power = p.d. × current
= 230 × 1.21 [1 mark for working] = 278 W [1 mark]
Total power = motor power + heater power
= 278 + 920 = 1198 W, which is approx. 1.2 kW [1 mark]

AS Answers

2) a) Energy supplied = p.d. × current × time
= 12 × 48 × 2 [1 mark for working] = 1152 J [1 mark]

b) Energy lost = current^2 × resistance × time
= 48^2 × 0.01 × 2 [1 mark for working] = 46 J [1 mark]

Page 33 — E.m.f. and Internal Resistance

1) a) Total resistance = $\mathbf{R} + \mathbf{r} = 4 + 0.8 = 4.8\ \Omega$ [1 mark]
$\mathbf{I}$ = e.m.f./total resistance = 24/4.8 = 5 A [1 mark]

b) $\mathbf{V} = \varepsilon - \mathbf{Ir} = 24 - 5 \times 0.8 = 20$ V [1 mark]

2) a) $\varepsilon = \mathbf{I}(\mathbf{R} + \mathbf{r})$, so $\mathbf{r} = \varepsilon/\mathbf{I} - \mathbf{R}$ [1 mark]
$\mathbf{r} = 500/(50 \times 10^{-3}) - 10 = 9990\ \Omega$ [1 mark]

b) This is a very high internal resistance [1 mark], so only small currents can be drawn, reducing the risk to the user [1 mark]

Page 35 — Conservation of Energy & Charge in Circuits

1) a) Resistance of parallel resistors: $1/\mathbf{R}_{parallel} = 1/6 + 1/3 = 1/2$
so $\mathbf{R}_{parallel} = 2\ \Omega$ [1 mark]
Total resistance: $\mathbf{R}_{total} = 4 + \mathbf{R}_{parallel} = 4 + 2 = 6\ \Omega$ [1 mark]

b) $\mathbf{V} = \mathbf{IR}$, so rearranging $\mathbf{I}_3 = \mathbf{V} / \mathbf{R}_{total}$ [1 mark]
$\mathbf{I}_3 = 12 / 6 = 2$ A [1 mark]

c) $\mathbf{V} = \mathbf{IR} = 2 \times 4 = 8$ V [1 mark]

d) E.m.f. = sum of p.d.s in circuit, so $12 = 8 + \mathbf{V}_{parallel}$
$\mathbf{V}_{parallel} = 12 - 8 = 4$ V [1 mark]

e) Current = p.d. / resistance
$\mathbf{I}_1 = 4 / 3 = 1.33$ A [1 mark] $\mathbf{I}_2 = 4 / 6 = 0.67$ A [1 mark]

Page 37 — The Potential Divider

1) Parallel circuit, so p.d. across both sets of resistors is 12 V.

i) $\mathbf{V}_{AB} = ½ \times 12 = 6$ V [1 mark]

ii) $\mathbf{V}_{AC} = 2/3 \times 12 = 8$ V [1 mark]

iii) $\mathbf{V}_{BC} = \mathbf{V}_{AC} - \mathbf{V}_{AB} = 8 - 6 = 2$ V [1 mark]

2) a) $\mathbf{V}_{AB} = 50/80 \times 12 = 7.5$ V [1 mark]
(ignore the 10 Ω — no current flows that way)

b) Total resistance of the parallel circuit:
$1/\mathbf{R}_T = 1/50 + 1/(10 + 40) = 1/25$
$\mathbf{R}_T = 25\Omega$ [1 mark]
p.d. over the whole parallel arrangement = 25/55 × 12 = 5.45 V [1 mark]
p.d. across AB = 40/50 × 5.45 = 4.36 V [1 mark]
current through 40 Ω resistor = $\mathbf{V}/\mathbf{R}$ = 4.36/40 = 0.11 A [1 mark]

Page 39 — Alternating Current

1)

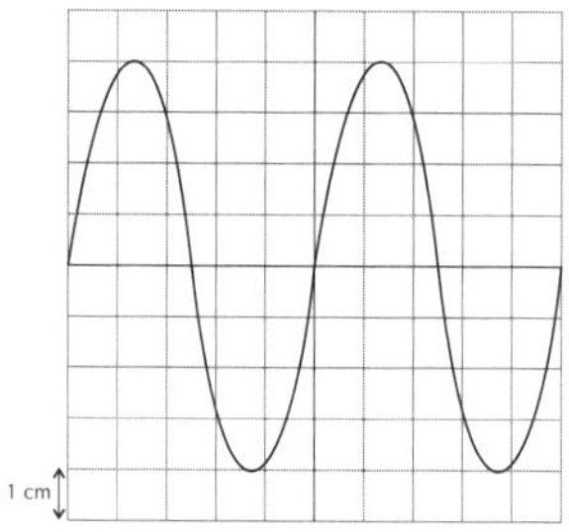

$\mathbf{T} = 1 / \mathbf{f} = 1/100 = 0.01$ s (10 ms) [1 mark]
Length of wave on screen = $\mathbf{T}$ / time base
= 10 ms / 2 ms cm^{-1} = 5 cm.
So, one wave occupies 5 cm horizontally and two waves take up the whole screen. [1 mark]
Height of wave on screen = peak voltage / voltage sensitivity
= 2 V / 0.5 V cm^{-1} = 4 cm.
Therefore the wave peaks 4 cm above the zero line and reaches its lowest value 4 cm below the zero line [1 mark]
[Plus 1 mark for drawing the diagram correctly]

2) a)

A d.c. signal produces a horizontal straight line. [1 mark]
Trace is a straight line 2 divs above the zero line. [1 mark]

b)

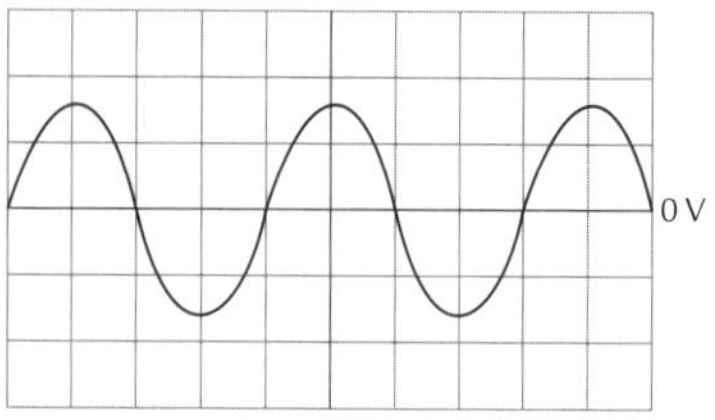

$\mathbf{f}$ = 50 Hz so $\mathbf{T}$ = 1 / 50 = 20 ms. [1 mark]
20 ms / 5 ms div^{-1} = 4 divs, one wave occupies 4 divisions [1 mark]
The height is 1½ divs above and below, vertically. [1 mark]

3) a) $\mathbf{T} = 1/2500 = 4 \times 10^{-4}$ s = 0.4 ms [1 mark]

b) One complete wave must occupy 2 divisions if five waves are seen on the whole screen. [1 mark]
2 divisions represents a time of 0.4 ms. So the time-base setting is 0.2 ms per division or 2×10^{-4} s per division [1 mark]

Unit 2: Section 1 — Mechanics

Page 41 — Scalars and Vectors

1) Start by drawing a diagram:

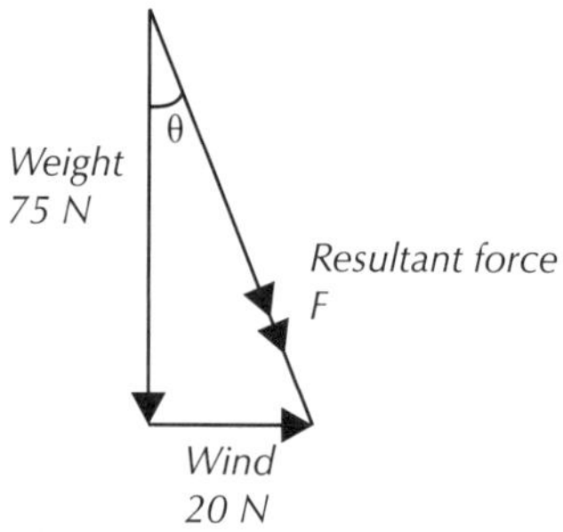

$\mathbf{F}^2 = 20^2 + 75^2 = 6025$
So $\mathbf{F} = 77.6$ N
$\tan\boldsymbol{\theta} = 20 / 75 = 0.267$
So $\boldsymbol{\theta} = 14.9°$
The resultant force on the rock is 77.6 N [1 mark] at an angle of 14.9° [1 mark] to the vertical.

Make sure you know which angle you're finding — and label it on your diagram.

2) Again, start by drawing a diagram:

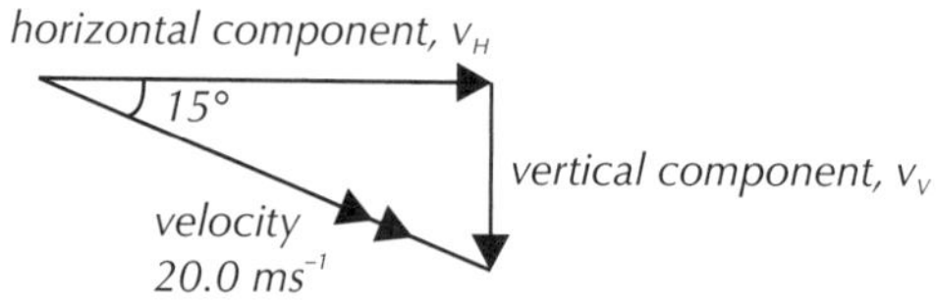

horizontal component $v_H = 20 \cos 15° = 19.3$ ms^{-1} [1 mark]
vertical component $v_V = 20 \sin 15° = 5.2$ ms^{-1} downwards [1 mark]

Always draw a diagram.

AS Answers

Page 43 — Motion with Constant Acceleration

1)a) $\mathbf{a} = -9.81\ ms^{-2}$, $\mathbf{t} = 5\ s$, $\mathbf{u} = 0\ ms^{-1}$, $\mathbf{v} = ?$

use : $\mathbf{v} = \mathbf{u} + \mathbf{at}$

$\mathbf{v} = 0 + 5 \times -9.81$ *[1 mark for either step of working]*

$\mathbf{v} = -49.05\ ms^{-1}$ *[1 mark]*

NB: It's negative because she's falling downwards and we took upwards as the positive direction.

b) Use: $\mathbf{s} = \left(\frac{\mathbf{u}+\mathbf{v}}{2}\right)\mathbf{t}$ or $\mathbf{s} = \mathbf{ut} + \frac{1}{2}\mathbf{at}^2$ *[1 mark for either]*

$\mathbf{s} = \frac{-49.05}{2} \times 5$ $\qquad$ $\mathbf{s} = 0 + \frac{1}{2} \times -9.81 \times 5^2$

$\mathbf{s} = -122.625\ m$ $\qquad$ $\mathbf{s} = -122.625\ m$

So she fell 122.625 m *[1 mark for answer]*

2)a) $\mathbf{v} = 0\ ms^{-1}$, $\mathbf{t} = 3.2\ s$, $\mathbf{s} = 40\ m$, $\mathbf{u} = ?$

use: $\mathbf{s} = \left(\frac{\mathbf{u}+\mathbf{v}}{2}\right)\mathbf{t}$ *[1 mark]*

$40 = 3.2\mathbf{u} \div 2$

$\mathbf{u} = \frac{80}{3.2} = 25\ ms^{-1}$ *[1 mark]*

b) use: $\mathbf{v}^2 = \mathbf{u}^2 + 2\mathbf{as}$ *[1 mark]*

$0 = 25^2 + 80\mathbf{a}$

$-80\mathbf{a} = 625$

$\mathbf{a} = -7.8\ ms^{-2}$ *[1 mark]*

3)a) Take upstream as negative: $\mathbf{v} = 5\ ms^{-1}$, $\mathbf{a} = 6\ ms^{-2}$, $\mathbf{s} = 1.2\ m$, $\mathbf{u} = ?$

use: $\mathbf{v}^2 = \mathbf{u}^2 + 2\mathbf{as}$ *[1 mark]*

$5^2 = \mathbf{u}^2 + 2 \times 6 \times 1.2$

$\mathbf{u}^2 = 25 - 14.4 = 10.6$

$\mathbf{u} = -3.26\ ms^{-1}$ *[1 mark]*

b) From furthest point: $\mathbf{u} = 0\ ms^{-1}$, $\mathbf{a} = 6\ ms^{-2}$, $\mathbf{v} = 5\ ms^{-1}$, $\mathbf{s} = ?$

use: $\mathbf{v}^2 = \mathbf{u}^2 + 2\mathbf{as}$ *[1 mark]*

$5^2 = 0 + 2 \times 6 \times \mathbf{s}$

$\mathbf{s} = 25 \div 12 = 2.08\ m$ *[1 mark]*

Page 45 — Free Fall and Projectile Motion

1)a) You only need to worry about the vertical motion of the stone.

$\mathbf{u} = 0\ ms^{-1}$, $\mathbf{s} = -560\ m$, $\mathbf{a} = -\mathbf{g} = -9.81\ ms^{-2}$, $\mathbf{t} = ?$

You need to find $\mathbf{t}$, so use: $\mathbf{s} = \mathbf{ut} + \frac{1}{2}\mathbf{at}^2$ *[1 mark]*

$-560 = 0 + \frac{1}{2} \times -9.81 \times \mathbf{t}^2$

$t = \sqrt{\frac{2\times(-560)}{-9.81}} = 10.7\,s\ (1\ d.p.) = 11\ s$ (to the nearest second)

[1 mark]

b) You know that in the horizontal direction:

$\mathbf{v} = 20\ ms^{-1}$, $\mathbf{t} = 10.7\ s$, $\mathbf{a} = 0$, $\mathbf{s} = ?$

So use velocity = $\frac{distance}{time}$, $\mathbf{v} = \frac{\mathbf{s}}{\mathbf{t}}$ *[1 mark]*

$\mathbf{s} = \mathbf{v} \times \mathbf{t} = 20 \times 10.7 = 214\ m$ (to the nearest metre) *[1 mark]*

2) You know that for the arrow's vertical motion (taking upwards as the positive direction):

$\mathbf{a} = -9.81\ ms^{-2}$, $\mathbf{u} = 30\ ms^{-1}$ and the arrow will be at its highest point just before it starts falling back towards the ground, so $\mathbf{v} = 0\ m/s$.

$\mathbf{s}$ = the distance travelled from the arrow's firing point

So use $\mathbf{v}^2 = \mathbf{u}^2 + 2\mathbf{as}$ *[1 mark]*

$0 = 30^2 + 2 \times -9.81 \times \mathbf{s}$

$900 = 2 \times 9.81\mathbf{s}$

$\mathbf{s} = \frac{900}{2 \times 9.81} = 45.9\ m$ *[1 mark]*

So the maximum distance reached from the ground = 45.9 + 1 = 47 m (to the nearest metre). *[1 mark]*

Page 47 — Displacement-Time Graphs

1) Split graph into four sections:

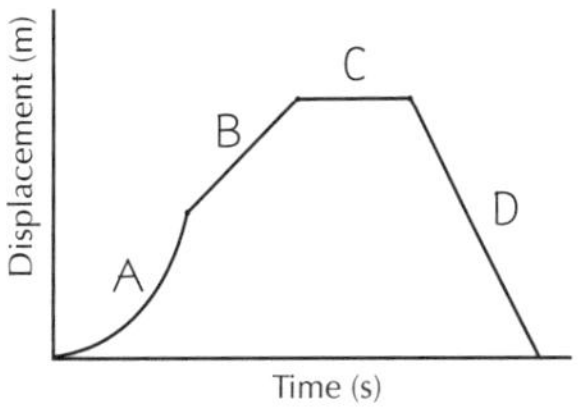

A: acceleration *[1 mark]*
B: constant velocity *[1 mark]*
C: stationary *[1 mark]*
D: constant velocity in opposite direction to A and B *[1 mark]*

2)a)

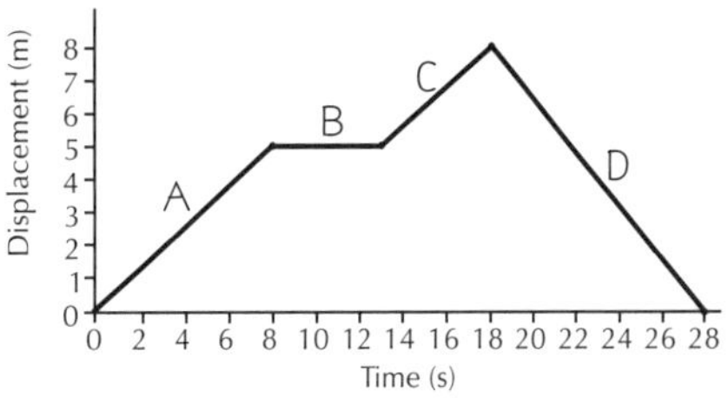

[4 marks — 1 mark for each section correctly drawn]

b) At A: $\mathbf{v} = \frac{\text{displacement}}{\text{time}} = \frac{5}{8} = 0.625\ ms^{-1}$

At B: $\mathbf{v} = 0$

At C: $\mathbf{v} = \frac{\text{displacement}}{\text{time}} = \frac{3}{5} = 0.6\ ms^{-1}$

At D: $\mathbf{v} = \frac{\text{displacement}}{\text{time}} = \frac{-8}{10} = -0.8\ ms^{-1}$

[2 marks for all correct or just 1 mark for 2 or 3 correct]

Page 49 — Velocity-Time Graphs

1)a)

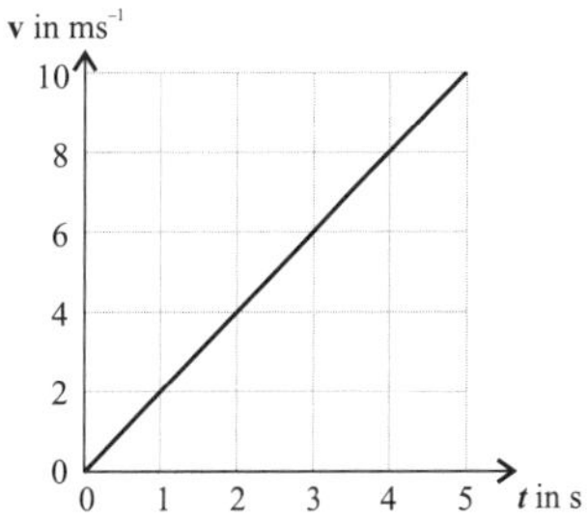

[2 marks]

b) use $\mathbf{s} = \mathbf{ut} + \frac{1}{2}\mathbf{at}^2$ *[1 mark]*

($\mathbf{t} = 1$, $\mathbf{s} = 1$) ($\mathbf{t} = 2$, $\mathbf{s} = 4$) ($\mathbf{t} = 3$, $\mathbf{s} = 9$) ($\mathbf{t} = 4$, $\mathbf{s} = 16$) ($\mathbf{t} = 5$, $\mathbf{s} = 25$)

[2 marks for all correct or 1 mark for at least 3 pairs of values right]

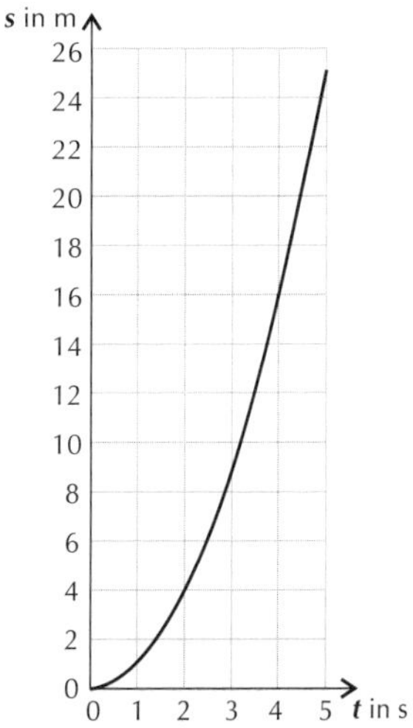

[2 marks]

AS Answers

c) E.g. another way to calculate displacement is to find the area under the velocity-time graph. [1 mark]
E.g. total displacement = ½ × 5 × 10 = 25 m [1 mark]

Page 51 — Mass, Weight and Centre of Gravity

1) a) Density is a measure of 'compactness' of a material — its mass per unit volume. [1 mark]

b) $\rho = \frac{\mathbf{m}}{\mathbf{V}}$ [1 mark]

V of cylinder = $\pi \mathbf{r}^2 \mathbf{h} = \pi \times 4^2 \times 6 = 301.6\ cm^3$ [1 mark]
$\rho = 820 \div 301.6 = 2.72\ g\,cm^{-3}$ [1 mark]

c) $\mathbf{V} = 5 \times 5 \times 5 = 125\ cm^3$
$\mathbf{m} = \rho \times \mathbf{V} = 2.7 \times 125 = 340\ g$ [1 mark]

2) Experiment:
Hang the object freely from a point. Hang a plumb bob from the same point, and use it to draw a vertical line down the object. [1 mark] Repeat for a different point and find the point of intersection. [1 mark]
The centre of gravity is halfway through the thickness of the object (by symmetry) at the point of intersection.
Identifying and reducing error, e.g.:
Source: the object and/or plumb line might move slightly while you're drawing the vertical line [1 mark]
Reduced by: hang the object from a third point to confirm the position of the point of intersection [1 mark]

Page 53 — Forces

1)

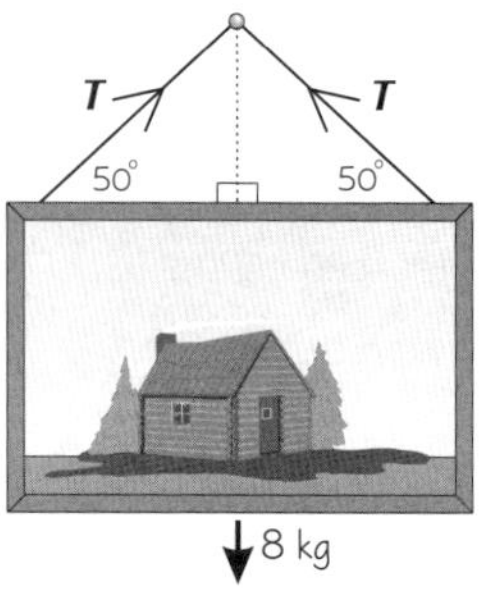

Weight = vertical component of tension × 2
$8 \times 9.81 = 2\mathbf{T} \sin 50°$ [1 mark]
$78.48 = 0.766 \times 2\mathbf{T}$
$102.45 = 2\mathbf{T} \Rightarrow \mathbf{T} = 51.2\ N$ [1 mark]

2)

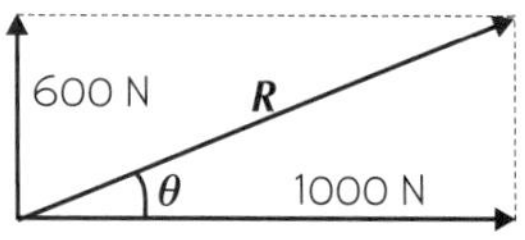

By Pythagoras: $\mathbf{R} = \sqrt{1000^2 + 600^2} = 1166\ N$ [1 mark]

$\tan \boldsymbol{\theta} = \frac{600}{1000}$, so $\boldsymbol{\theta} = \tan^{-1} 0.6 = 31.0°$ [1 mark]

Page 55 — Moments and Torques

1) Torque = Force × distance [1 mark]
$60 = 0.4\mathbf{F}$, so $\mathbf{F} = 150\ N$ [1 mark]

2)

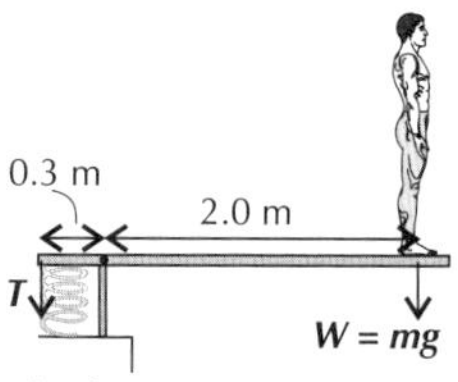

clockwise moment = anticlockwise moment
$\mathbf{W} \times 2.0 = \mathbf{T} \times 0.3$ [1 mark for either line of working]
$60 \times 9.81 \times 2.0 = \mathbf{T} \times 0.3 \Rightarrow \mathbf{T} = 3924\ N$ [1 mark]

Page 57 — Newton's Laws of Motion

1)

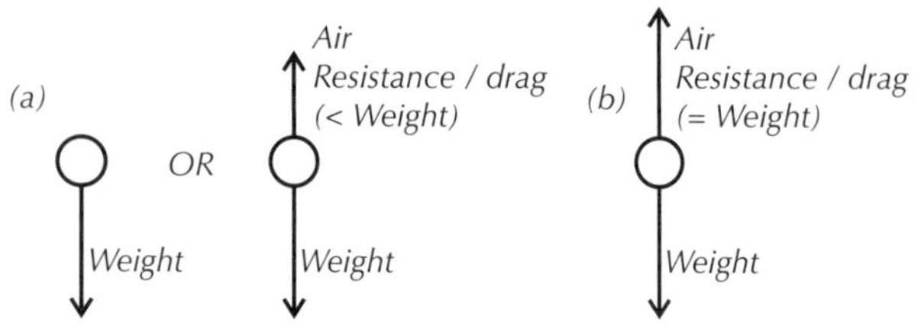

[1 mark for each diagram]

2) a) Force perpendicular to river flow = 500 – 100 = 400 N [1 mark]
Force parallel to river flow = 300 N

Resultant force = $\sqrt{400^2 + 300^2} = 500\ N$ [1 mark]

b) $\mathbf{a} = \mathbf{F}/\mathbf{m}$ (from $\mathbf{F} = \mathbf{ma}$) [1 mark]
$= 500/250 = 2\ ms^{-2}$ [1 mark]

3) a) The resultant force acting on it [1 mark] and its mass. [1 mark]

b) Michael is able to exert a greater force than Tom.
Michael is lighter than Tom. [1 mark each for 2 sensible points]

c) The only force acting on each of them is their weight = $\mathbf{m}g$ [1 mark]. Since $\mathbf{F} = \mathbf{ma}$, this gives $\mathbf{ma} = \mathbf{m}g$, or $\mathbf{a} = g$ [1 mark]. Their acceleration doesn't depend on their mass — it's the same for both of them — so they reach the water at the same time. [1 mark]

Page 59 — Drag and Terminal Velocity

1) a) The velocity increases at a steady rate, which means the acceleration is constant [1 mark]. Constant acceleration means there must be no atmospheric resistance (atmospheric resistance would increase with velocity, leading to a decrease in acceleration). So there must be no atmosphere. [1 mark]

b)

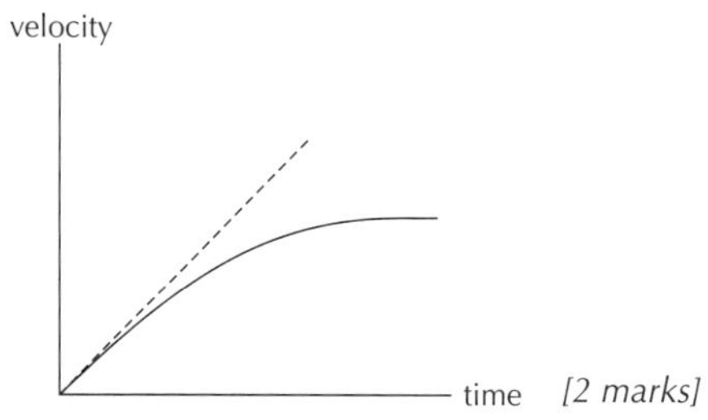

[2 marks]

Your graph must be a smooth curve which levels out. It must NOT go down at the end.

c) (The graph becomes less steep)
because the acceleration is decreasing [1 mark]
because air resistance increases with speed [1 mark]
(The graph levels out)
because air resistance has become equal to weight [1 mark]

If the question says 'explain', you won't get marks for just describing what the graph shows — you have to say why it is that shape.

Page 61 — Work and Power

1) a)

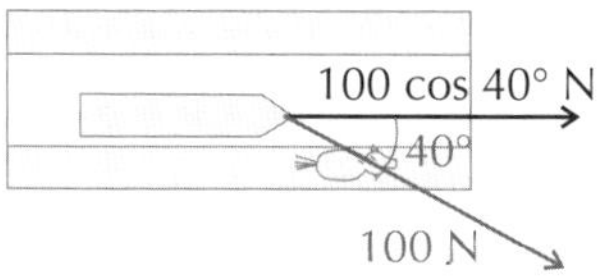

Force in direction of travel = 100 cos40° = 76.6 N [1 mark]
$\mathbf{W} = \mathbf{Fs} = 76.6 \times 1500 = 114\,900\ J$ [1 mark]

b) Use $\mathbf{P} = \mathbf{Fv}$ [1 mark]
$= 100 \cos 40° \times 0.8 = 61.3\ W$ [1 mark]

2) a) Use $\mathbf{W} = \mathbf{Fs}$ [1 mark]
$= 20 \times 9.81 \times 3 = 588.6\ J$ [1 mark]

b) Use $\mathbf{P} = \mathbf{Fv}$ [1 mark]
$= 20 \times 9.81 \times 0.25 = 49.05\ W$ [1 mark]

AS Answers

Page 63 — Conservation of Energy

1) a) Use $E_k = \frac{1}{2}mv^2$ and $E_p = mgh$ [1 mark]
$\frac{1}{2}mv^2 = mgh$
$\frac{1}{2}v^2 = gh$
$v^2 = 2gh = 2 \times 9.81 \times 2 = 39.24$ [1 mark]
$v = 6.26\ ms^{-1}$ [1 mark]
'No friction' allows you to say that the changes in kinetic and potential energy will be the same.

b) 2 m — no friction means the kinetic energy will all change back into potential energy, so he will rise back up to the same height as he started. [1 mark]

c) Put in some more energy by actively 'skating'. [1 mark]

2) a) If there's no air resistance, $E_k = E_p = mgh$ [1 mark]
$E_k = 0.02 \times 9.81 \times 8 = 1.57\ J$ [1 mark]

b) If the ball rebounds to 6.5 m, it has gravitational potential energy:
$E_p = mgh = 0.02 \times 9.81 \times 6.5 = 1.28\ J$ [1 mark]
So $1.57 - 1.28 = 0.29\ J$ is converted to other forms [1 mark]

Unit 2: Section 2 — Materials

Page 65 — Hooke's Law

1) a) Force is proportional to extension.
The force is 1.5 times as great, so the extension will also be 1.5 times the original value.
Extension = 1.5×4.0 mm = 6.0 mm [1 mark]

b) $F = ke$ and so $k = F/e$ [1 mark]
$k = 10 \div 4.0 \times 10^{-3}$ = 2500 Nm^{-1} or 2.5 Nmm^{-1} [1 mark]
There is one mark for rearranging the equation and another for getting the right numerical answer.

c) One mark for any sensible point e.g.
The string now stretches much further for small increases in force.
When the string is loosened it is longer than at the start. [1 mark]

2) The rubber band does not obey Hooke's law [1 mark] because when the force is doubled from 2.5 N to 5 N, the extension increases by a factor of 2.3. [1 mark]

Page 67 — Stress and Strain

1) a) Area = $\pi d^2/4$ or πr^2.
So area = $\pi \times (1 \times 10^{-3})^2/4 = 7.85 \times 10^{-7}\ m^2$ [1 mark]

b) Stress = force/area = $300/(7.85 \times 10^{-7})$
$= 3.82 \times 10^8\ Nm^{-2}$ or Pa [1 mark]

c) Strain = extension/length = $4 \times 10^{-3}/2.00 = 2 \times 10^{-3}$ [1 mark]

2) a) $F = ke$ and so rearranging $k = F/e$ [1 mark]
So $k = 50/(3.0 \times 10^{-3}) = 1.67 \times 10^4\ Nm^{-1}$ [1 mark]

b) Elastic strain energy = $\frac{1}{2}Fe$
Giving the elastic strain energy as
$\frac{1}{2} \times 50 \times 3 \times 10^{-3} = 7.5 \times 10^{-2}\ J$ [1 mark]

3) Elastic strain energy,
$E = \frac{1}{2}ke^2 = \frac{1}{2} \times 40.8 \times 0.05^2 = 0.051\ J$ [1 mark]
To find maximum speed, assume all this energy is converted to kinetic energy in the ball. $E_{kinetic} = E$ [1 mark]
$E = \frac{1}{2}mv^2$, so rearranging, $v^2 = 2E/m$ [1 mark]
$v^2 = (2 \times 0.051)/0.012 = 8.5$, so $v = 2.92\ ms^{-1}$ [1 mark]

Page 69 — The Young Modulus

1) a) Cross-sectional area = $\pi d^2/4$ or πr^2.
So the cross-sectional area = $\pi \times (0.6 \times 10^{-3})^2/4 = 2.83 \times 10^{-7}\ m^2$ [1 mark]

b) Stress = force/area = $80/(2.83 \times 10^{-7}) = 2.83 \times 10^8\ Nm^{-2}$ [1 mark]

c) Strain = extension/length = $3.6 \times 10^{-3}/2.5 = 1.44 \times 10^{-3}$ [1 mark]

d) The Young modulus for steel = stress/strain
$= 2.83 \times 10^8/(1.44 \times 10^{-3}) = 2.0 \times 10^{11}\ Nm^{-2}$ [1 mark]

2) a) The Young modulus, E = stress/strain and so strain = stress/E [1 mark]
Strain on copper = $2.6 \times 10^8/1.3 \times 10^{11} = 2 \times 10^{-3}$ [1 mark]
There's one mark for rearranging the equation and another for using it.

b) Stress = force/area and so area = force/stress
Area of the wire = $100/(2.6 \times 10^8) = 3.85 \times 10^{-7}\ m^2$ [1 mark]

c) Strain energy per unit volume = $\frac{1}{2}$ × stress × strain
$= \frac{1}{2} \times 2.6 \times 10^8 \times 2 \times 10^{-3} = 2.6 \times 10^5\ Jm^{-3}$ [1 mark]
Give the mark if answer is consistent with the value calculated for strain in part a).

Page 71 — Interpreting Stress-Strain Graphs

1) a) Liable to break suddenly without deforming plastically. [1 mark]

b)

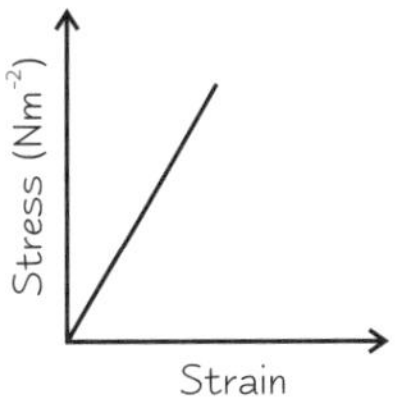

[1 mark for correctly labelled axes, 1 mark for straight line through the origin]

c) When stress is applied to hardened steel, tiny cracks at the material's surface get bigger until the material fractures. [1 mark]
The cracks are able to grow because the steel has a rigid structure. [1 mark]

2) a) E = stress ÷ strain [1 mark]
$= 800 \div 0.2 = 4000\ MNm^{-2}$ (or $4.0 \times 10^9\ Nm^{-2}$) [1 mark]

b) Energy = area under the graph [1 mark]
$= (800 \times 0.2) \div 2 = 80$ MJ (or 8.0×10^7 J) [1 mark]

Unit 2: Section 3 — Waves

Page 73 — The Nature of Waves

1) a) Use $v = \lambda f$ and $f = 1/T$
So $v = \lambda/T$, giving $\lambda = vT$ [1 mark]
$\lambda = 3\ ms^{-1} \times 6\ s = 18\ m$ [1 mark]
The vertical movement of the buoy is irrelevant to this part of the question.

b) The trough to peak distance is twice the amplitude, so the amplitude is 0.6 m [1 mark]

Page 75 — Longitudinal and Transverse Waves

1) [This question could equally well be answered using diagrams.]
For ordinary light, the EM field vibrates in all planes at right angles to the direction of travel. [1 mark]
Iceland spar acts as a polariser. [1 mark]
When light is shone through the first disc, it only allows through vibrations in one particular plane, so the light emerges less bright. [1 mark]
As the two crystals are rotated relative to each other there comes a point when the allowed planes are at right angles to each other. [1 mark]
So all the light is blocked. [1 mark]
Try to remember to say that for light and other EM waves it's the electric and magnetic fields that vibrate.

2) E.g. Polarising filters are used in photography to remove unwanted reflections [1 mark]. Light is partially polarised when it reflects, so putting a polarising filter over the lens at 90 degrees to the plane of polarisation will block most of the reflected light. [1 mark].

AS Answers

Page 77 — The Electromagnetic Spectrum

1) At the same speed. [1 mark] Both are electromagnetic waves and hence travel at **c** in a vacuum. [1 mark]

2) a) Medical X-rays [1 mark] rely on the fact that X-rays penetrate the body well but are blocked by bone. [1 mark] OR
Security scanners at airports [1 mark] rely on the fact that X-rays penetrate suitcases and clothes but are blocked by metal e.g. of a weapon. [1 mark]

b) The main difference between gamma rays and X-rays is that gamma rays arise from nuclear decay [1 mark] but X-rays are generated when metals are bombarded with electrons. [1 mark]

3) Any of: unshielded microwaves, excess heat, damage to eyes from too bright light, sunburn or skin cancer from UV, cancer or eye damage due to ionisation by X-rays or gamma rays.
[1 mark for the type of EM wave, 1 mark for the danger to health]

Page 79 — Refractive Index

1) a) $n_{diamond} = c / v_{diamond} = (3 \times 10^8) / (1.24 \times 10^8) = 2.42$ [1 mark]

b) $n_{air} \sin i = n_{diamond} \sin r$, $n_{air} = 1$
So, $n_{diamond} = \sin i / \sin r$ [1 mark]

$$\sin r = \frac{\sin i}{n_{diamond}} = \frac{\sin 50}{2.42} = 0.317 \Rightarrow r = 18.5° \text{ [1 mark]}$$

You can assume the refractive index of air is 1, and don't forget to write the degree sign in your answer.

2) a) When the light is pointing steeply upwards some of it is refracted and some reflected — the beam emerging from the surface is the refracted part. [1 mark]
However when the beam hits the surface at more than the critical angle (to the normal to the boundary) refraction does not occur. All the beam is totally internally reflected to light the tank, hence its brightness. [1 mark]

b) The critical angle is 90° – 41.25° = 48.75°. [1 mark]
$n_{water} = 1 / \sin C = 1 / \sin 48.75° = 1 / 0.752 = 1.330$ [1 mark]
The question talks about the angle between the light beam and the floor of the aquarium. This angle is 90° minus the incident angle — measured from a normal to the surface of the water.

Page 81 — Superposition and Coherence

1) a) The frequencies and wavelengths of the two sources must be equal [1 mark] and the phase difference must be constant. [1 mark]

b) Interference will only be noticeable if the amplitudes of the two waves are approximately equal. [1 mark]

2) a) 180° (or 180° + 360n°). [1 mark]

b) The displacements and velocities of the two points are equal in size [1 mark] but in opposite directions. [1 mark]

Page 83 — Stationary Waves

1) a)

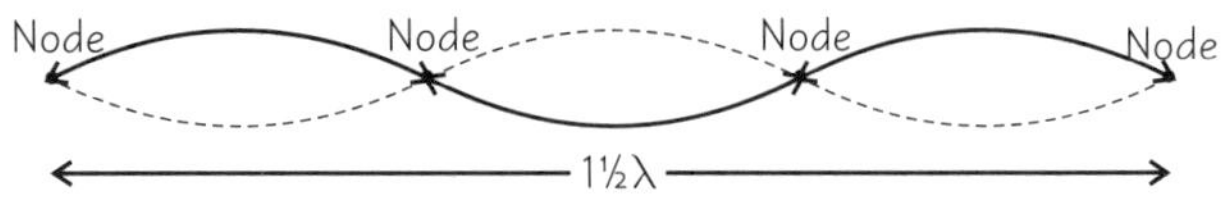

[1 mark for the correct shape, 1 mark for labelling the length]

b) For a string vibrating at three times the fundamental frequency, length = 3**λ** / 2
1.2 m = 3**λ** / 2
λ = 0.8 m [1 mark]

c) When the string forms a standing wave, its amplitude varies from a maximum at the antinodes to zero at the nodes. [1 mark] In a progressive wave all the points have the same amplitude. [1 mark]

Page 85 — Diffraction

1) When a wavefront meets an obstacle, the waves will diffract round the corners of the obstacle. When the obstacle is much bigger than the wavelength, little diffraction occurs. In this case, the mountain is much bigger than the wavelength of short-wave radio. So the "shadow" where you cannot pick up short wave is very long. [1 mark]

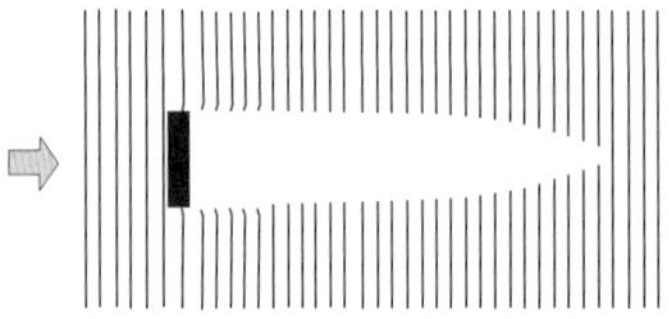

[1 mark]

When the obstacle is comparable in size to the wavelength, as it is for the long-wave radio waves, more diffraction occurs. The wavefront re-forms after a shorter distance, leaving a shorter "shadow". [1 mark]

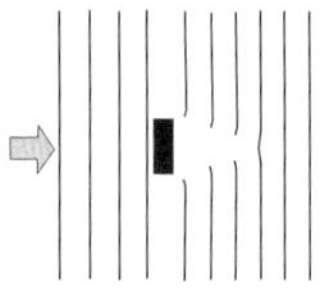

[1 mark]

Page 87 — Two-Source Interference

1) a)

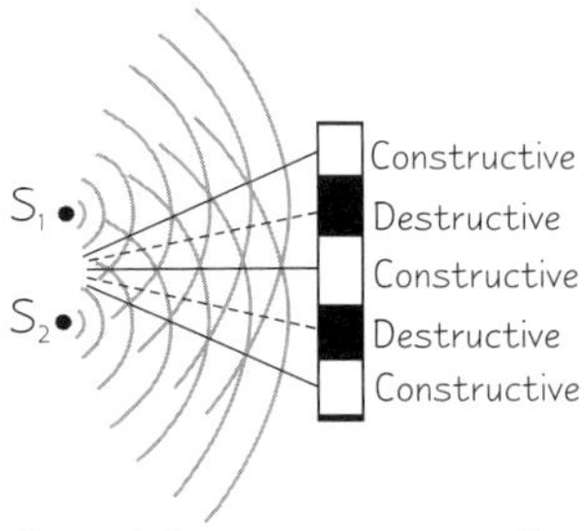

[1 mark for correct constructive interference patterns, 1 mark for correct destructive interference patterns]

b) Light waves from separate sources are not coherent, as light is emitted in random bursts of energy. To get coherent light the two sets of waves must emerge from one source. [1 mark] A laser is used because it emits coherent light that is all of one wavelength. [1 mark]

2) a) $\lambda = v / f = 330 / 1320 = 0.25$ m. [1 mark]

b) Separation = $X = \lambda D / d$ [1 mark]
= 0.25 m × 7 m / 1.5 m = 1.17 m. [1 mark]

Page 89 — Diffraction Gratings

1) a) Use $\sin \theta = n\lambda / d$
For the first order, $n = 1$
So, $\sin \theta = \lambda / d$ [1 mark]
No need to actually work out d. The number of lines per metre is 1 / d. So you can simply multiply the wavelength by that.
$\sin \theta = 600 \times 10^{-9} \times 4 \times 10^5 = 0.24$
$\theta = 13.9°$ [1 mark]
For the second order, $n = 2$ and $\sin \theta = 2\lambda / d$. [1 mark]
You already have a value for λ / d. Just double it to get sin θ for the second order.
$\sin \theta = 0.48$
$\theta = 28.7°$ [1 mark]

b) No. Putting $n = 5$ into the equation gives a value of $\sin \theta$ of 1.2, which is impossible. [1 mark]

2) $\sin \theta = n\lambda / d$, so for the 1st order maximum, $\sin \theta = \lambda / d$ [1 mark]
$\sin 14.2° = \lambda \times 3.7 \times 10^5$
$\lambda = 663$ nm (or 6.63×10^{-7} m) [1 mark].

A2 Answers

Unit 4: Section 1 — Further Mechanics

Page 95 — Momentum and Impulse

1) a) total momentum before collision = total momentum after [1 mark]
$(0.6 \times 5) + 0 = (0.6 \times -2.4) + 2v$
$3 + 1.44 = 2v$ [1 mark for working] $\Rightarrow v = 2.22\ ms^{-1}$ [1 mark]

b) Kinetic energy before collision = $\frac{1}{2} \times 0.6 \times 5^2 + \frac{1}{2} \times 2 \times 0^2 = 7.5$ J
Kinetic energy after the collision = $\frac{1}{2} \times 0.6 \times 2.4^2 + \frac{1}{2} \times 2 \times 2.22^2$
$= 1.728 + 4.9284 = 6.6564$ J [1 mark]
The kinetic energy of the two balls is greater before the collision than after (i.e. it's not conserved) [1 mark], so the collision must be inelastic [1 mark].

2) momentum before = momentum after [1 mark]
$\Rightarrow (0.7 \times 0.3) + 0 = 1.1v$
$0.21 = 1.1v$ [1 mark for working] $\Rightarrow v = 0.19\ ms^{-1}$ [1 mark]

Page 97 — Circular Motion

1) a) $\omega = \frac{\theta}{t}$ [1 mark] so $\omega = \frac{2\pi}{3.2 \times 10^7} = 2.0 \times 10^{-7}\ rad\,s^{-1}$ [1 mark]

b) $v = r\omega$ [1 mark] $= 1.5 \times 10^{11} \times 2.0 \times 10^{-7}$
$= 30\ kms^{-1}$ [1 mark]

c) $F = m\omega^2 r$ [1 mark] $= 6.0 \times 10^{24} \times (2.0 \times 10^{-7})^2 \times 1.5 \times 10^{11}$
$= 3.6 \times 10^{22}$ N [1 mark]
The answers to b) and c) use the rounded value of ω calculated in a).

d) The gravitational force between the Sun and the Earth [1 mark]

2) a) Gravity pulling down on the water at the top of the swing gives a centripetal acceleration of $9.81\ ms^{-2}$ [1 mark]. If the circular motion of the water needs a centripetal acceleration of less than $9.81\ ms^{-2}$, gravity will pull it in too tight a circle. The water will fall out of the bucket.

Since $a = \omega^2 r$, $\omega^2 = \frac{a}{r} = \frac{9.81}{1}$, so $\omega = 3.1\ rad\,s^{-1}$ [1 mark]

$\omega = 2\pi f$, so $f = \frac{\omega}{2\pi} = 0.5\ rev\,s^{-1}$ [1 mark]

b) Centripetal force = $m\omega^2 r = 10 \times 5^2 \times 1 = 250$ N [1 mark].
This force is provided by both the tension in the rope, T, and gravity: $T + (10 \times 9.81) = 250$.
So $T = 250 - (10 \times 9.81) = 152$ N [1 mark].

Page 99 — Simple Harmonic Motion

1) a) Simple harmonic motion is an oscillation in which an object always accelerates towards a fixed point [1 mark] with an acceleration directly proportional to its displacement from that point [1 mark]. [The SHM equation would get you the marks if you defined all the variables.]

b) The acceleration of a falling bouncy ball is due to gravity. This acceleration is constant, so the motion is not SHM. [1 mark].

2) a) Maximum velocity = $(2\pi f)A = 2\pi \times 1.5 \times 0.05 = 0.47\ ms^{-1}$ [1 mark].

b) Stopclock started when object released, so $x = A\cos(2\pi ft)$ [1 mark]
$x = 0.05 \times \cos(2\pi \times 1.5 \times 0.1) = 0.05 \times \cos(0.94)$
$= 0.029$ m [1 mark].

c) $x = A\cos(2\pi ft) \Rightarrow 0.01 = 0.05 \times \cos(2\pi \times 1.5t)$.
So $0.2 = \cos(3\pi t) \Rightarrow \cos^{-1}(0.2) = 3\pi t$.
$3\pi t = 1.37 \Rightarrow t = 0.15$ s.
[1 mark for working, 1 mark for correct answer]

Page 101 — Simple Harmonic Oscillators

1) a) Extension of spring = 0.20 – 0.10 = 0.10 m [1 mark]. Hooke's Law gives $k = \frac{\text{force}}{\text{extension}}$, so $k = \frac{0.10 \times 9.8}{0.10} = 9.8\ Nm^{-1}$ [1 mark].

b) $T = 2\pi\sqrt{\frac{m}{k}} \Rightarrow T = 2\pi \times \sqrt{\frac{0.10}{9.8}} = 2\pi \times \sqrt{0.01} = 0.63$ s [1 mark].

c) $m \propto T^2$ so if T is doubled, T^2 is quadrupled and m is quadrupled [1 mark]. So mass needed = $4 \times 0.10 = 0.40$ kg [1 mark].

2) E.g. $5T_{\text{short pendulum}} = 3T_{\text{long pendulum}}$, and $T = 2\pi\sqrt{\frac{l}{g}}$ [1 mark]. Let length of long pendulum = l. So $5\left(2\pi\sqrt{\frac{0.20}{g}}\right) = 3\left(2\pi\sqrt{\frac{l}{g}}\right)$ [1 mark].

Dividing by 2π gives $5 \times \sqrt{\frac{0.20}{g}} = 3 \times \sqrt{\frac{l}{g}}$. Squaring and simplifying gives $5 = 9l$ so length of long pendulum = 5/9 = 0.56 m [1 mark].

Page 103 — Free and Forced Vibrations

1) a) When a system is forced to vibrate at a frequency that's close to, or the same as its natural frequency [1 mark] and oscillates with a much larger than usual amplitude [1 mark].

b) [1 mark] for a peak at the natural frequency, [1 mark] for a sharp peak.

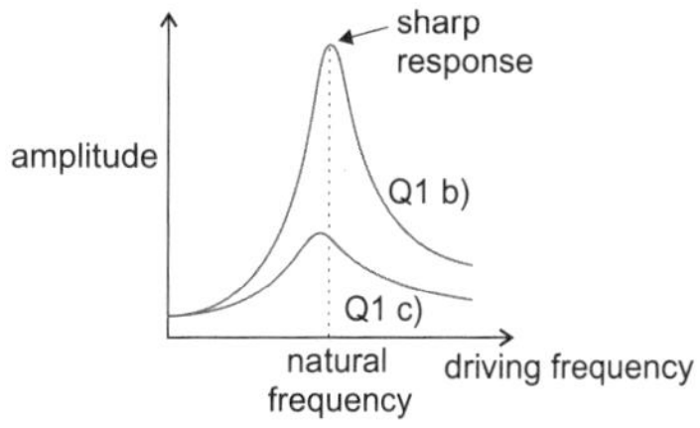

c) See graph. [1 mark] for a smaller peak at the natural frequency [the peak will actually be slightly to the left of the natural frequency, but you'll get the mark if the peak is at the same frequency in the diagram].

2) a) A system is critically damped if it returns to rest in the shortest time possible [1 mark] when it's displaced from equilibrium and released.

b) e.g. suspension in a car [1 mark].

Unit 4: Section 2 — Gravitation and Electric Fields

Page 105 — Gravitational Fields

1) a) $g = \frac{GM}{r^2} \Rightarrow M = \frac{gr^2}{G} = \frac{9.81 \times (6400 \times 1000)^2}{6.67 \times 10^{-11}}$ [1 mark]
$= 6.02 \times 10^{24}$ kg [1 mark]

b) $F = \frac{GMm}{r^2} = \frac{6.67 \times 10^{-11} \times 1.99 \times 10^{30} \times 6.02 \times 10^{24}}{(1.5 \times 10^{11})^2}$ [1 mark]
$= 3.6 \times 10^{22}$ N [1 mark]

2) a) $g = \frac{GM}{r^2} = \frac{6.67 \times 10^{-11} \times 7.35 \times 10^{22}}{(1740 \times 1000)^2} = 1.62\ Nkg^{-1}$ [1 mark]

b) $F = mg = 25 \times 1.62 = 40.5$ N [1 mark]

Page 107 — Gravitational Fields

1) a) $T = \sqrt{\frac{4\pi^2 r^3}{GM}} = \sqrt{\frac{4\pi^2 \times [(6400 + 200) \times 1000]^3}{6.67 \times 10^{-11} \times 5.98 \times 10^{24}}}$ [1 mark]
= 5334 seconds OR 1.48 hours [1 mark]

b) $v = \sqrt{\frac{GM}{r}} = \sqrt{\frac{6.67 \times 10^{-11} \times 5.98 \times 10^{24}}{(6400 + 200) \times 1000}}$ [1 mark]
$= 7774\ ms^{-1} \approx 7.77\ kms^{-1}$

A2 Answers

2) Period = 24 hours = 24 × 60 × 60 = 86400 s *[1 mark]*

$$T=\sqrt{\frac{4\pi^2 r^3}{GM}} \Rightarrow r=\sqrt[3]{\frac{T^2GM}{4\pi^2}}=\sqrt[3]{\frac{86400^2\times 6.67\times 10^{-11}\times 5.98\times 10^{24}}{4\pi^2}}$$

$r = 4.23 \times 10^7$ m $= 4.23 \times 10^4$ km *[1 mark]*

Height above Earth $= 4.23 \times 10^4 - 6.4 \times 10^3 = 35900$ km *[1 mark]*

Page 109 — Electric Fields

1) Charge on alpha particle, $Q_1 = +2e = 2 \times 1.6 \times 10^{-19} = 3.2 \times 10^{-19}$ C

Charge on gold nucleus, $Q_2 = +79e = 79 \times 1.6 \times 10^{-19}$ C $= 1.3 \times 10^{-17}$ C *[1 mark]*

$$F=\frac{1}{4\pi\varepsilon_0}\frac{Q_1Q_2}{r^2}=\frac{1}{4\pi\varepsilon_0}\frac{3.2\times10^{-19}\times1.3\times10^{-17}}{\left(5\times10^{-12}\right)^2} \text{ [1 mark]}$$

$= 1.5 \times 10^{-3}$ N *[1 mark]* away from the gold nucleus *[1 mark]*

2) a) $E = V/d = 1500/(4.5 \times 10^{-3}) = 3.3 \times 10^5$ *[1 mark]* Vm^{-1} *[1 mark]*

The field is perpendicular to the plates. *[1 mark]*

b) $d = 2 \times (4.5 \times 10^{-3}) = 9.0 \times 10^{-3}$ m *[1 mark]*

$E = V/d \Rightarrow V = Ed = 3.3 \times 10^5 \times 9 \times 10^{-3} = 3000$ V *[1 mark]*

Page 111 — Electric Fields

1)

$$F_g=-\frac{Gm_1m_2}{r^2}=\frac{6.67\times10^{-11}\times(9.11\times10^{-31})^2}{(8\times10^{-10})^2}=-8.65\times10^{-53}\text{N}$$

$$F_e=\frac{1}{4\pi\varepsilon_0}\frac{Q_1Q_2}{r^2}=\frac{1}{4\pi\varepsilon_0}\frac{(1.60\times10^{-19})^2}{(8\times10^{-10})^2}=3.60\times10^{-10}\text{N [1 mark]}$$

The electric force on each electron is much larger than the gravitational force, by a factor of over 10^{40} *[1 mark]*. The gravitational force is attractive, while the electric force is repulsive *[1 mark]*.

2) a) Oil drop is stationary, so $mg = F_e = Vq/d$ *[1 mark]*

$\Rightarrow q = mgd/V = (1.5 \times 10^{-14} \times 9.81 \times 0.03)/5000$ *[1 mark]*

$= 8.8 \times 10^{-19}$ C *[1 mark]*

b) The drop would accelerate towards the positive lower plate *[1 mark]*.

Unit 4: Section 3 — Capacitance

Page 113 — Capacitors

1) a) Capacitance is the amount of energy stored per volt. *[1 mark]*

b) i) Capacitance $=\frac{Q}{V}=$ gradient of line $=\frac{660\ \mu C}{3\ V}=220\ \mu F$.

[1 mark for 'gradient', 1 mark for correct answer.]

ii) Charge stored $= Q =$ area $= 15 \times 10^{-6} \times 66 = 990\ \mu C$.

[1 mark for 'area', 1 mark for correct answer.]

2) a) E.g.

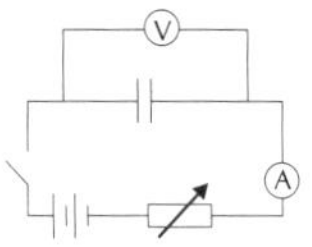

[2 marks, or 1 mark for 3 components correct]

b) Close the switch and start a stopwatch. Adjust the variable resistor to keep the current constant as the capacitor charges *[1 mark]*. Record the p.d. across the capacitor at regular intervals until it equals the battery voltage *[1 mark]*. Plot a graph of charge stored (using $Q = It$) against voltage, then find the area under it. This is the energy stored by the capacitor *[1 mark]*.

3) a) $E=\frac{1}{2}CV^2$ *[1 mark]* $=\frac{1}{2}\times0.5\times12^2=36$ J *[1 mark]*

b) $Q = CV$ *[1 mark]* $= 0.5 \times 12 = 6$ C *[1 mark]*

Page 115 — Charging and Discharging

1) a) The charge falls to 37% after RC seconds *[1 mark]*, so $t = 1000 \times 2.5 \times 10^{-4} = 0.25$ seconds *[1 mark]*

b) $Q=Q_0e^{-\frac{t}{RC}}$ *[1 mark]*, so after 0.7 seconds:

$Q=Q_0e^{-\frac{0.7}{0.25}}=Q_0\times0.06$ *[1 mark]*.

6% of the initial charge is left on the capacitor after 0.7 s *[1 mark]*.

c) i) The total charge stored doubles: V is proportional to Q *[1 mark]*.

ii) None: this is a fixed property of the capacitor *[1 mark]*.

iii) None: charging time depends only on the capacitance of the capacitor and the resistance of the circuit, which don't change *[1 mark]*.

Unit 4: Section 4 — Magnetic Fields

Page 117 — Magnetic Fields

1) a) $F = BIl = 2 \times 10^{-5} \times 3 \times 0.04$ *[1 mark]* $= 2.4 \times 10^{-6}$ N *[1 mark]*

b) The force is zero *[1 mark]* because the two magnetic fields are perpendicular, so do not affect each other *[1 mark]*.

Page 119 — Electromagnetic Induction

1) a) $\phi = BA$ *[1 mark]* $= 2 \times 10^{-3} \times 0.23 = 4.6 \times 10^{-4}$ Wb *[1 mark]*

b) $\Phi = BAN$ *[1 mark]* $= 2 \times 10^{-3} \times 0.23 \times 150 = 0.069$ Wb *[1 mark]*

c) $$V=\frac{\Delta\Phi}{\Delta t}=\frac{(B_{start}-B_{end})AN}{\Delta t}=\frac{\left(2\times10^{-3}-1.5\times10^{-3}\right)(0.23\times150)}{2.5}=6.9\times10^{-3}\text{ V}$$

[3 marks available, one for each stage of the workings]

2) a) $V = Blv$ *[1 mark]* $= 60 \times 10^{-6} \times 30 \times 100 = 0.18$ V *[1 mark]*

b)

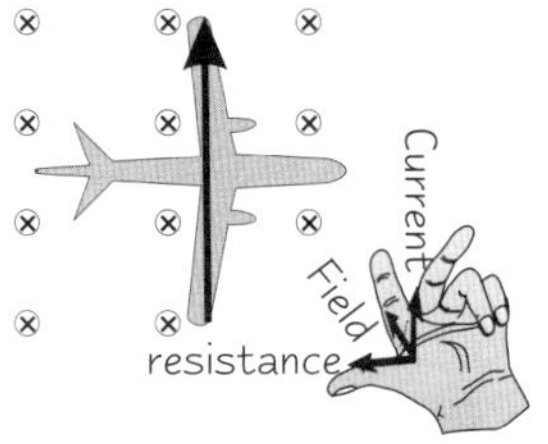

[1 mark]

Page 121 — Transformers and Alternators

1) a) $\frac{V_p}{V_s}=\frac{N_p}{N_s}$ *[1 mark]* so, $N_s=\frac{45\times150}{9}=750$ turns *[1 mark]*

b) $\frac{V_p}{V_s}=\frac{I_s}{I_p}$ *[1 mark]* so, $I_s=\frac{V_pI_p}{V_s}=\frac{9\times1.5}{45}=0.3$ A *[1 mark]*

c) efficiency $=\frac{V_sI_s}{V_pI_p}$ *[1 mark]* $=\frac{10.8}{9\times1.5}=0.8$ (i.e. 80%) *[1 mark]*

2) a) $\Phi = BAN\cos\theta$ *[1 mark]* $= 0.9 \times 0.01 \times 500 \times \cos 60^\circ = 2.25$ Wb *[1 mark]*

b) Peak e.m.f. when $\sin\omega t = \pm1$, giving: $\varepsilon = \pm BAN\omega$ *[1 mark]*

So, peak e.m.f. is: $\varepsilon = \pm\ 0.9 \times 0.01 \times 500 \times 40\pi$ *[1 mark]*

$\varepsilon = \pm565$ V *[1 mark]*

A2 Answers

Unit 5: Section 1 — Radioactivity and Nuclear Energy

Page 123 — Scattering to Determine Structure

1) a) The majority of alpha particles are not scattered because the nucleus is a very small part of the whole atom and so the probability of an alpha particle getting near it is small [1 mark]. Most alpha particles pass undeflected through the empty space around the nucleus [1 mark].

b) Alpha particles and atomic nuclei are both positively charged [1 mark]. If an alpha particle travels close to a nucleus, there will be a significant electrostatic force of repulsion between them [1 mark]. This force deflects the alpha particle from its original path. [1 mark]

2) a) All particles have wave-like properties, with an associated wavelength [1 mark]. If the wavelength of a beam of particles is similar to the atomic spacing of the material it's passing through, the beam will produce a diffraction pattern. [1 mark]

b) Electrons are not affected by the strong nuclear force. [1 mark]

c) Maximum diffraction occurs when the nucleus is the same size as the wavelength of the electrons [1 mark]. Larger nuclei cause less diffraction for the same electron energy [1 mark].

Page 125 — Nuclear Radius and Density

1) a) Rearrange $\mathbf{r} = \mathbf{r}_0\mathbf{A}^{1/3}$ [1 mark], then substitute the given values of $\mathbf{r}$ and $\mathbf{A}$:

$$r_0 = \frac{\mathbf{r}}{\mathbf{A}^{1/3}} = \frac{3.2\times10^{-15}}{12^{1/3}} = 1.40\times10^{-15}\,\text{m} \quad \text{[1 mark]}$$

b) For a radium nucleus with $\mathbf{A}$ = 226:

$$\mathbf{r} = \mathbf{r}_0\mathbf{A}^{1/3} = 1.4\times10^{-15}\times226^{1/3} = 8.53\times10^{-15}\ \text{m} \quad \text{[1 mark]}$$

c) Volume = $\frac{4}{3}\pi \mathbf{r}^3 = \frac{4}{3}\pi(8.53\times10^{-15})^3 = 2.6\times10^{-42}\ \text{m}^3$ [1 mark]

So density (ρ) = $\frac{\mathbf{m}}{\mathbf{v}} = \frac{3.75\times10^{-25}}{2.6\times10^{-42}} = 1.44\times10^{17}\ \text{kgm}^{-3}$ [1 mark]

2) The mass density of a gold nucleus is much larger than the mass density of a gold atom [1 mark]. This implies that the majority of a gold atom's mass is contained in the nucleus [1 mark].
The nucleus is small compared to the size of the atom [1 mark].
There must be a lot of nearly empty space inside each atom [1 mark].

Page 127 — Radioactive Emissions

1) Place different materials between the source and detector and measure the amount of radiation getting through [1 mark]:

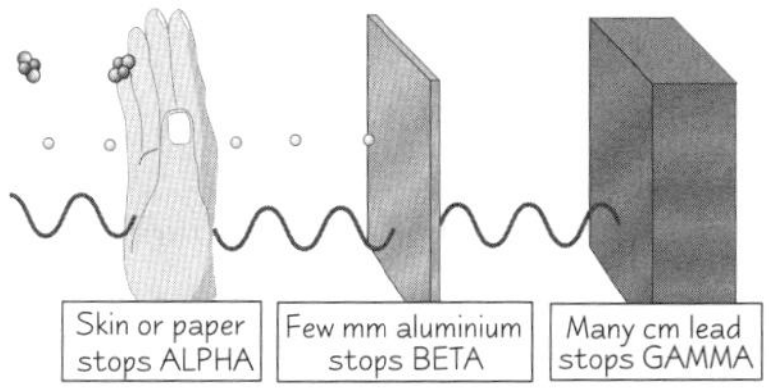

[1 mark for each material stopping correct radiation]

2) $\mathbf{I} \propto \text{count rate} \propto \frac{1}{\mathbf{x}^2}$ [1 mark]

The G-M tube is 4 times the original distance from the source.

$\frac{1}{4^2} = \frac{1}{16}$, so the count rate at 40 cm will be 1/16th that at 10 cm [1 mark].

The count rate at 40 cm will be $\frac{240}{16} = 15$ counts s^{-1} [1 mark].

Page 129 — Exponential Law of Decay

1) Any one of: You can't say which atom/nucleus in a sample will decay next. / You can only estimate the fraction of nuclei that will decay or the probability an atom/nucleus will decay in a given time. / You cannot say exactly how many atoms will decay in a given time. [1 mark]

2) a) Activity, $\mathbf{A}$ = measured activity – background activity
= 750 – 50 = 700 Bq [1 mark]
$\mathbf{A} = \lambda\mathbf{N} \Rightarrow 700 = 50\,000\,\lambda$ [1 mark]
So $\lambda = 0.014\ s^{-1}$ [1 mark]

b) $\mathbf{T}_{\frac{1}{2}} = \frac{\ln 2}{\lambda} = \frac{0.693}{0.014} = 49.5$ seconds

[1 mark for half-life equation, 1 mark for correct half-life]

c) $\mathbf{N} = \mathbf{N}_0\mathbf{e}^{-\lambda t} = 50\,000 \times \mathbf{e}^{-0.014\times300} = 750$
[1 mark for decay equation, 1 mark for correct number of atoms]

Page 131 — Nuclear Decay

1) a) $^{226}_{88}\mathbf{Ra} \rightarrow {}^{222}_{86}\mathbf{Rn} + {}^{4}_{2}\alpha$
[3 marks available — 1 mark for alpha particle, 1 mark each for proton and nucleon number of radon]

b) $^{40}_{19}\mathbf{K} \rightarrow {}^{40}_{20}\mathbf{Ca} + {}^{0}_{-1}\beta$
[3 marks available — 1 mark for beta particle, 1 mark each for proton and nucleon number of calcium]

2) Mass defect = $(6.695\times10^{-27}) - (6.645\times10^{-27}) = 5.0\times10^{-29}$ kg [1 mark]. Using the equation $\mathbf{E} = \mathbf{mc}^2$ [1 mark],
$\mathbf{E} = (5.0\times10^{-29})\times(3\times10^8)^2 = 4.5\times10^{-12}$ J [1 mark]

Page 133 — Binding Energy

1) a) There are 6 protons and 8 neutrons, so the mass of individual parts = (6 × 1.007276) + (8 × 1.008665) = 14.112976 u [1 mark]
Mass of $^{14}_{6}C$ nucleus = 13.999948 u

so, mass defect = 14.112976 – 13.999948 = 0.113028 u [1 mark]
Converting this into kg gives mass defect = 1.88×10^{-28} kg [1 mark]

b) $\mathbf{E} = \mathbf{mc}^2 = (1.88\times10^{-28})\times(3\times10^8)^2 = 1.69\times10^{-11}$ J [1 mark]

1 MeV = 1.6×10^{-13} J, so energy = $\frac{1.69\times10^{-11}}{1.6\times10^{-13}} = 106$ MeV [1 mark]

2) a) Fusion [1 mark]

b) The increase in binding energy per nucleon is about 0.86 MeV [1 mark]. There are 2 nucleons in ^2H, so the increase in binding energy is about 1.72 MeV — so about 1.7 MeV is released (ignoring the positron) [1 mark].

Page 135 — Nuclear Fission and Fusion

1) a) Nuclear fission can be induced by neutrons and produces more neutrons during the process [1 mark]. This means that each fission reaction induces more fission reactions, resulting in a ongoing chain of reactions [1 mark].

b) For example, control rods limit the rate of fission by absorbing neutrons [1 mark]. The number of neutrons absorbed by the rods is controlled by varying the amount they are inserted into the reactor [1 mark]. A suitable material for the control rods is boron [1 mark].

c) In an emergency shut-down, the control rods are released into the reactor [1 mark]. The control rods absorb the neutrons, and stop the reaction as quickly as possible [1 mark].

2) Advantages (any two of e.g.): the nuclear reactor itself doesn't produce any waste gases that could be harmful to the environment e.g. sulfur dioxide (leading to acid rain) or carbon dioxide [1 mark]. It can be used to supply a continuous supply of electricity unlike some renewable sources [1 mark].
Disadvantages (any two of e.g.): problems with the reactor getting out of control [1 mark], risks of radiation from radioactive waste produced [1 mark], the long half-life of nuclear waste [1 mark].

A2 Answers

Unit 5: Section 2 — Thermal Physics

Page 137 — Ideal Gases

1) a) i) Number of moles $= \frac{\text{mass of gas}}{\text{molar mass}} = \frac{0.014}{0.028} = 0.5$ [1 mark]

ii) Number of molecules = number of moles × Avogadro's constant $= 0.5 \times 6.02 \times 10^{23} = 3.01 \times 10^{23}$ [1 mark]

b) $pV = nRT$, so $p = \frac{nRT}{V}$ [1 mark] $= \frac{0.5 \times 8.31 \times 300}{0.01} = 125\ 000$ Pa [1 mark]

c) The pressure would also halve [1 mark] because it is proportional to the number of molecules — $pV = NkT$ [1 mark].

2) At ground level, $\frac{pV}{T} = \frac{1 \times 10^5 \times 10}{293} = 3410\ JK^{-1}$ [1 mark]

pV/T is constant, so higher up $pV/T = 3410\ JK^{-1}$ [1 mark]

Higher up, $p = \frac{3410 \times T}{V} = \frac{3410 \times 260}{25} = 35\ 500$ Pa [1 mark]

Page 139 — The Pressure of an Ideal Gas

1) a) $pV = \frac{1}{3}Nm\overline{c^2}$ Rearrange the equation:

$\overline{c^2} = \frac{3pV}{Nm}$ [1 mark] $= \frac{3 \times 1 \times 10^5 \times 7 \times 10^{-5}}{2 \times 10^{22} \times 6.6 \times 10^{-27}} = 159\ 091\ (ms^{-1})^2$ [1 mark]

b) r.m.s. speed $= \sqrt{\overline{c^2}} = \sqrt{159\ 091} = 399\ ms^{-1}$ [1 mark]

c) pV is proportional to T, so doubling T will double pV. [1 mark]

r.m.s. speed $= \sqrt{\overline{c^2}} = \sqrt{\frac{3pV}{Nm}}$, so doubling pV will increase the r.m.s. speed by a factor of $\sqrt{2}$.

r.m.s. speed $= 399 \times \sqrt{2} = 564\ ms^{-1}$ [1 mark]

Page 141 — Energy and Temperature

1) a) $\frac{1}{2}m\overline{c^2} = \frac{3kT}{2}$ Rearranging gives: $\overline{c^2} = \frac{3kT}{m}$ [1 mark]

m = mass of 1 mole $\div N_A = 2.8 \times 10^{-2} \div 6.02 \times 10^{23} = 4.65 \times 10^{-26}$ kg [1 mark]

$\overline{c^2} = \frac{3 \times 1.38 \times 10^{-23} \times 300}{4.65 \times 10^{-26}} = 2.67 \times 10^5\ m^2s^{-2}$ [1 mark]

Typical speed = r.m.s. speed $= \sqrt{2.67 \times 10^5} = 517\ ms^{-1}$ [1 mark]

b) Gas molecules move at different speeds because they have different amounts of energy [1 mark].
The molecules have different amounts of energy because they constantly collide and transfer energy between themselves [1 mark].

2) a) Time = distance ÷ speed = 8.0 m ÷ 400 = 0.02 s [1 mark]

b) Although the particles move at 400 ms^{-1} on average, they frequently collide with other particles [1 mark]. So their motion in any one direction is limited and they only slowly move from one end of the room to the other [1 mark].

3) Electrical energy supplied: $\Delta Q = VI\Delta t = 12 \times 7.5 \times 180 = 16200$ J [1 mark]

The temperature rise is 12.7 – 4.5 = 8.2 °C

$c = \frac{\Delta Q}{m\Delta\theta}$ [1 mark] $= 16200 \div (2 \times 8.2) = 988\ Jkg^{-1}°C^{-1}$ [1 mark]

You need the right unit for the third mark — J kg^{-1} K^{-1} would be right too.

4) $\Delta Q = ml = 2.26 \times 10^6 \times 0.5 = 1.13 \times 10^6$ J [1 mark]
3 kW means 3000 J in 1s, so time $= 1.13 \times 10^6 / 3000$ [1 mark] = 377 s [1 mark]

Unit 5: Option A — Astrophysics and Cosmology

Page 144 — Optical Telescopes

1) a)

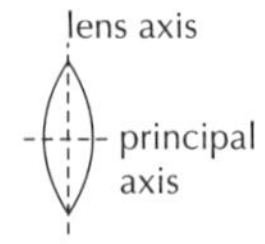

The principal focus is where rays parallel to the principal axis converge [1 mark]. The focal length is the distance between the lens axis and the principal focus [1 mark].

b) $\frac{1}{u} + \frac{1}{v} = \frac{1}{f}$ [1 mark]. So $\frac{1}{0.2} + \frac{1}{v} = \frac{1}{0.15}$ [1 mark].

$\frac{1}{v} = \frac{1}{0.15} - \frac{1}{0.2} = \frac{5}{3} \Rightarrow v = 0.6m$ [1 mark].

c) Using lens equation: $\frac{1}{v} = \frac{1}{0.15} - \frac{1}{0.10} = -\frac{10}{3} \Rightarrow v = -0.3m$ [1 mark]
The sign of v is negative, indicating that the image is a virtual image on the same side of the lens as the object [1 mark].

2) a) Separation of lenses needs to be $f_o + f_e = 5.0 + 0.10 = 5.1$ m [1 mark].

b) Angular magnification = angle subtended by image at eye / angle subtended by object at unaided eye [1 mark].
$M = f_o/f_e = 5.0 / 0.10 = 50$ [1 mark].

3) a) When light strikes a pixel on a CCD, electrons [1 mark] are liberated from the silicon and are stored in a potential well [1 mark]. Once the exposure has been taken, electrons are shunted along the potential wells [1 mark] and emerge in sequence at the output, where they can be measured.

b) CCDs have a quantum efficiency greater than 70% [1 mark] whereas photographic emulsion only has a 4% efficiency.
So fewer photons are needed for an image and fainter objects can be detected [1 mark].
The output is in electronic form and can be processed digitally by computers, making it easier for the images to be enhanced [1 mark].

Page 147 — Non-Optical Telescopes

1) The collecting power of the telescope is proportional to the area of the objective dish or mirror [1 mark].
Radio telescopes tend to have larger dishes than UV telescopes, so radio telescopes tend to have greater collecting powers [1 mark].
Resolving power depends on the wavelength of the radiation and the diameter of the dish [1 mark].
Since UV radiation has a much, much smaller wavelength than radio, UV telescopes have a greater resolving power. [1 mark]

2) a) The telescope emits infrared radiation, which masks the infrared it is trying to detect [1 mark]. The colder the telescope, the less infrared it emits [1 mark].

b) They are set up at high altitude in dry places [1 mark].

3) a) On high altitude aeroplanes / weather balloons [1 mark], to get above the level of the atmosphere that absorbs the radiation [1 mark].

b) A UV telescope uses a single parabolic mirror, whereas an X-ray telescope uses a series of nested 'grazing' mirrors [1 mark].
This is because UV reflects in the same way as visible light [1 mark] but X-rays can only be reflected at very shallow angles / would be absorbed by a parabolic mirror [1 mark].

4) a) power $\propto$ diameter^2 [1 mark].

b) $\frac{\text{power of Arecibo}}{\text{power of Lovell}} = \frac{300^2}{76^2}$ [1 mark].
Ratio = 15.6:1 [1 mark].

A2 Answers

Page 149 — Distances and Magnitude

1) The absolute magnitude is the apparent magnitude [1 mark] that the object would have if it were 10 parsecs [1 mark] away from Earth.

2) Distance to Sun in parsecs = $1/(2 \times 10^5) = 5 \times 10^{-6}$ pc [1 mark].
$\mathbf{m} - \mathbf{M} = 5 \lg (\mathbf{d}/10)$ [1 mark]
$\Rightarrow -27 - \mathbf{M} = 5 \lg (5 \times 10^{-6}/10)$ [1 mark]
$\Rightarrow -27 - \mathbf{M} = 5 \lg (5 \times 10^{-7}) \Rightarrow -27 - \mathbf{M} = -31.5$
$\Rightarrow \mathbf{M} = 4.5$ [1 mark].

3) a) Sirius is the brighter of the two [1 mark].
b) $\mathbf{m} - \mathbf{M} = 5 \lg (\mathbf{d}/10)$ [1 mark]
$\Rightarrow -0.72 - (-5.5) = 5 \lg (\mathbf{d}/10)$ [1 mark]
$\Rightarrow 4.78 = 5 \lg (\mathbf{d}/10)$
$\Rightarrow \lg (\mathbf{d}/10) = 0.956$
$\Rightarrow \mathbf{d}/10 = 10^{0.956}$ So $\mathbf{d} = 90$ pc [1 mark].

Page 151 — Stars as Black Bodies

1) a) According to Wein's displacement law $\lambda_{max} \times \mathbf{T} = 0.0029$, so for this star $\lambda_{max} = 0.0029 \div 4000 = 7.25 \times 10^{-7}$ m [1 mark].
Curve Y peaks at around 0.7 μm ($= 7 \times 10^{-7}$ m), so could represent the star [1 mark].
b) $\mathbf{L} = \sigma \mathbf{A} \mathbf{T}^4$, so $3.9 \times 10^{26} = 5.67 \times 10^{-8} \times \mathbf{A} \times 4000^4$ [1 mark], which gives $\mathbf{A} = 2.7 \times 10^{19}$ m^2 [1 mark].

2) $\lambda_{max} \times \mathbf{T} = 0.0029$ [1 mark].
So $\mathbf{T} = 0.0029/(436 \times 10^{-9}) \approx 6650$ K [1 mark].
$\mathbf{L} = \sigma \mathbf{A} \mathbf{T}^4$ [1 mark]. So $2.3 \times 10^{27} = 5.67 \times 10^{-8} \times \mathbf{A} \times 6650^4$, which gives $\mathbf{A} = 2.1 \times 10^{19}$ m^2 [1 mark].

Page 153 — Spectral Classes and the H-R Diagram

1) a) To get strong Balmer lines, the majority of the electrons need to be at the $\mathbf{n} = 2$ level [1 mark]. At low temperatures, few electrons have enough energy to be at the $\mathbf{n} = 2$ level [1 mark].
At very high temperatures, most electrons are at $\mathbf{n} = 3$ or above, leading to weak Balmer lines [1 mark].
b) Spectral classes A [1 mark] and B [1 mark]
c) Class F stars are white [1 mark], have a temperature of 6000 – 7500 K [1 mark] and show prominent absorption lines from metal ions [1 mark].

2) Molecules are only present in the lowest temperature stars [1 mark]. At higher temperatures molecules are broken up into individual atoms [1 mark].

3)

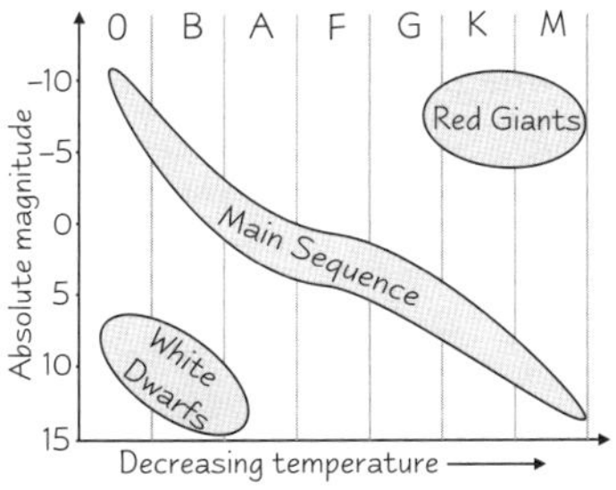

[5 marks maximum, 1 mark each for correctly labelled axes, 1 mark each for 'Main Sequence', 'White Dwarfs' and 'Red Giants'.]

Page 155 — Stellar Evolution

1) High mass stars spend less time on the main sequence than low mass stars [1 mark].
As red giants, low mass stars only fuse hydrogen and/or helium but the highest mass stars fuse nuclei up to iron [1 mark].
Lower mass stars eject their atmospheres to become white dwarfs [1 mark], but high mass stars explode in supernovae [1 mark] to leave neutron stars [1 mark] or black holes [1 mark].

2) a) The Schwarzschild radius is the distance [1 mark] from the centre of a black hole to where the escape velocity is the speed of light [1 mark].
b) $\mathbf{R_s} = \frac{2\mathbf{GM}}{\mathbf{c}^2} = \frac{2 \times 6.67 \times 10^{-11} \times 6 \times 10^{30}}{(3 \times 10^8)^2}$ [1 mark] ≈ 8.9 km [1 mark]

Page 157 — The Doppler Effect and Redshift

1) a) Object A is moving towards us [1 mark].
b) Object B is part of a binary star system (or is being orbited by a planet) [1 mark] with a period of two weeks [1 mark].
c) Find the velocity of object C using $\frac{\Delta\lambda}{\lambda} = \frac{\mathbf{v}}{\mathbf{c}}$, so $\mathbf{v} = \mathbf{c}\frac{\Delta\lambda}{\lambda}$. [1 mark]

$$\mathbf{v} = 3.0 \times 10^8 \times \frac{(667.83 \times 10^{-9} - 656.28 \times 10^{-9})}{656.28 \times 10^{-9}} = 5.28 \times 10^6 \text{ ms}^{-1}$$

So object C is moving away from us [1 mark] at 5.28×10^6 ms^{-1} [1 mark].

2) a) Their spectrum shows an enormous redshift [1 mark].
b) Intensity is proportional to 1/distance^2 [1 mark].
So, e.g., if a quasar is 500 000 times further away than, but just as bright as, a star in the Milky Way it must be 500 000^2 times brighter than the star [1 mark].
c) A supermassive black hole [1 mark] surrounded by a doughnut-shaped mass of whirling gas [1 mark].

Page 159 — The Big Bang Model of the Universe

1) a) $\mathbf{v} = \mathbf{H_0 d}$ [1 mark] where $\mathbf{v}$ is recessional velocity (in kms^{-1}), $\mathbf{d}$ is distance (in Mpc) and $\mathbf{H_0}$ is Hubble's constant in (kms^{-1}Mpc^{-1}). [1 mark]
b) Hubble's law suggests that the Universe originated with the Big Bang [1 mark] and has been expanding ever since. [1 mark]
c) i) $\mathbf{H_0} = \mathbf{v} \div \mathbf{d} = 50$ kms$^{-1} \div 1$ Mpc^{-1}.
50 kms^{-1} = 50×10^3 ms^{-1} and 1 Mpc^{-1} = 3.09×10^{22} m
So, $H_0 = 50 \times 10^3$ ms$^{-1} \div 3.09 \times 10^{22}$ m $= 1.62 \times 10^{-18}$ s^{-1}
[1 mark for the correct value, 1 mark for the correct unit]
ii) $\mathbf{t} = 1/\mathbf{H_0}$ [1 mark] $= 1/1.62 \times 10^{-18}$
$= 6.18 \times 10^{17}$ s $\approx 2 \times 10^{10}$ years [1 mark]
The observable Universe has a radius of ~20 billion light years. [1 mark]

2) a) $\mathbf{z} \approx \mathbf{v}/\mathbf{c}$ [1 mark]
so $\mathbf{v} \approx 0.37 \times 3.0 \times 10^8 \approx 1.1 \times 10^8$ ms^{-1} [1 mark]
b) $\mathbf{d} = \mathbf{v}/\mathbf{H_0} \approx 1.1 \times 10^8 / 2.4 \times 10^{-18} = 4.6 \times 10^{25}$ m [1 mark]
$= 4.6 \times 10^{25} / 9.5 \times 10^{15}$ ly = 4.9 billion ly [1 mark]
c) $\mathbf{z} = \mathbf{v}/\mathbf{c}$ is only valid if $\mathbf{v} << \mathbf{c}$ — it isn't in this case [1 mark].

3) The cosmic background radiation is microwave radiation [1 mark] showing a perfect black body spectrum [1 mark] of a temperature of about 3 K [1 mark].
It is very nearly isotropic and homogeneous [1 mark].
It suggests that the ancient Universe was very hot, producing lots of electromagnetic radiation [1 mark] and that its expansion has stretched the radiation into the microwave region [1 mark].
[1 mark for quality of written communication.]

Unit 5: Option B — Medical Physics

Page 161 — Physics of the Eye

1) a) The distance will be the focal length of the lens [1 mark]
$\mathbf{v} = \mathbf{f} = 1$/power $= 1/60 = 0.017$ m [1 mark]
b) $1/\mathbf{u} + 1/\mathbf{v} = 1/\mathbf{f}$, $\mathbf{u} = 0.3$ m, $\mathbf{v} = 1/60$ m [1 mark].
$1/\mathbf{f} = 63.3$ D [1 mark]. So the extra power needed = 3.3 D [1 mark].

A2 Answers

2)a) Light enters the eye through the cornea, which focuses the light [1 mark]. It then passes through the aqueous humour, then through the iris, which controls the amount of light entering the eye [1 mark], to reach the lens.
The lens acts as a fine focus [1 mark] so that light travels through the vitreous humour and is focused on the retina [1 mark].
b) Spatial resolution is greatest at the yellow spot on the retina where the photoreceptors are most densely packed [1 mark].
Away from the yellow spot, spatial resolution decreases with receptor density, and because of an increase in the number of receptors per nerve cell [1 mark].

Page 163 — Defects of Vision

1) Focal length of diverging lens needs to be –4 m [1 mark].
Power = $1/\mathbf{f}$ = –0.25 D [1 mark for value, 1 mark for negative sign]

2) Lens equation $1/\mathbf{u} + 1/\mathbf{v} = 1/\mathbf{f}$ [1 mark].
When $\mathbf{u}$ = 0.25 m, $\mathbf{v}$ = –2 m ⇒ $1/\mathbf{f} = 1/0.25 - 1/2 = 3.5$.
Power = +3.5 D [1 mark for value, 1 mark for sign]

3)a) Cylindrical lenses [1 mark].
b)

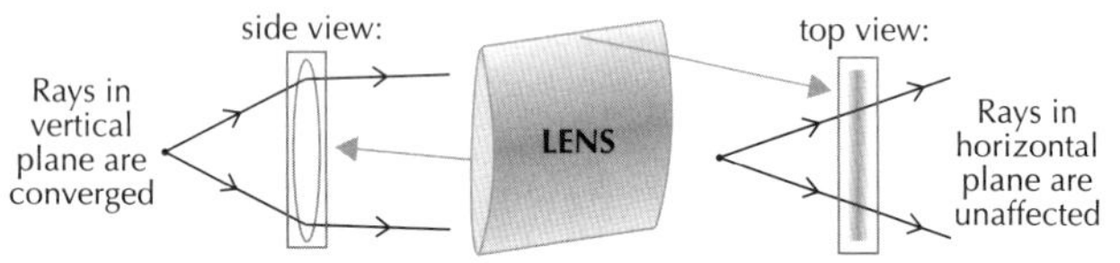

[2 marks maximum, 1 mark for each correctly drawn axis]

Page 165 — Physics of the Ear

1)a) The pinna concentrates the sound energy entering the ear into the auditory canal, increasing its intensity [1 mark].
b) Sound energy entering the ear causes the tympanic membrane (eardrum) to vibrate [1 mark].
The vibrations are transmitted through the middle ear by the malleus, incus and stapes in turn [1 mark].
The stapes is connected to the oval window, so causes it to vibrate [1 mark].
c) The amplitude of a sound is proportional to the square root of its intensity, and the intensity is inversely proportional to the area [1 mark].
This means that the amplitude is inversely proportional to the square root of the area [1 mark], so if the area is decreased by a factor of 14, the amplitude is increased by a factor of $\sqrt{14}$ ≈ 3.74 [1 mark].
d) Pressure waves in the cochlea cause the basilar membrane to vibrate [1 mark], which causes hair cells on the membrane to trigger electrical impulses [1 mark].
e) Different regions of the basilar membrane have different natural frequencies [1 mark].
When the frequency of a sound wave matches the natural frequency of a part of the membrane, that part resonates, causing the hair cells in that area to trigger impulses, so different frequencies trigger different nerve cells [1 mark].

Page 167 — Intensity and Loudness

1)a) 1×10^{-12} Wm^{-2} [1 mark]

b) $\mathbf{IL} = 10\log\left(\frac{\mathbf{I}}{\mathbf{I}_0}\right)$ [1 mark] $= 10\log\left(\frac{0.94}{1\times10^{-12}}\right) = 119.7$ dB [1 mark]

c) The ear is most sensitive at about 3000 Hz, so the siren will sound as loud as possible [1 mark].

2) The patient has suffered hearing loss at all frequencies, but the loss is worst at high frequencies [1 mark]. If the patient's hearing had been damaged by excessive noise, you would expect to see a peak at a particular frequency [1 mark]. This isn't present, so the patient's hearing loss is more likely to be age-related [1 mark].

Page 169 — Physics of the Heart

1)a) Dead skin cells and hairs are removed [1 mark]. Conductive gel is used to give a good electrical contact [1 mark].
b) The P wave corresponds to depolarisation and contraction of the atria [1 mark]. The QRS wave corresponds to depolarisation and contraction of the ventricles [1 mark] and repolarisation and relaxation of the atria [1 mark].

2) A membrane is initially polarised so that the outside is positively charged and the inside is negatively charged [1 mark].
When the membrane is stimulated, it becomes permeable to sodium ions [1 mark].
The ions move through the membrane and into the cell, depolarising the system [1 mark] and then polarising it the other way. In repolarisation, the membrane becomes impermeable to sodium, but very permeable to potassium ions [1 mark].
The potassium ions move through the membrane to reverse the polarisation, and the Na-K pump moves sodium ions out of the cell to restore equilibrium [1 mark].
[1 mark for quality of written communication]

3) The sinoatrial node produces ~70 electrical pulses a minute [1 mark]. These make the atria contract [1 mark].
They then pass to the atrioventricular node, which delays the pulses [1 mark] then passes them to the ventricles to make them contract shortly after [1 mark].

Page 171 — X-ray Production

1)

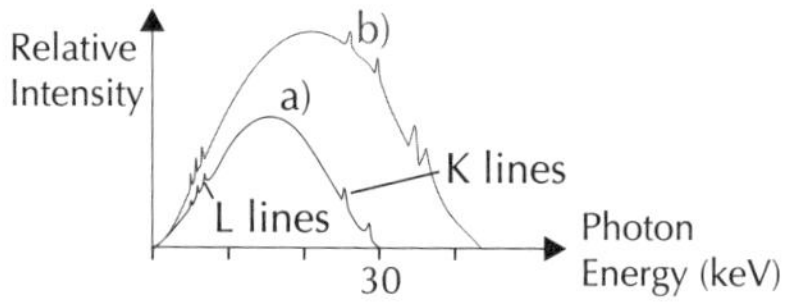

a) See graph [1 mark for shape of graph, 1 mark for 30 keV maximum energy and 1 mark for correct labelling of line spectrum]
b) See graph [1 mark for higher intensity and higher maximum energy and 1 mark for a few extra lines in the line spectrum]
c) Find the energy of each electron using $\mathbf{E} = \mathbf{QV}$:
$\mathbf{E} = 1.6 \times 10^{-19} \times 30 \times 10^3 = 4.8 \times 10^{-15}$ J [1 mark]
Kinetic energy = $\frac{1}{2}\mathbf{mv}^2 = 4.8 \times 10^{-15}$ [1 mark]
So, $\mathbf{v}^2 = (2 \times 4.8 \times 10^{-15}) \div (9.11 \times 10^{-31}) = 1.054 \times 10^{16}$ [1 mark]
$\mathbf{v} = 1.03 \times 10^8$ ms^{-1} [1 mark]

Page 173 — X-ray and MRI Imaging

1)a) Half-value thickness is the thickness of material required to reduce the intensity of an X-ray beam to half its original value [1 mark].

b) $\mu = \frac{\ln 2}{\mathbf{x}_{\frac{1}{2}}} = \frac{\ln 2}{3} = 0.23\,\text{mm}^{-1}$ [1 mark], $\mathbf{I} = \mathbf{I}_0 \mathbf{e}^{-\mu x} \Rightarrow \frac{\mathbf{I}}{\mathbf{I}_0} = \mathbf{e}^{-\mu x}$ [1 mark].

So, $0.01 = \mathbf{e}^{-0.23\mathbf{x}} \Rightarrow \ln(0.01) = -0.23\mathbf{x}$ [1 mark]
$\mathbf{x}$ = 20mm [1 mark].

2) The patient lies in the centre of a large magnet, which produces a magnetic field [1 mark]. The magnetic field aligns hydrogen protons in the patient's body [1 mark].
Radio frequency coils are used to transmit radio waves, which cause the aligned protons to absorb energy [1 mark]. When the radio waves stop the protons emit the stored energy as radio waves, which are recorded by the scanner [1 mark].
A computer analyses the received radio waves to produce an image of the patient's body [1 mark].
[1 mark for quality of written communication]

3) Advantages, e.g. there are no known side effects / doesn't use ionising radiation / an image can be made for any slice in any orientation of the body / images of soft tissue are higher quality than using other techniques (CT, X-ray) / contrast can be weighted to investigate different situations.

A2 Answers

Disadvantages, e.g. other techniques give better quality images of bony structures / people can suffer claustrophobia inside the scanner / MRI cannot be used on people with pacemakers/surgical implants/other metal in their bodies / MRI scanners are very expensive. [1 mark for each point explained, to a total of 6 marks]

Page 175 — Ultrasound Imaging

1) a) $\alpha = \left(\frac{Z_{tissue} - Z_{air}}{Z_{tissue} + Z_{air}}\right)^2 = \left(\frac{1630 - 0.430}{1630 + 0.430}\right)^2$ *[1 mark]*, $\alpha = 0.999$ *[1 mark]*

b) *From part a), 0.1% enters the body when no gel is used [1 mark].*

$\alpha = \left(\frac{Z_{tissue} - Z_{gel}}{Z_{tissue} + Z_{gel}}\right)^2 = \left(\frac{1630 - 1500}{1630 + 1500}\right)^2 = 0.002$ *[1 mark], so 99.8% of the ultrasound is transmitted [1 mark]. Ratio is ~1000 [1 mark].*

2) a) $Z = \rho v$ *[1 mark]*,
$v = (1.63 \times 10^6)/(1.09 \times 10^3) = 1495 ms^{-1} \approx 1.50 kms^{-1}$ *[1 mark].*

b) *A pulse from the far side of the head travels an extra 2**d** cm, where **d** is the diameter of the head [1 mark].*
Time taken to travel this distance $= 2.4 \times 50 = 120\ \mu s$ *[1 mark].*
Distance = speed × time, so $2d = 1500 \times 120 \times 10^{-6} = 0.18$ m *[1 mark]. So* $d = 9$ cm *[1 mark].*

Page 177— Endoscopy

1) a) $\sin \theta_c = n_2/n_1 = 1.30/1.35$ *[1 mark]*, $\theta = 74.4°$ *[1 mark].*

b) *When the angle of incidence is greater than or equal to the critical angle, the beam of light will undergo total internal reflection [1 mark]. If the angle of incidence falls below the critical angle, then some light will be lost [1 mark].*

2) *A coherent fibre-optic bundle consists of a large number of very thin fibres [1 mark], arranged in the same way at either end of the bundle [1 mark].*
Lots of thin fibres are used to increase the resolution of the image [1 mark].
The relative positions of the fibres have to remain constant or the image would be jumbled up [1 mark].

Unit 5: Option D — Turning Points in Physics

Page 179 — Charge/Mass Ratio of the Electron

1) a) *1000 eV [1 mark]*

b) $1000\ eV \times 1.6 \times 10^{-19}\ J/eV = 1.6 \times 10^{-16}\ J$ *[1 mark]*

c) *Kinetic energy* $= \frac{1}{2}mv^2 = 1.6 \times 10^{-16}\ J$ *[1 mark]*
$v^2 = (2 \times 1.6 \times 10^{-16}) \div (9.1 \times 10^{-31}) = 3.5 \times 10^{14}$
$\Rightarrow v = 1.9 \times 10^7\ ms^{-1}$ *[1 mark]*
Divide by 3.0×10^8*: 6.3% of the speed of light [1 mark]*

2) *[Your answer will depend on which experiment you describe, e.g.]*
Electrons are accelerated using an electron gun [1 mark].
A magnetic field [1 mark] exerts a centripetal force [1 mark] on the electrons, making them trace a circular path.
By measuring the radius of this path and equating the magnetic and centripetal forces [1 mark] you can calculate e/m.
[1 mark for quality of written communication]

Page 181 — Millikan's Oil-Drop Experiment

1) a) *The forces acting on the drop are its weight, acting downwards [1 mark] and the equally sized force due to the electric field, acting upwards [1 mark].*

b) *Weight = electric force, so* $mg = \frac{QV}{d}$, *and* $Q = \frac{mgd}{V}$ *[1 mark].*

$Q = \frac{1.63 \times 10^{-14} \times 9.81 \times 3.00 \times 10^{-2}}{5000} = 9.59 \times 10^{-19} C$ *[1 mark]*

Divide by e: $9.59 \times 10^{-19} \div 1.6 \times 10^{-19} = 6 \Rightarrow Q = 6e$ *[1 mark]*

c) *The forces on the oil drop as it falls are its weight and the viscous force from the air [1 mark].*
As the oil drop accelerates, the viscous force increases until it equals the oil drop's weight [1 mark].
At this point, there is no resultant force on the oil drop, so it stops accelerating, but continues to fall at terminal velocity [1 mark].

d) *At terminal velocity,* $mg = 6\pi\eta rv$*. Rearranging,* $v = \frac{mg}{6\pi\eta r}$ *[1 mark]*

Find the radius of the oil drop, using mass = volume × density:

$m = \frac{4}{3}\pi r^3 \rho$. *So,* $r^3 = \frac{3m}{4\pi\rho} = \frac{3 \times 1.63 \times 10^{-14}}{4 \times \pi \times 880} = 4.42 \times 10^{-18}$

and $r = 1.64 \times 10^{-6}$ m *[1 mark].*

So, $v = \frac{1.63 \times 10^{-14} \times 9.81}{6 \times \pi \times 1.84 \times 10^{-5} \times 1.64 \times 10^{-6}} = 2.81 \times 10^{-4} ms^{-1}$ *[1 mark]*

Page 183 — Light — Newton vs Huygens

1) *Light consists of particles [1 mark]. The theory was based on Newton's laws of motion with the straight-line motion of light as evidence [1 mark].*

2) *Most scientists in the 18th century supported Newton's corpuscular theory [1 mark]. He said that light was made up of particles that obey his laws of motion [1 mark].*
In Huygens' wave theory, light is a wave [1 mark]. This is supported by the diffraction and interference seen in Young's double-slit experiment [1 mark].
In the second half of the 19th century, Maxwell described light as an electromagnetic wave [1 mark] consisting of oscillating electric and magnetic fields.
[1 mark for quality of written communication]

Page 185 — The Photoelectric Effect

1) *Electrons in the metal absorb energy from the UV light and leave the surface [1 mark], causing the plate to become positively charged [1 mark].*

2) a) $\phi = hf - (\frac{1}{2}mv^2)_{max} = hc/\lambda - (\frac{1}{2}mv^2)_{max}$ *[1 mark]*
$= ((6.63 \times 10^{-34} \times 3.00 \times 10^8)/(0.5 \times 10^{-6})) - 2.0 \times 10^{-19}\ J$
[1 mark] $= 1.98 \times 10^{-19}\ J$ *[1 mark]*

b) *Electrons will only be emitted if the energy they gain from photons in the beam of light is greater than the work function energy [1 mark].*
The energy supplied by the beam of light is:
$E = hc/\lambda = (6.63 \times 10^{-34} \times 3.00 \times 10^8)/(1.5 \times 10^{-6}) = 1.33 \times 10^{-19}\ J$
[1 mark], which is less than the work function calculated in (a), so will not supply enough energy for electrons to be released [1 mark].

3) $\phi = 2.2\ eV = 2.2 \times 1.6 \times 10^{-19}\ J = 3.52 \times 10^{-19}\ J$
$(\frac{1}{2}mv^2)_{max} = hf - \phi = (hc/\lambda) - \phi$ *[1 mark]*
$= [6.63 \times 10^{-34} \times (3.00 \times 10^8/350 \times 10^{-9})] - 3.52 \times 10^{-19}$
$= 2.16 \times 10^{-19}\ J$ *[1 mark]*
Stopping potential, $V_s = (\frac{1}{2}mv^2)_{max}/e$ *[1 mark]*
$\Rightarrow V_s = 2.16 \times 10^{-19}/1.6 \times 10^{-19}\ C = 1.35\ V$ *[1 mark]*

Page 187 — Wave-Particle Duality

1) a) i) *Velocity is given by* $\frac{1}{2}mv^2 = eV$ *[1 mark]*
$\Rightarrow v^2 = 2eV/m$
$\Rightarrow v = 1.3 \times 10^7\ ms^{-1}$ *[1 mark]*

ii) *de Broglie* $\lambda = h/mv$ *[1 mark]* $\Rightarrow \lambda = 5.5 \times 10^{-11}$ m *[1 mark]*

b) *This is in the X-ray region of the EM spectrum [1 mark].*

2) a) *A stream of electrons is accelerated towards the sample using an electron gun [1 mark]. The beam of electrons is focused onto the sample using a magnetic field [1 mark].*
The parts of the beam that pass through the sample are projected onto a screen to form an image of the sample [1 mark].

A2 Answers

b) *To resolve detail around the size of an atom, the electron wavelength needs to be around 0.1 nm [1 mark]. The relationship between anode voltage and electron wavelength is given by* $\lambda = \frac{h}{\sqrt{2meV}}$, *which rearranges to give* $V = \frac{h^2}{2me\lambda^2}$ *[1 mark].*

Substituting $m = 9.1 \times 10^{-31}$ *kg,* $e = 1.6 \times 10^{-19}$ *C,* $l = 0.1 \times 10^{-9}$ *m*

gives: $V = \frac{(6.63 \times 10^{-34})^2}{2 \times 9.1 \times 10^{-31} \times 1.6 \times 10^{-19} \times (0.1 \times 10^{-9})^2} = 151\ V$,
showing that the minimum anode voltage ≈ 150 V [1 mark].

Page 189 — The Speed of Light and Relativity

1) a) *The interference pattern would move/be shifted [1 mark].*
b) *The speed of light has the same value for all observers [1 mark]. It is impossible to detect absolute motion [1 mark].*

2) a) *An inertial reference frame is one in which Newton's 1st law is obeyed [1 mark], e.g. a train carriage moving at constant speed along a straight track (or any other relevant example) [1 mark].*
b) *The speed of light is unaffected by the motion of the observer [1 mark] or the motion of the light source [1 mark].*

Page 191 — Special Relativity

1) $t = \frac{t_0}{\sqrt{1 - \frac{v^2}{c^2}}}$ *[1 mark] and* $t_0 = 20 \times 10^{-9}$ *s [1 mark]*

$t = \frac{20 \times 10^{-9}}{\sqrt{1 - \frac{(0.995c)^2}{c^2}}} = 200$ *ns or* 2×10^{-7} *s [1 mark]*

2) *Your description must include:*
A diagram or statement showing relative motion [1 mark].
An event of a specified duration in one reference frame [1 mark].
Measurement of the time interval by a moving observer [1 mark].
Time interval for "external" observer greater than time interval for the "stationary" observer or equivalent [1 mark].

3) a) $m = m_0 \div \sqrt{1 - \frac{v^2}{c^2}} = 1.67 \times 10^{-27} \div \sqrt{1 - \frac{2.8^2}{3.0^2}} = 4.65 \times 10^{-27}$ kg
[1 mark]
b) $E = mc^2 = 4.65 \times 10^{-27} \times (3 \times 10^8)^2 = 4.2 \times 10^{-10}$ *J [1 mark]*

Index

Index

Index

Index